Project Management, Planning and Control

Dedicated to my two sons
Guy and Marc

Project Management, Planning and Control

Managing Engineering, Construction and Manufacturing Projects to PMI, APM and BSI Standards

8th Edition

EUR. ING. Albert Lester
CEng, FICE, FIMechE, FIStructE, Hon FAPM

Butterworth-Heinemann
An imprint of Elsevier

Butterworth-Heinemann is an imprint of Elsevier
The Boulevard, Langford Lane, Kidlington, Oxford OX5 1GB, United Kingdom
50 Hampshire Street, 5th Floor, Cambridge, MA 02139, United States

Notices
Knowledge and best practice in this field are constantly changing. As new research and experience broaden our understanding, changes in research methods, professional practices, or medical treatment may become necessary.

Practitioners and researchers must always rely on their own experience and knowledge in evaluating and using any information, methods, compounds, or experiments described herein. In using such information or methods they should be mindful of their own safety and the safety of others, including parties for whom they have a professional responsibility.

To the fullest extent of the law, neither the Publisher nor the authors, contributors, or editors, assume any liability for any injury and/or damage to persons or property as a matter of products liability, negligence or otherwise, or from any use or operation of any methods, products, instructions, or ideas contained in the material herein.

Library of Congress Cataloging-in-Publication Data
A catalog record for this book is available from the Library of Congress

British Library Cataloguing-in-Publication Data
A catalogue record for this book is available from the British Library

ISBN: 978-0-12-824339-8

For information on all Butterworth-Heinemann publications visit our website at https://www.elsevier.com/books-and-journals

Publisher: Matthew Deans
Acquisitions Editor: Brian Guerin
Editorial Project Manager: Chiara Giglio
Production Project Manager: Nirmala Arumugam
Cover Designer: Miles Hitchen

Typeset by TNQ Technologies

Contents

Foreword to the 1st edition

A key word in the title of this book is 'control'. This word, in the context of manage-ment, implies the observation of performance in relation to plan and the swift taking of corrective action when the performance is inadequate. In contrast to many other publications which purport to deal with the subject, the mechanism of control permeates the procedures that Mr Lester advocates. In some chapters, such as that on Manual and Computer Analysis, it is there by implication. In others, such as that on Cost Control, it is there in specific terms.

The book, in short, deals with real problems and their real solutions. I recommend it therefore both to students who seek to understand the subject and to managers who wish to sharpen their performance.

by Geoffrey Trimble[†]
Professor of Construction Management
University of Technology, Loughborough, United Kingdom

[†] Deceased author.

Foreword to the 8th edition

The eighth edition of Albert Lester's publication says it all. The widespread acceptance and critical acclaim of his overall grasp of the basic principles of 'project management' confirms the demand for this book.

My supervisory involvement with Albert over many years in different organizations and eras has given me the opportunity of seeing him employ his procedures in real time and enjoy satisfactory results.

The control of cost, resources and time spent is fundamental, but systems are tested to the limit when an unexpected event occurs. This is where Albert's understanding of the sophisticated methods shines through and how we can handle these 'unexpected' events which can interfere to the severe detriment of a project's success if not dealt with efficiently.

Quoting an old Egyptian saying 'It is the leading camels that hold up the train it is the last camel that gets beaten' indicates that little has changed over the years. Criticism for delays in modern projects is often misdirected by the media and misunderstood by the general public. Late changes in design, afterthoughts by the owner or developer, which need to be incorporated, have to be handled effectively if the end objectives are to be attained.

This is where 'Project Management, Planning and Control' comes into its own in helping to equip managers with the tools to operate effectively.

W.T.J. Davies
CEng FIMech.E FIChem.E

Preface to the 8th edition

The shortest distance between two points is a straight line.

Euclid

The longest distance between two points is a shortcut.

Lester

As in previous editions, the basic principles of network analysis have been retained despite the unfortunate trend of entering the scheduling data directly into the computer.

I firmly believe that a manually drafted network at which all the departmental managers take part is the key to a successful project schedule.

In order to show readers that the principles, methods and techniques given in this book actually work and have been successfully used by me over a period of 30 years, I have included in this 8th edition case studies of two large design-and-construction projects, both of which were completed within days of the anticipated completion date. In both cases, the scheduling was done using critical path networks and the point I wanted to emphasize was that while the analysis and presentation could be performed by any of the latest computer programs, the drafting of the network and the formulation of the logic should always be done manually by a team of experts in the various disciplines, led by the project manager and assisted by the appointed planning engineer.

While this new edition incorporates all the topics of the last (7th) edition, some chapters and appendices have been brought up to date in line with the latest developments.

Chapter 7, Stakeholder management, has been expanded to discuss more deeply the relationship between the project manager and the different stakeholders.

In Chapter 42, the section on Adjudication has been updated and now includes the major changes in the Local Democracy Economic Development and Construction Act in 2011 and the Low Value Dispute (LVD) Model Adjudication Procedure 2020.

The updated Chapter 43, Governance of project management, takes into account the new and revised guidance since 2016 and notes the relevance of this topic to the Strategic Report now required of UK companies. The reference to the updated publication 'Directing Change' has been added as this places greater emphasis on culture and ethics. This contribution by the original author, David Shannon, is greatly appreciated.

Chapter 51, Primavera P2, has been updated by Arnaud Morvan of Oracle to reflect the latest software revision, and the risk analysis has now been carried out before the baseline review and final report.

Three new chapters and three new appendices have been added.

Chapter 53, Virtual Design and Construction (VDC), by Dale Dutton of InEight Inc. explains the processes developed by this software provider to meet the needs of designers and constructors in adopting and operating BIM.

Chapter 54, Sustainability, has been added to reflect the current realization that serious changes are necessary in the way we deliver projects to protect the environment and conserve the natural resources of this planet. The whole philosophy of mass-producing consumer items has to change. Consideration must be given to easy repair to prolong the life of the product, as well as maximizing its reusable and recyclable components.

The new Chapter 55, Project assurance, reflects the need to monitor and report progress and performance on an ongoing basis to top management.

A number of chapters have been updated including Chapter 36 (Health, safety and environment). As I researched the latest publications on project management standards by the leading authorities on project management, I was surprised and appalled to find that only scant, if any, references have been made to safety. I have therefore listed in this chapter a list of project disasters caused by failures in safety systems or procedures to emphasize the importance of safety.

The new Appendix 2, Artificial Intelligence (AI) by Graham Collins, which also includes Big Data, has been added as this topic has now been incorporated into many new projects. The chapter also describes how Big Data was used in fighting the COVID-19 pandemic.

Appendix 3 is a case study of a Portland cement plant capable of an output of 1000 tons per day. In this study, I have broken with convention by writing it in the first person in order to convey some of the feelings that were aroused by some of the episodes during the construction period. It is hoped that the reader will find this interesting.

Appendix 4 is a case study which describes the construction and commissioning of phase 1 of the Teesside oil terminal which was at the time the largest petrochemical construction project in Europe.

Appendix 9 (a repeat of Appendix 7 of the 7th edition) is a syllabus summary of bullet points selected from the APM Body of Knowledge (BoK) 6th edition. Although APM has now published a 7th edition of the BoK, I have retained the syllabus summary based on its 6th edition because I consider it to be more appropriate and helpful for potential examinees.

Albert Lester

Acknowledgements

The author and publishers acknowledge with thanks all the individuals and organizations whose contributions were vital in the preparation of this book.

Particular acknowledgement is given to the following contributors:

Arnaud Morvan and Oracle/Milestone Ltd, for providing the description of their highly regarded Primavera P6 computer software package.

David Shannon of Oxford Project Management, for writing the chapter on project governance.

Clive Robinson and Trimble Solutions Corporation, for contributing the description and procedures for BIM.

Graham Collins from UCL, for providing the description of Agile Project Management and the new section on Artificial Intelligence (AI) and Big Data.

Dale A. Dutton from InEight Inc, for contributing the chapter on his company's VDC process package.

The author would also like to thank the following for their help and cooperation:

The Association for Project Management (APM) for permission to reproduce excerpts from their publications *A guide to Conducting Integrated Baseline Reviews, Directing Change, A guide to governance of project management* and *The APM Body of Knowledge*

CIMAX/RPC for their approval of the case study of the South Ferriby Cement Plant.

Conoco/Phillips for supplying the aerial photographs of the Teesside Terminal during the construction phase.

The National Economic Development Office for permission to reproduce the relevant section of their report 'Engineering Construction Performance Mechanical & Electrical Engineering Construction, EDC, NEDO December 1976'.

Foster Wheeler Power Products Limited for assistance in preparing the text and manuscripts and permission to utilize the network diagrams of some of their contracts

Mr Tony Benning, my co-author of *Procurement in the Process Industry*, for permission to include certain texts from that book.

British Standards Institution for permission to reproduce extracts from BS 6079-1-10 (Project management life cycle) and BS5499-10-2006 (Safety signs).

British Standards can be obtained in PDF or hard copy formats from the BSI online shop:

www.bsigroup.com/Shop or by contacting BSI Customer Services for hardcopies only: Tel: +44 (0)20 8996 9001, Email: cservices@bsigroup.com.

A. P. Watt for permission to quote the first verse of Rudyard Kipling's poem, *The Elephant's Child*.

Daimler Chrysler for permission to use their diagram of the Mercedes-Benz 190 car.

The Automobile Association for the diagram of a typical motor car engine.

Mrs Mary Willis for her agreement to use some of the diagrams in the chapters on Risk and Quality management.

Jane Walker and University College London for permission to include diagrams in the chapters on project context, leadership and negotiations.

Last, but by no means least, I must thank the Elsevier project editor, Chiara Giglio, for her help in getting the various manuscripts to a stage from which they could be sent for final collation and publication.

Project definition

1

Chapter outline

Project definition

Many people and organizations have defined what a project is, or should be, but probably the most authoritative definition is that given in BS 6079-2:2000 *Project Management Vocabulary*, which states that a project is:

> *A unique process, consisting of a set of co-ordinated and controlled activities with start and finish dates, undertaken to achieve an objective conforming to specific requirements, including constraints of time, cost and resources.*

The next question that can be asked is 'Why does one need project management?' What is the difference between project management and management of any other business or enterprise? Why has project management taken off so dramatically in the last 20 years?

The answer is that project management is essentially management of change, while running a functional or ongoing business is managing a continuum or 'business-as-usual'.

Project management is not applicable to running a factory making sausage pies, but it will be the right system when there is a requirement to relocate the factory, build an extension, or produce a different product requiring new machinery, skills, staff training, and even marketing techniques.

It is immediately apparent therefore that there is a fundamental difference between project management and functional or line management where the purpose of management is to continue the ongoing operation with as little disruption (or change) as possible. This is reflected in the characteristics of the two types of managers. While the project manager thrives on and is *proactive* to change, the line manager is *reactive* to change and hates disruption. In practice, this often creates friction and organizational problems when a change has to be introduced.

Project Management, Planning and Control. https://doi.org/10.1016/B978-0-12-824339-8.00001-8

Projects may be undertaken either to generate revenue, such as introducing methods for improving cash flow, or be capital projects that require additional expenditure and resources to introduce a change to the capital base of the organization. It is to this latter type of project that the techniques and methods described in this book can be most easily applied.

Fig. 1.1 shows the types of operations suitable for a project type of organization which are best managed as a functional or 'business-as-usual' organization.

Both types of operations have to be managed, but only the ones in column (A) require project-management skills.

It must be emphasized that the suitability of an operation being run as a project is independent of size. Project-management techniques are equally suitable for building a cathedral or a garden shed. Moving house, a very common project for many people, lends itself as effectively to project-management techniques such as tender analysis and network analysis as relocating a major government department from the capital city to another town. There just is no upper or lower limit to projects!

As stated in the definition, a project has a definite starting and finishing point and must meet certain specified objectives.

Broadly these objectives, which are usually defined as part of the business case and set out in the project brief, must meet three fundamental criteria:

1. The project must be completed on time.
2. The project must be accomplished within the budgeted cost.
3. The project must meet the prescribed quality requirements.

These criteria can be graphically represented by the well-known project triangle (Fig. 1.2). Some organizations like to substitute the word 'quality' with 'performance', but the principle is the same — the operational requirements of the project must be met, and met safely.

The order of priority given to any of these criteria is dependent not only on the industry but also on the individual project. For example, in designing and constructing an aircraft, motor car or railway carriage, safety must be paramount. The end product may cost more than budgeted or it may be late in going into service, and certain quality requirements in terms of comfort may have to be sacrificed, but under no circumstances can safety be compromised. Airplanes, cars and railways *must* be safe under all operating conditions.

(A) Project organisation	(B) Functional or line organisation
Building a house	Manufacturing bricks
Designing a car	Mass-producing cars
Organising a party	Serving the drinks
Setting up a filing system	Doing the filing
Setting up retail cash points	Selling goods & operating tills
Building a process plant	Producing sausages
Introducing a new computer system	Operating credit control procedures

Figure 1.1 Organization comparison.

Figure 1.2 Project triangle.

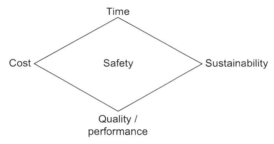

Figure 1.3 Project diamond.

The following (rather obvious) examples show where different priorities on the project triangle (or diamond) apply.

'However, because Sustainability can now be classified as the fourth criterion, the Project Management Triangle should now be replaced by the Project Management Rhomboid, where the negotiable criteria of Time, Cost, Quality/Performance and Sustainability are shown at the corners of the diagram while the non-negotiable Safety is still in the centre. See Fig. 1.3'. This is discussed more fully in Chapter 53, 'Sustainability'.

Time-bound project

A scoreboard for a prestigious tennis tournament must be finished in time for the opening match, even if it costs more than anticipated and the display of some secondary information, such as the speed of the service, has to be abandoned. In other words, cost and performance may have to be sacrificed to meet the unalterable starting date of the tournament. (In practice, the increased cost may well be a matter of further negotiation and the temporarily delayed display can usually be added later during the non-playing hours.)

Cost-bound project

A local authority housing development may have to curtail the number of housing units and may even overrun the original construction programme, but the project cost cannot be exceeded, because the housing grant allocated by the central government for this type of development was frozen at a fixed sum. Another solution to this problem would be to reduce the specification of the internal fittings instead of reducing the number of units.

Performance (quality)-bound project

An armaments manufacturer has been contracted to design and manufacture a new type of rocket launcher to meet the client's performance specification in terms of range, accuracy and rate of fire. Even if the delivery has to be delayed to carry out more tests and the cost has increased, the specification must be met. Again, if the weapons were required during a war, the specification might be relaxed to get the equipment into the field as quickly as possible.

Safety-bound project

Apart from the obvious examples of public transport given previously, safety is a factor that is required by law and enshrined in the Health and Safety at Work Act.

Not only must safe practices be built into every project, but constant monitoring is an essential element of a safety policy. To that extent, it could be argued that *all* projects are safety-bound, because, if it became evident after an accident that safety was sacrificed for speed or profitability, some or all of the project stakeholders could find themselves in real trouble, even in jail. This is true for almost every industry, especially agriculture, food/drink production and preparation, pharmaceuticals, chemicals, toy manufacture, aircraft production, motor vehicle manufacture and, of course, building and construction.

A serious accident that may kill or injure people will not only cause anguish among the relatives but, while not necessarily terminating the project, could very well destroy the company. For this reason, the 'S' symbol when shown in the middle of the project-management triangle gives more emphasis on its importance (see Fig. 1.2).

While the other three criteria (cost, time and quality/performance) can be juggled by the project manager to suit the changing requirements and environment of a project, safety cannot, under any circumstances, be compromised. As any project manager knows, the duration (time) may be reduced by increasing resources (cost), and cost may be saved by sacrificing quality or performance, but any diminution of safety can quickly lead to disaster, death and even the closure of an organization. The catastrophic explosions on the Piper Alpha gas platform in the North Sea in July 1988 killed 167 men and cost millions of dollars to Occidental and its insurers, and the explosion at the Buncefield, England, oil depot in 2009 caused massive destruction of its surroundings and huge costs to Total Oil Co. Additionally, the explosion on its Texas City refinery in March 2005, which killed 15 men and injured 170, and the blowout of the Deepwater Horizon drilling rig in the Gulf of Mexico in April 2010, causing 11 fatalities, have seriously damaged the reputation of BP and resulted in a considerable drop in its share price. In the transport industry, the series of railway accidents in 2000 resulted in the winding up of British Rail and subsequently one of its main contractors. More recently, Toyota had to recall millions of cars to rectify an unsafe breaking and control system, after which Mr Toyoda, the Chairman of the company, publicly stated that Toyota's first priority is safety, the second is quality, and the third is volume (quantity). These occurrences clearly show that safety must head the list of priorities for any project or organization.

The priorities of the other three criteria can of course change with the political climate or the commercial needs of the client, even within the life cycle of the project, and therefore the project manager has to constantly evaluate these changes to determine the new priorities. Ideally, all the main criteria should be met (and indeed this is the case for many well-run projects), but there are times when the project manager, with the agreement of the sponsor or client, has to make difficult decisions to satisfy the best interests of most, if not all, the stakeholders.

However, the examples given earlier highlight the importance of ensuring a safe operating environment, even at the expense of the other criteria. It is important to note that while a project manager can be reprimanded or dismissed for not meeting any of the three 'corner criteria', the one transgression for which a project manager can actually be jailed is not complying with the provisions of the Health and Safety regulations.

If one were to list the four project-management criteria in the order of their importance, the sequence would be safety, performance, time and cost, which can be remembered using the acronym SAPETICO. The rationale for this order is as follows:

If the project is not safe, it can cost lives and/or destroy the constructor and other stakeholders.

If the performance is not acceptable, the project will have been a waste of time and money.

If the project is not on time, it can still be a success but may have caused a financial loss.

Even if the cost exceeds the budget, the project can still be viable, as extra money can usually be found. The most famous (or infamous) example is the Sydney Opera House, which was so much over budget that the extra money had to be raised via a New South Wales State lottery but is now celebrated as a great Sydney landmark.

Project management

2

Chapter outline

It is obvious that project management is not new. Noah must have managed one of the earliest recorded projects in the Bible — the building of the ark. He may not have completed it within the budget, but he certainly had to finish it by a specified time — before the flood — and it must have met his performance criteria, as it successfully accommodated a pair of all the animals.

There are many published definitions of *project management* (see BS 6079 and ISO 21,500), but the following definition covers all the important ingredients:

> *The planning, monitoring, and control of all aspects of a project and the motivation of all those involved in it, in order to achieve the project objectives within agreed criteria of time, cost, and performance.*

Whilst this definition includes the fundamental criteria of time, cost and performance, the operative word, as far as the management aspect is concerned, is *motivation*. A project will not be successful unless all (or at least most) of the participants are not only competent but also motivated to produce a satisfactory outcome.

To achieve this, a number of methods, procedures and techniques have been developed, which, together with the general management and people skills, enable the project manager to meet the set criteria of time cost and performance/quality in the most effective ways.

Many textbooks divide the skills required in project management into *hard* skills (or topics) and *soft* skills. This division is not exact as some of the skills are clearly interdependent. Furthermore, it depends on the type of organization, type and size of project, authority given to a project manager and which of the listed topics are in his or her remit for a particular project. For example, in many large construction companies, the project manager is not permitted to get involved in industrial (site) disputes as these are more effectively resolved by specialist industrial relations managers who are conversant with the current labour laws, national or local labour agreements and site conditions.

Project Management, Planning and Control. https://doi.org/10.1016/B978-0-12-824339-8.00002-X

The hard skills cover such subjects as business case, cost control, change management, project life cycles, work breakdown structures, project organization, network analysis, earned value analysis, risk management, quality management, estimating, tender analysis and procurement.

The soft topics include health and safety, stakeholder analysis, team building, leadership, communications, information management, negotiation, conflict management, dispute resolutions, value management, configuration management, financial management, marketing and sales and law.

A quick inspection of the two types of topics shows that the hard subjects are largely required only for managing specific projects, while the soft ones can be classified as general management and are more or less necessary for any type of business operation whether running a design office, factory, retail outlet, financial services institution, charity, public service organization, national or local government or virtually any type of commercial undertaking.

A number of organizations, such as APM, PMI, ISO, OGC and licensees of PRINCE (project in a controlled environment), have recommended and advanced their own methodology for project management, but by and large the differences are on emphasis or sequence of certain topics. For example, PRINCE requires the resources to be determined before the commencement of the time scheduling and the establishment of the completion date, while in the construction industry the completion date or schedule is often stipulated by the customer and the contractor has to provide (or recruit) whatever resources (labour, plant, equipment or finance) are necessary to meet the specified objectives and complete the project on time.

Project manager

A *project manager* can be defined as follows:

> *The individual or body with authority, accountability and responsibility for managing a project to achieve specific objectives (BS 6079-2:2000).*

Few organizations will have problems with the earlier definition, but unfortunately in many instances, while the responsibility and accountability are vested in the project manager, the authority given to him or her is either severely restricted or non-existent. The reasons for this may be the reluctance of a department (usually one responsible for the accounts) to relinquish financial control or it is perceived that the project manager does not have sufficient experience to handle certain tasks, such as control of expenditure. There may indeed be good reasons for these restrictions which depend on the size and type of project, the size and type of the organization and of course the personality and experience of the project manager, but if the project manager is supposed to be in effect the managing director of the project (as one large construction organization liked to put it), he or she must have control over costs and expenditure, albeit within specified and agreed limits.

Apart from the conventional responsibilities for time, cost and performance/quality, the project manager must ensure that all the safety requirements and safety procedures are complied with. For this reason, the word *safety* has been inserted into the project management triangle to reflect the importance of ensuring that various important health and safety requirements are met. Serious accidents not only have personal tragic consequences, but they can also destroy a project or indeed a business overnight. Lack of attention to safety is just bad business, as any oil, airline, bus or railroad companies can confirm.

Project manager's charter

Because the terms of engagement of a project manager are sometimes difficult to define in a few words, some organizations issue a *project manager's charter*, which sets out the responsibilities and limits of authority of the project manager. This makes it clear to the project manager what his or her areas of accountability are, and if this document is included in the project management plan, all stakeholders will be fully aware of the role the project manager will have in this particular project.

The project manager's charter is project-specific and will have to be amended for every manager as well as the type, size, complexity or importance of a project (see Fig. 2.1).

Project office

The project manager needs to be supported on large projects, either by one or more assistant project managers (one of whom can act as deputy) or a specially created *project office*. The main duties of such a project office are to establish a uniform organizational approach for systems, processes and procedures, carry out the relevant configuration management functions, disseminate project instructions and other information and collect, retrieve or chase information required by the project manager on a regular or ad hoc basis. Such an office can assist greatly in the seamless integration of all the project systems and would also prepare programmes, schedules, progress reports, cost analyses, quality reports and a host of other useful tasks that would otherwise have to be carried out by the project manager. In addition, the project office can also be required to service the requirements of a programme or portfolio manager, in which case it will probably have its own office manager responsible for the onerous task of satisfying the different and often conflicting priorities set by the various projects managers (see also Chapter 10).

PROJECT MANAGER'S CHARTER

1. **Project Manager:**

   Name: _____

   Appointment/Position: _____

   Date of Appointment: _____

2. **Project Title:** _____

3. **Responsibility and Authority given to the Project Manager:**

   The above named Project Manager has been given the authority, responsibility and accountability for_____

4. **Project Goals and Deliverables are:**

   a: _____

   b: _____

   c: _____

5. **The Project will be reviewed.**

   _____

6. **Financial Authority:**

   The Project Manager's delegated financial powers are: _____

   _____

7. **Intramural Resources:**

   The following resources have been/are to be made available:

8. **Trade-offs:**

   a: Cost: _____ %

   b: Time: _____ days/weeks.

   c: Performance: _____

9. **Charter Review: No charter review is expected to take place for the duration of this project unless it becomes clear that the PM cannot fulfil his/her duties or a reassessment of the trade-offs is required.**

10. **Approved:**

    Sponsor/Client/Customer/Programme Manager: _____

    Project Manager: _____

    Line Manager: _____

11. **Distribution:**

    a: Sponsor; b: Programme Manager; c: Line Manager

Figure 2.1 Project manager's charter.

Further reading

Burke, R. (2011). *Advanced project management*. Burke Publishing.

Cleland, D. I. (2006). *Global project management handbook*. McGraw-Hill.

Gordon, J., & Lockyer, K. (2005). *Project management & project network techniques* (7th ed.). Prentice Hall.

Heldman, K. Project management jump start. Syber.

Kerzner, H. (2009). *Project management: A managerial approach*. Wiley.

O'Connell, F. (2010). *What you need to know about project management*. Wiley.

Rad, P. F., & Levin, G. (2002). *The advanced project management office*. St Lucie Press.

Taylor, P. (2011). *Leading successful PMOs*. Gower.

Turner, J. R. (2008). *The Gower handbook of project management* (4th ed.). Gower.

Programme and portfolio management

Chapter outline

Programme management can be defined as 'The coordinated management of a group of related projects to ensure the best use of resources in delivering the projects to the specified time, cost and quality/performance criteria'.

A number of organizations and authorities have coined different definitions, but the operative word in any definition is *related*. Unless the various projects are related to a common objective, the collection of projects would be termed a 'portfolio' rather than a 'programme'.

A programme manager could therefore be defined as '*The individual to whom responsibility has been assigned for the overall management of the time, cost and performance aspects of a group of related projects and the motivation of those involved*'.

Again, different organizations have different definitions for the role of the programme manager or portfolio manager. In some companies, he or she would be called manager of projects or operations manager or operations director, etc., but it is generally understood that the programme manager's role is to coordinate the individual projects that are linked to a common objective. Whatever the definition, it is the programme manager who has the overall picture of the organization's project commitments.

Many organizations carrying out a number of projects have limited resources. It is the responsibility of the programme manager to allocate these resources in the most cost-effective manner, taking into consideration the various project milestones and deadlines as well as the usual cost restrictions. It is the programme manager who may have to obtain further authority to engage any external resources as necessary and decide on their disposition.

As an example, the construction of a large cruise ship would be run by a programme manager who coordinates many (often very large) projects such as the ship's hull, propulsion system and engines, control systems, catering system and interior design. One of the associated projects might even include recruitment and training of the crew.

A manager responsible for diverse projects such as the design, supply and installation of a computerized supermarket check-out and stock-control system, an electronic scoreboard for a cricket ground, or a cheque-handling system for a bank would be a portfolio manager, because although all the projects require computer systems, they are for different clients at different locations and are independent of each other. Despite

Project Management, Planning and Control. https://doi.org/10.1016/B978-0-12-824339-8.00003-1

this diversity of the projects, the portfolio manager, like the project manager, still has the responsibility to set priorities, maximize the efficient use of the organization's resources and monitor and control the costs, schedule and performance of each project.

As with project management, programme management and the way programmes are managed depend primarily on the type of organization carrying out the programme. There are two main types of organizations:

- Client organizations
- Contracting organizations

In a client-type organization, the projects or programmes will probably not be the main source of income and may well constitute or require a major change in the management structure and culture. New resources may have to be found and managers involved in the normal running of the business may have to be consulted, educated and finally convinced of the virtues, not only of the project itself but also of the ways it has to be managed.

The programme manager in such an organization has to ensure that the project fits into the corporate strategy and meets the organization's objectives. He or she has to ensure that established project management procedures, starting with the business case through implementation and ending with disposal, are correctly employed. In other words, the full life-cycle systems using all the 'soft' techniques to create a project environment have to be in place in an organization that may well be set up for 'business-as-usual', employing only well-established line-management techniques. In addition, the programme manager has to monitor all projects to ensure that they meet the strategic objectives of the organization as well as fulfilling the more obvious requirements of being performed safely, minimizing and controlling risks at the same time meeting the cost, time and performance criteria for every project.

Programme management can, however, mean more than coordinating a number of related projects. The prioritization of the projects themselves, not just the required resources, can be a function of programme management. It is the programme manager who decides which project, or which type of project, is the best investment and which one is the most cost-effective one to start. It may even be advantageous to merge two or more small projects into one larger project, if they have sufficient synergy or if certain resources or facilities can be shared.

Another function of programme management is to monitor the performance of the projects that are part of the programme and check that the expected deliverables have produced the specified benefits, whether to the parent organization or the client. This could take several days or months depending on the project, but unless it is possible to measure these benefits, it is not possible to assess the success of the project or, indeed say, whether the whole exercise is worthwhile. It can be seen therefore that it is just as important for the programme manager to set up the monitoring and close-out reporting system for the end of a project, as the planning and control systems for the start.

In a contracting organization, such a culture change will either not be necessary, as the organization will already be set up on a project basis, or the change to a project-oriented company will be easier because the delivery of projects is after all the 'raison d'être' of the organization. Programme management in a contracting organization is therefore more of the coordination of the related or overlapping projects covering such topics as resource management, cost management and procurement, and ensuring conformity with standard company systems and procedures. The cost, time and performance/quality criteria therefore relate more to the obligations of the contractor (apart from performance) than those of the client.

The life cycles of projects in a contracting organization usually start after the feasibility study has been carried out and finishes when the project is handed over to the client for the operational phase. There are clearly instances when these life-cycle terminal points occur earlier or later, but a contractor is rarely concerned with whether or not the strategic or business objectives of the client have been met.

Portfolio management

The APM Body of Knowledge defines portfolio management as follows:

The selection and management of all of an organisations projects, programmes and related operational activities taking into account resource restraints.

Portfolio management, which can be regarded as a subset of corporate management, is very similar to programme management, but the projects in the programme manager's portfolio, though not necessarily related, are still required to meet an organization's objectives. Furthermore, portfolios (unlike projects or programmes) do not necessarily have a defined start and finish date. Indeed portfolios can be regarded as a rolling set of programmes monitored in a continuous life cycle from the strategic planning stage to the delivery of the programme. In a large organization, a portfolio manager may be in charge of several programme managers, whilst in a smaller company he or she may be in direct control of a number of project managers.

Companies do not have unlimited resources, so the portfolio manager has to prioritize the deployment of these resources for competing projects, each of which has to be assessed in terms of the following:

1. Profitability and cost/benefit
2. Return on investment
3. Cash flow
4. Risks
5. Prestige
6. Importance of the client
7. Company strategy and objectives

Portfolio management therefore involves the identification of these project attributes and the subsequent analysis, prioritization, balancing, monitoring and reporting of progress of each project or in the case of large organizations, each programme. As each project develops, different pressures and resource requirements appear, often as a result of contractual changes or the need to rectify errors or omissions. Unforeseen environmental issues may require immediate remedial action to comply with health and safety requirements, and there is always the danger of unexpected resignations of key members of one of the project teams.

A portfolio manager must therefore possess the ability to reassign resources, both human and material (such as office equipment, construction plant and bulk materials), in an effective and economical manner, often in emergency or other stressful situations, always taking into account the cost/benefit calculations, the performance and sustainability criteria and the overall strategic objectives of the organization.

The difference between programme management and portfolio management is that in the former the projects being managed are related in some form, while in the latter, the projects may or may not be related. For example, the projects controlled by a portfolio manager may be as diverse as an update of the company's IT system to the development of a commercial building or shopping centre.

The portfolio manager will normally be part of the senior management team which determines which projects go ahead and which should be shelved, not started or even abandoned.

Clearly, the degree of detailed involvement in the individual projects by the portfolio manager must therefore be limited, as no one can be an expert in everything. Instead, the portfolio manager has to ensure that the projects under his control meet the corporate ethical and quality standards as well as the basic criteria of cost, time, performance and the last, but not the least, safety.

As with programme management, the order of priority of the various projects must be established at an early stage, but as circumstances change (often outside the control of the manager or even the organization) the priorities will have to be adjusted to suit the latest overall strategy or the resources (often financial) of the organization.

Further reading

APM. (2007). *APM introduction to programme management*. APM.
Bartlett, J. (2010). *Managing programmes of business change*. Project Manager Today.
OGC. (2010). *An executive guide to portfolio management*. The Stationary Office.
PMI. (2008). *The standard for portfolio management*. PMI.
Reiss, G. (2006). *The Gower handbook of programme management*. Gower.
Sanwal, A. (2007). *Optimising corporate portfolio management*. Wiley.
Thiry, M. (2010). *Programme management*. Gower.
Venning, C. (2007). *Managing portfolios of change with MSP for programmes and prince for projects*. The Stationary Office.

Project context (project environment)

Chapter outline

Projects are influenced by a multitude of factors which can be external or internal to the organization responsible for its management and execution. The important thing for the project manager is to recognize what these factors are and how they impact the project during various phases from inception to final handover, or even disposal.

These external or internal influences are known as the *project context* or *project environment*. The external factors making up this environment are the client or customer, various external consultants, contractors, suppliers, competitors, politicians, national and local government agencies, public utilities, pressure groups, the end users and even the general public. Internal influences include the organization's management, the project team, internal departments (technical and financial) and possibly the shareholders.

Fig. 4.1 illustrates the project surrounded by its external environment.

All these influences are neatly encapsulated by the acronym PESTLE, which stands for the following:

- Political
- Economic
- Social
- Technical
- Legal
- Environmental

A detailed discussion of these areas of influence is given in the following.

Political

Two types of politics have to be considered here.

Project Management, Planning and Control. https://doi.org/10.1016/B978-0-12-824339-8.00004-3

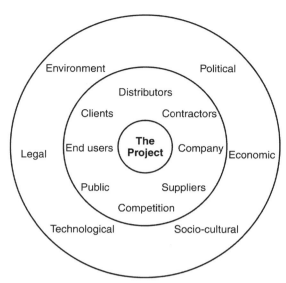

Figure 4.1 The project environment.

First, there are the internal politics that inevitably occur in all organizations whether governmental, commercial, industrial or academic, and which manifest itself in the opinions and attitudes of the different stakeholders in these organizations. These stakeholder's relationships to the project can vary from the very supportive to the downright antagonistic, but depending on their field of influence, they must be considered and managed. Even within an apparently cohesive project, team jealousies and personal vested interests can have a disruptive influence that the project manager has to recognize and diffuse.

The fact that a project relies on clients, consultants, contractors (with their numerous subcontractors), material and service suppliers, statutory authorities and, of course, the end user, all of which may have their own agenda and preferences, gives some idea of the potential political problems that may occur.

Second, there are the external politics, over which neither the sponsor nor the project manager may have much, if any, control. Any project that has international ramifications is potentially subject to disruption due to the national or international political situation. In the middle of a project, the government may change and impose additional import, export or exchange restrictions, impose penal working conditions or even cancel the contracts altogether. For overseas construction contracts in countries with inherently unstable economies or governments, sudden coups or revolutions may require the whole construction team to be evacuated at short notice. Such a situation should be envisaged, evaluated and planned for as a part of the political risk assessment when the project is first considered.

Even on a less dramatic level, the political interplay between national and local government, lobbyists and pressure groups has to be taken into consideration, as can be appreciated when the project consists of a road bypass, reservoir, power station or airport extension.

Economic

Here again, there are two levels of influence: internal or microeconomic, and external or macroeconomic.

The internal economics relates to the viability of the project and the soundness of the business case. Unless there is a net gain, whether financial or non-financial, such as required by prestige, environmental, social service or national security considerations, there is no point in even considering embarking on a project. It is therefore vital that financial models and proven accountancy techniques are applied during the evaluation phase to ensure the economic viability of the project. These tests must be applied at regular intervals throughout the life of a project to check that with the inevitable changes that may be required, it is still worthwhile to proceed. The decision to abort the whole project at any stage after the design stage is clearly not taken lightly, but once the economic argument has been lost, it may be a better option in the end. A typical example is the case of an oil-fired power station that had to be mothballed over halfway through construction, when the price of fuel oil rose above the level at which power generation was no longer economic. It is not uncommon for projects to be shelved when the cost of financing the work has to be increased and the resulting interest payments exceed the foreseeable revenues.

The external economics, often related to the political climate, can have a serious influence on the project. Higher interest or exchange rates, and additional taxes on labour, materials or the end product, can seriously affect the viability of the project. A manufacturer may abandon the construction of a factory in its home country and transfer the project abroad if just one of these factors changes enough to make such a move economically viable. Again, changes to fiscal and interest movements must be constantly monitored so that representations can be made to government or the project be curtailed. Other factors that can affect a project are tariff barriers, interstate taxes, temporary embargoes, shipping restrictions such as only being permitted to use Conference Line vessels and special licences.

Social (or sociological)

Many projects, and indeed most of the construction projects, inevitably affect the community of the areas they are carried out in. It is therefore vital to inform the residents in the affected areas as early as possible of the intent, purpose and benefits of the project to the organization and community.

This may require a public relations campaign to be initiated, which includes meetings, exploratory discussions, consultations at various levels and possible trade-offs. This is particularly important when public funding from central or local government is involved or when public spaces and access facilities are affected. A typical example of a trade-off is when a developer wishes to build a shopping centre, the local authority may demand that it includes a recreation area or leisure park for free use by the public.

Some projects cannot even be started without first being subjected to a public enquiry, environmental impact assessment, route surveys or lengthy planning procedures. There are always pressure groups that have special interests in a particular project, and it is vital that they are given the opportunity to state their case while at the same time informing them of the positive and often less desirable implications. The ability to listen to their point of views and give sympathetic attention to their grievances is essential, but as it is almost impossible to satisfy all the parties, compromises may be necessary. The last things a project manager wants are the constant demonstrations and disruptions while the project is being carried out.

On another level, the whole object of the project may be to enhance the environment and facilities of the community, in which case the involvement of local organizations can be very helpful in focussing on areas which give the maximum benefit, and avoiding pitfalls which only people with local knowledge are aware of. A useful method to ensure local involvement is to set up advisory committees or even invite a local representative to be a part of the project management team.

Technical

It goes without saying that, unless the project is technically sound, it will end in failure. Whether the project involves rolling out a new financial service product or building a power station, the technology must be in place or be developed as the work proceeds. The mechanisms by which these technical requirements are implemented have to be firmed up at a very early stage after a rigorous risk assessment of all the realistically available options. Each option may then be subjected to a separate feasibility study and investment appraisal. Alternatives to be considered may include the following:

- Should in-house or external design, manufacture or installation be used?
- Should existing facilities be used or should new ones be acquired?
- Should one's own management team be used or should specialist project managers be appointed?
- Should existing components (or documents) be incorporated?
- What is the anticipated life of the end product (deliverable) and how soon must it be updated?
- Are materials available on a long-term basis and what alternatives can be substituted?
- What is the nature and size of the market and can this market be expanded?

These and many more technical questions have to be asked and assessed before a decision can be made to proceed with the project. The financial implications of these factors can then be fed into the overall investment appraisal, which includes the commercial and financing, and environmental considerations.

Legal

One of the fundamental requirements of a contract, and by implication a project, is that it should be legal. In other words, if it is illegal in a certain country to build a brewery, little protection can be expected from the law.

The relationships between the contracting parties must be confirmed by a legally binding contract that complies with the laws (and preferably customs) of the participating organizations. The documents themselves have to be legally acceptable and equitable, and unfair and unreasonable clauses must be eliminated.

Where suppliers of materials, equipment or services are based in countries other than the main contracting parties, the laws of those countries have to be complied with in order to minimize future problems regarding deliveries and payments.

In the event of disputes, the law under which the contract is administered and adjudicated must be written into the contract together with the location of the court for litigation.

Generally, project managers are strongly advised always to take legal advice from specialists in contract law and especially, where applicable, in international law.

The project context includes the established conditions of contract and other standard forms and documents used by industry, and can also include all the legal, political and commercial requirements stipulated by international bodies as well as national and local governments in their project management procedures and procurement practices.

Environmental

Some of the environmental aspects of a project have already been alluded to under 'Social', from which it became apparent that environmental-impact assessments are highly desirable where they are not already mandatory.

The location of the project clearly has an enormous influence on the cost and completion time. The same type of plant or factory can be constructed in the United Kingdom, in the Sahara desert, in China, or even on an offshore platform, but the problems, costs and construction times can be very different. The following considerations must therefore be taken into account when deciding to carry out a project in a particular area of the world:

- Temperature (daytime and night time) in different seasons.
- Rainy seasons (monsoon).
- Tornado or typhoon seasons.
- Access by road, rail, water or air.
- Ground conditions and earthquake zones.
- Possible ground contamination.
- Nearby rivers and lakes.
- Is the project onshore or offshore?
- Tides and storm conditions.
- Nearby quarries for raw materials.
- Does the project involve the use of radioactive materials?

Most countries now have strict legislation to prevent or restrict emissions of polluting substances whether solids, liquids or gases. In addition, noise restrictions may apply at various times and cultural or religious laws may prohibit work at specified times or on special days in the year.

The following list is a very small sample of over 15,000 web pages covering European Economic Community (EEC) directives and gives some idea of the regulations that may have to be followed when carrying out a project.

EC	85/337/EEC	Environmental impact assessment
Directive	97/11/EEC	Assessment of effects on certain public and private projects
	92/43/EEC	Chapter 4 Environment
	86/278/EEC	Protection of the environment
	90/313/EEC	Sustainable development
	90/679/EEC	Substances hazardous to health
	79/409/EEC	Conservation of natural habitats
	96/82/EEC	Control of major accident hazards
	91/156/EEC	Control of pollution
	87/217/EEC	Air pollution
	89/427/EEC	Air pollution
	80/779/EEC	Air quality limit values
	75/442/EEC	Ozone-depleting substances
	89/427/EEC	Quality limit of sulphur dioxide
	80/1268/EEC	Fuel and CO_2 emissions
	91/698/EEC	Hazardous waste
	78/659/EEC	Quality standard of water
	80/68/EEC	Groundwater directive
	80/778/EEC	Spring waters
	89/336/EEC	Noise emissions

Further reading

Longdin, I. (2009). *Legal aspects of purchasing and supply chain management* (3rd ed.). Liverpool Academic.
Wright. (2004). *The law for project managers*. Gower.

Business case

Chapter outline

Before embarking on a project, it is clearly necessary to understand that there will be a benefit either in terms of money or service, or both. The document that sets out the main advantages and parameters of the project is called the *business case* and is (or should be) produced by either the client or the sponsor of the project who in effect becomes the owner of the document.

A business case in effect outlines the 'why' and 'what' of the project as well as making the financial case by including the investment appraisal.

As with all documents, a clear procedure for developing the business case is highly desirable, and the following headings provide some indication of the subjects to be included:

1. Why is the project required?
2. What are we trying to achieve?
3. What are the deliverables?
4. What is the anticipated cost?
5. How long will it take to complete?
6. What quality standards must be achieved?
7. What are the performance criteria?
8. What are key performance indicators (KPIs)?
9. What are the main risks?
10. What are the success criteria?
11. Who are the main stakeholders?

In addition, any known information, such as location, key personnel, and resource requirements, should be included so that the recipients, usually a board of directors, are in a position to accept or reject the case for carrying out the project.

The project sponsor

It is clear that the business case has to be prepared before the project can be started. Indeed, the business case is the first document to be submitted to the directorate of an organization to enable this body to discuss the purpose and virtues of the project

Project Management, Planning and Control. https://doi.org/10.1016/B978-0-12-824339-8.00005-5

before making any financial commitments. Therefore, it follows that the person responsible for producing the business case cannot normally be the project manager, but must be someone who has a direct interest in the project going ahead. This person, who is often a director of the client's organization with a special brief to oversee the project, is the *project sponsor* (sometimes also known as the project champion).

The role of the project sponsor is far greater than being the initiator or champion of the project. Even after the project has started, the sponsor's role is to:

1. Monitor the performance of the project manager
2. Constantly ensure that the project's objectives and main criteria are met
3. Ensure that the project is run effectively as well as efficiently
4. Assess the need and viability of variations and agree to their implementation
5. Assist in smoothing out difficulties with other stakeholders
6. Support the project by ensuring sufficient resources (especially financial) are available
7. Act as business leader and top-level advocate to the company board
8. Ensure that the perceived benefits of the project are realized

Depending on the value, size and complexity of the project, the sponsor is a key player who, as a leader and mentor, can greatly assist the project manager to meet all the project's objectives and KPIs.

Requirements management

As has been explained previously, the two main components of a business case are 'what' is required and 'why' it is required. *Requirements management* is concerned with the 'what'.

Clients, end users and indeed most of the stakeholders have their own requirements on what they expect from the project, even if the main objectives were agreed upon. Requirements management is concerned with eliciting, capturing, collating, assessing, analysing, testing, prioritizing, organizing and documenting of all these different requirements. Many of these may of course be the common needs of a number of stakeholders and will therefore be high on the priority list, but it is the project manager who is responsible for deciding on the viability or desirability of a particular requirement and to agree with the stakeholder on whether it should or could be incorporated, taking into account the cost, time and performance factors associated with the requirement. Once agreed, these requirements become the benchmark against which the success of the project is measured.

Ideally, all the requirements should be incorporated as clear deliverables in the objectives enshrined in the business case and confirmed by the project manager in the project-management plan. However, it is always possible that one or more stakeholders may wish to change these requirements either just before or even after the project scope has been agreed and finalized. The effect of such a change of requirement will have to be carefully examined by the project manager, who must take into account any cost implication, effects on the project programme, changes to the procedures and processes needed to incorporate the new requirement and the environmental impact in its widest sense.

In such a situation, the project manager must immediately advise all the relevant stakeholders of the additional cost, time and performance implications and obtain their approval before amending the objectives, scope and cost of the project.

If the change of requirements is requested after the official start of the project, that is, after the cost and time criteria have been agreed, the new requirements will be subject to the normal project (or contract)-change procedure and configuration management described elsewhere.

To log and control the requirement documents during the life of the project, a simple 'reporting matrix', as shown in the following, will be helpful.

No.	Document (requirement)	Prepared by	Information from	Sent or copied to	Issued date

Testing and periodic reviews of the various requirements will establish their viability and ultimate effect on the outcome of the project. The following are some of the major characteristics that should be examined as part of this testing process:

- Feasibility, operability and time constraints
- Functionality, performance and quality requirements and reliability
- Compliance with health and safety regulations and local by-laws
- Buildability, delivery (transportability), storage and security
- Environmental and sociological impact
- Labour, staffing, outsourcing and training requirements

There may be occasions where the project manager is approached by a stakeholder, or even the client, to incorporate a 'minor' requirement 'as a favour'. The dangers of agreeing to such a request without following the normal change-management procedures are self-apparent. A small request can soon escalate into a large change once all the ramifications and spin-off effects have become apparent which leads to the all too common 'scope creep'. All changes to requirements, however small, must be treated as official and handled accordingly. It may of course be politically expedient not to charge a client for any additional requirement, but this is a commercial decision taken by senior management for reasons of creating goodwill, obtaining possible future contracts or succumbing to political pressure.

Further reading

Gambles, I. (2009). *Making the business case*. Gower.
West, D. (2010). *Project sponsorship*. Gower.

Investment appraisal

Chapter outline

The investment appraisal, which is a part of the business case, will, if properly structured, improve the decision-making process regarding the desirability or viability of the project. It should examine all the realistic options before making a firm recommendation for the proposed case. The investment appraisal must also include a cost/benefit analysis and take into account all the relevant factors such as the following:

- Capital costs, operating costs and overhead costs
- Support and training costs
- Dismantling and disposal costs
- Expected residual value (if any)
- Any cost savings that the project will bring
- Any benefits that cannot be expressed in monetary terms

To enable the comparison to be made of some of the options, the payback, return on capital, net present value (NPV) and anticipated profit must be calculated. In other words, the project viability must be established.

Project viability

Return on investment

The simplest way to ascertain whether the investment in a project is viable is to calculate the *return on investment* (ROI).

Project Management, Planning and Control. https://doi.org/10.1016/B978-0-12-824339-8.00006-7

If a project investment is £10,000 and gives a return of £2000 per year over 7 years,

$$\text{The average return/year} = \frac{(7 \times £2000) - £10,000}{7}$$

$$= \frac{£4000}{7} = £571.4$$

The ROI, usually given as a percentage, is the average return over the period considered × 100, divided by the original investment, that is,

$$\text{Return on investment} = \frac{\text{average return} \times 100}{\text{investment}}$$

$$= \frac{£571.4 \times 100}{£10,000} = £5.71$$

This calculation does not, however, take into account the cash flow of the investment, which in a real situation may vary year by year.

Net present value

As the value of money varies with time due to the interest it could earn if invested in a bank or other institution, the actual cash flow must be taken into account to obtain a realistic measure of the profitability of the investment.

If £100 were invested in a bank earning an interest of 6%:

The value in 1 year would be £100 × 1.05 = £106.

The value in 2 years would be £100 × 1.06 × 1.06 = £112.36.

The value in 3 years would be £100 × 1.06 × 1.06 × 1.06 = £119.10.

It can be seen therefore that, today, to obtain £119.10 in 3 years it would cost £100. In other words, the present value (PV) of £119.10 is £100.

Another way of finding the PV of £119.10 is to divide it by 1.06 × 1.06 × 1.06 or 1.191 or

$$\frac{£119.10}{1.06 \times 1.06 \times 1.06} = \frac{£119.10}{1.191} = £100.$$

If instead of dividing the £119.10 by 1.191, it is multiplied by the inverse of 1.191, one obtains the same answer, since

$$£\frac{119.10 \times 1}{1.191} = £119.10 \times 0.840 = £100$$

The 0.840 is called the *discount factor* or present-value factor and can be quickly found from discount factor tables, a sample of which is given in Fig. 6.1.

Discount Rate

Period	1%	2%	3%	4%	5%	6%	7%	8%	9%	10%	11%	12%	13%	14%	15%	Period
1	0.9901	0.9804	0.9709	0.9615	0.9524	0.9434	0.9346	0.9259	0.9174	0.9091	0.9009	0.8929	0.8850	0.8772	0.8696	1
2	0.9803	0.9612	0.9426	0.9246	0.9070	0.8900	0.8734	0.8573	0.8417	0.8264	0.8116	0.7972	0.7831	0.7695	0.7561	2
3	0.9705	0.9423	0.9151	0.8890	0.8638	0.8396	0.8163	0.7938	0.7722	0.7513	0.7312	0.7118	0.6931	0.6750	0.6575	3
4	0.9610	0.9238	0.8885	0.8548	0.8227	0.7921	0.7629	0.7350	0.7084	0.6830	0.6587	0.6355	0.6133	0.5921	0.5718	4
5	0.9515	0.9057	0.8626	0.8219	0.7835	0.7473	0.7130	0.6806	0.6499	0.6209	0.5935	0.5674	0.5428	0.5194	0.4972	5
6	0.9420	0.8880	0.8375	0.7903	0.7462	0.7050	0.6663	0.6302	0.5963	0.5645	0.5346	0.5066	0.4803	0.4556	0.4323	6
7	0.9327	0.8706	0.8131	0.7599	0.7107	0.6651	0.6227	0.5835	0.5470	0.5132	0.4817	0.4523	0.4251	0.3996	0.3759	7
8	0.9235	0.8535	0.7894	0.7307	0.6768	0.6274	0.5820	0.5403	0.5019	0.4665	0.4339	0.4039	0.3762	0.3506	0.3269	8
9	0.9143	0.8368	0.7664	0.7026	0.6446	0.5919	0.5439	0.5002	0.4604	0.4241	0.3909	0.3606	0.3329	0.3075	0.2843	9
10	0.9053	0.8203	0.7441	0.6756	0.6139	0.5584	0.5083	0.4632	0.4224	0.3855	0.3522	0.3220	0.2946	0.2697	0.2472	10
11	0.8963	0.8043	0.7224	0.6496	0.5847	0.5268	0.4751	0.4289	0.3875	0.3505	0.3173	0.2875	0.2607	0.2366	0.2149	11
12	0.8874	0.7885	0.7014	0.6246	0.5568	0.4970	0.4440	0.3971	0.3555	0.3186	0.2858	0.2567	0.2307	0.2076	0.1869	12
13	0.8787	0.7730	0.6810	0.6006	0.5303	0.4688	0.4150	0.3677	0.3262	0.2897	0.2575	0.2292	0.2042	0.1821	0.1625	13
14	0.8700	0.7579	0.6611	0.5775	0.5051	0.4423	0.3878	0.3405	0.2992	0.2633	0.2320	0.2046	0.1807	0.1597	0.1413	14
15	0.8613	0.7430	0.6419	0.5553	0.4810	0.4173	0.3624	0.3152	0.2745	0.2394	0.2090	0.1827	0.1599	0.1401	0.1229	15
16	0.8528	0.7284	0.6232	0.5339	0.4581	0.3936	0.3387	0.2919	0.2519	0.2176	0.1883	0.1631	0.1415	0.1229	0.1069	16
17	0.8444	0.7142	0.6050	0.5134	0.4363	0.3714	0.3166	0.2703	0.2311	0.1978	0.1696	0.1456	0.1252	0.1078	0.0929	17
18	0.8360	0.7002	0.5874	0.4936	0.4155	0.3503	0.2959	0.2502	0.2120	0.1799	0.1528	0.1300	0.1106	0.0946	0.0808	18
19	0.8277	0.6964	0.5703	0.4746	0.3957	0.3305	0.2765	0.2317	0.1945	0.1635	0.1377	0.1161	0.0981	0.0829	0.0703	19
20	0.8195	0.6730	0.5537	0.4564	0.3769	0.3118	0.2584	0.2145	0.1784	0.1486	0.1240	0.1037	0.0868	0.0728	0.0611	20
21	0.8114	0.6598	0.5375	0.4388	0.3589	0.2942	0.2415	0.1987	0.1637	0.1351	0.1117	0.0926	0.0768	0.0638	0.0531	21
22	0.8034	0.6468	0.5219	0.4220	0.3418	0.2775	0.2257	0.1839	0.1502	0.1228	0.1007	0.0826	0.0680	0.0560	0.0462	22
23	0.7954	0.6342	0.5087	0.4057	0.3256	0.2618	0.2109	0.1703	0.1378	0.1117	0.0907	0.0738	0.0601	0.0491	0.0402	23
24	0.7876	0.6217	0.4919	0.3901	0.3101	0.2470	0.1971	0.1577	0.1264	0.1015	0.0817	0.0659	0.0532	0.0431	0.0349	24
25	0.7798	0.6095	0.4776	0.3751	0.2953	0.2330	0.1842	0.1460	0.1160	0.0923	0.0736	0.0588	0.0471	0.0378	0.0304	25
26	0.7720	0.5976	0.4637	0.3607	0.2812	0.2198	0.1722	0.1352	0.1064	0.0839	0.0663	0.0525	0.0417	0.0331	0.0264	26
27	0.7644	0.5859	0.4502	0.3468	0.2678	0.2074	0.1609	0.1252	0.0976	0.0763	0.0597	0.0469	0.0369	0.0291	0.0230	27
28	0.7568	0.5744	0.4371	0.3336	0.2561	0.1956	0.1504	0.1159	0.0896	0.0693	0.0538	0.0419	0.0326	0.0255	0.0200	28
29	0.7493	0.5631	0.4243	0.3207	0.2429	0.1846	0.1406	0.1073	0.0822	0.0630	0.0485	0.0374	0.0289	0.0224	0.0174	29
30	0.7419	0.5521	0.4120	0.3083	0.2314	0.1741	0.1314	0.0994	0.0754	0.0573	0.0437	0.0334	0.0258	0.0196	0.0151	30

Figure 6.1 Discount factors.

It can be noticed from these tables that 0.840 is the PV factor for a 6% return after 3 years. The PV factor for a 6% return after 2 years is 0.890 or

$$\frac{1}{1.06 \times 1.06} = \frac{1}{1.1236} = 0.890$$

In the earlier example, the income was the same every year. In most of the projects, however, the projected annual net cash flow (income minus expenditure) will vary year by year, and to obtain a realistic assessment of the NPV of an investment, the net cash flow must be discounted separately for every year of the projected life.

The following example will make this clear.

Year	Income (£)	Discount rate (%)	Discount factor	NPV (£)
1	10,000	5	$1/1.05 = 0.9523$	$10,000 \times 0.9523 = 9523.8$
2	11,000	5	$1/1.05^2 = 0.9070$	$10,000 \times 0.9070 = 9070.3$
3	12,000	5	$1/1.05^3 = 0.8638$	$12,000 \times 0.8638 = 10,365.6$
4	12,000	5	$1/1.05^4 = 0.8227$	$12,000 \times 0.8227 = 9872.4$
Total	45,000			39,739.1

One of the main reasons for finding the NPV is to be able to compare the viability of competing projects or different repayment modes. Again, an example will demonstrate the point.

A company decides to invest £12,000 for a project which is expected to give a total return of £24,000 over 6 years. The discount rate is 8%.

There are two options of receiving the yearly income.

1. £6000 for years 1 and 2 = £12,000		2. £5000 for years 1, 2, 3 and 4 = £20,000	
£4000 for years 2 and 3 = £8000		£2000 for years 5 and 6 = £4000	
£2000 for years 5 and 6 = £4000			
Total	£24,000		£24,000

The DCF method will quickly establish the most profitable option to take as will be shown in the following table.

Year	Discount factor	Cash flow A (£)	NPV A (£)	Cash flow B (£)	NPV B (£)
1	$1/1.08 = 0.9259$	6000	5555.40	5000	4629.50
2	$1/1.08^2 = 0.8573$	6000	5143.80	5000	4286.50
3	$1/1.08^3 = 0.7938$	4000	3175.20	5000	3969.00
4	$1/1.08^4 = 0.7350$	4000	2940.00	5000	3675.00
5	$1/1.08^5 = 0.6806$	2000	1361.20	2000	1361.20
6	$1/1.08^6 = 0.6302$	2000	1260.40	2000	1260.40
Total		24,000	19,437.00	24,000	19,181.50

Clearly, A gives the better return, and after deducting the original investment of £12,000, the net discounted return for A = £7437.00 and for B = £7181.50.

The mathematical formula for calculating the NPV is as follows:

If NPV	=	Net present value
R	=	The interest rate
N	=	Number of years the project yields a return
B1, B2, B3, etc.	=	The annual net benefits for years 1, 2 and 3, etc.
NPV for year 1	=	$B1/(1 + r)$
For year 2	=	$B1/(1 + r) + B2/(1 + r)^2$
For year 3	=	$B1/(1 + r) + B2/(1 + r)^2 + B3/(1 + r)^3$ and so on

If the annual net benefit is the same for each year for n years, the formula becomes

$$NPV = B/(1 + r)^n.$$

As explained previously, the discount rate can vary year by year, so the rate relevant to the year for which it applies must be used when reading off the discount factor table.

Two other financial calculations need to be carried out to enable a realistic decision to be taken as to the viability of the project.

Payback

Payback is the time taken to recover the capital outlay of the project, having taken into account all the operating and overhead costs during this period. Usually, this is based on the undiscounted cash flow. Knowledge of the payback is particularly important when the capital must be recouped as quickly as possible, as would be the case in short-term projects or projects whose end products have a limited appeal due to changes in fashion, competitive pressures or alternative products. Payback is easily calculated by summating all the net incomes until the total equals the original investment (e.g. if the original investment is £600,000, and the net income is £75,000 per year for the next 10 years, the payback is £600,000/£75,000 = 8 years).

Internal rate of return

It has already been shown that the higher the discount rate (usually the cost of borrowing) of a project, the lower the NPV. Therefore, there must come a point at which the discount rate is such that the NPV becomes zero. At this point, the project ceases to be viable and the discount rate is the *internal rate of return* (IRR). In other words, it is the discount rate at which the NPV is 0.

While it is possible to calculate the IRR by trial and error, the easiest method is to draw a graph as shown in Fig. 6.2.

The horizontal axis is calibrated to give the discount rates from 0 to any chosen value, say 20%. The vertical axis represents the NPVs, above and below the horizontal axis denoted by (+) and (−).

By choosing two discount rates (one low and one high), two NPVs can be calculated for the same envisaged net cash flow. These NPVs (preferably one +ve and one −ve) are then plotted on the graph and joined by a straight line. Where this line cuts the horizontal axis, that is, where the NPV is zero, the IRR can be read off.

The basic formulae for the financial calculations are given in the following.

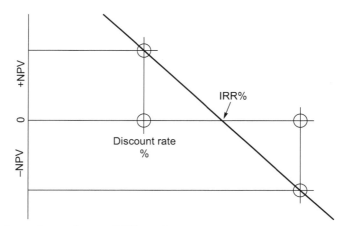

Figure 6.2 Internal rate of return (IRR) graph.

Investment appraisal definitions

NPV	=	Summation of PVs − original investment
Net income	=	Incoming money − outgoing money
Payback period	=	No. of years it takes for net income to equal original investment
Profit	=	Total net income − original investment
Average return/annum	=	$\dfrac{\text{total net income}}{\text{no. of years}}$
Return on investment (%)	=	$\dfrac{\text{average return} \times 100}{\text{investment}}$
	=	$\dfrac{\text{net income} \times 100}{\text{no. of years} \times \text{investment}}$
IRR	=	% Discount rate for NPV = 0

Cost/benefit analysis

Once the cost of the project has been determined, an analysis has to be carried out which compares these costs with the perceived benefits. The first cost/benefit analysis should be carried out as part of the business case investment appraisal, but in practice such an analysis should really be undertaken at the end of every phase of the life cycle to ensure that the project is still viable. The phase interfaces give management the opportunity to proceed with or, alternatively, abort the project if there is an unacceptable escalation in costs or a diminution of the benefits due to changes in market conditions, such as a reduction in demand caused by political, economic, climatic, demographic or a host of other reasons.

It is relatively easy to carry out a cost/benefit analysis where there is a tangible deliverable producing a predictable revenue stream. Provided there is an acceptable NPV, the project can usually go ahead. However, where the deliverables are intangible, such as better service, greater customer satisfaction, lower staff turnover, higher staff morale, etc., there may be considerable difficulty in quantifying the benefits. It will be necessary in such cases to run a series of tests and reviews and assess the results of interviews and staff reports.

Similarly, whilst the cost of redundancy payments can be easily calculated, the benefits in terms of lower staff costs over a number of years must be partially offset by lower production volume or poorer customer service. Where the benefits can only be realized over a number of years, a benefit profile curve as shown in Fig. 6.3 should be produced, making due allowance for the NPV of the savings.

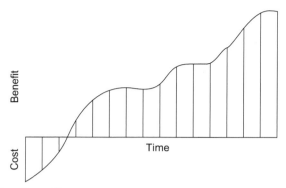

Figure 6.3 Cost/benefit profile.

Following is a list of some of the benefits that have to be considered, from which it will be apparent that some will be very difficult to quantify in monetary terms:

- Financial
- Statutory
- Economy
- Risk reduction
- Productivity
- Reliability
- Staff morale
- Cost reduction
- Safety
- Flexibility
- Quality
- Delivery
- Social

Further reading

Bradley, G. (2010). *Fundamentals of project realisation*. The Stationary Office.
Gatti, S. (2007). *Project finance in theory and practice*. Academic Press.
Goldsmith, L. (2005). *Project management accounting*. Wiley.
Khan, F., & Parra, R. (2003). *Financing large projects*. Pearson Education India.
Rogers, M. (2001). *Engineering project appraisal*. Blackwell Science.

Stakeholder management

Almost any person or organization with an interest in a project can be termed a *stakeholder*.

The type and interest of a stakeholder are of great importance to a project manager as they enable him or her to use these interests to the greatest benefit of the project. The process of listing, classifying and assessing the influence of these stakeholders is termed *stakeholder analysis*.

Stakeholders can be divided into two main groups:

1. Direct (or primary)
2. Indirect (or secondary)

Direct stakeholders

This group is made up, as the name implies, of all those directly associated or involved in the planning, administration or execution of the project. These include the client, project sponsor, project manager, members of the project team, technical and financial services providers, internal or external consultants, material and equipment suppliers, site personnel, contractors and subcontractors as well as end users. In other words, people or organizations directly involved in all or some of the various phases of the project are called direct stakeholders.

Indirect stakeholders

This group covers all those indirectly associated with the project, such as internal managers of the organization and support staff not directly involved in the project, including the HR department, accounts department, secretariat, senior management levels not directly responsible for the project and, last but not the least, the families of the project manager and team members.

A subsection of indirect stakeholders are those representing the regulatory authorities, such as national and local government, public utilities, licensing and inspecting organizations, technical institutions, professional bodies and personal interest groups such as stockholders, labour unions and pressure groups.

Each of these groups can contain the following:

* *Positive stakeholders* who support the aims and objectives of the project, and
* *Negative stakeholders* who do not support the project and do not wish for it to proceed.

Direct stakeholders mainly consist of positive stakeholders as they are the ones concerned with the design and implementation of the project with the object of completing it within the specified parameters of time, cost and quality/performance. They therefore include the sponsor, the project manager and the project design and construction/installation teams. This group could also have negative stakeholders such as employees of the end user, who would prefer to retain the existing facility because the new installation might result in relocation or even redundancy.

The indirect group probably contains the greatest number of potential negative stakeholders. These could include environmental pressure groups, trade (labour) unions, local residents' associations and even politicians (usually in opposition) who object to the project on principle or on environmental grounds.

Local residents' associations can be either positive or negative. For example, when it has been decided to build a bypass road around a town, the residents in the town may well be in favour to reduce the traffic congestion in the town centre, while residents in the outer villages, whose environment will be degraded by additional noise and pollution, will undoubtedly protest and will try to stop the road being constructed. It is these pressure groups who cause the greatest problems to the project manager.

In some situations, statutory/regulatory authorities or even government agencies, who have the power to issue or withhold permits, access, way leaves or other consents, can be considered as negative stakeholders.

Fig. 7.1 shows some of the types of people or organizations in the different groups and subgroups.

Although most negative stakeholders are clearly disruptive and tend to hamper progress, often in ingenious ways, they must nevertheless be given due consideration and afforded the opportunity to state their case. Whether it is possible to change their attitude by debate or argument depends on the strength of their convictions and the persuasiveness of the project supporters.

Positive stakeholders				Negative stakeholders	
Direct		Indirect		Indirect	
Internal	External	Internal	External	Internal	External
Sponsor	Client	Management	Stockholders	Disgruntled employees	Disgruntled end user
Project manager	Contractors Suppliers Consultants	Accounts dept HR dept Tech. depts Families	Banks Insurers Utilities Local authorities Government agencies		Pressure groups Unions Press (media) Competitors Politicians Residents' associations
Project team Project office					

Figure 7.1 Stakeholder groups.

Diplomacy and tact are essential when negotiating with potentially disruptive organizations, and it is highly advisable to enlist experts to participate in the discussion process. Most of the large organizations employ labour and public relation experts as well as lawyers well versed in methods for dealing with difficult stakeholders. Their services can be of enormous help to the project manager.

Therefore, it can be seen that for the project manager to be able to take advantage of the positive contributions of stakeholders and counter the negative ones most effectively, a detailed analysis must be carried out setting out the interests of each positive and negative stakeholder, the impact of these interests on the project, the probability of occurrence, particularly in the case of action by negative stakeholders and the actions, or reactions, to be taken.

Fig. 7.2 shows how this information can best be presented for analysis.

The Stakeholder column should contain the name of the organization and the main person or contact involved.

The Interest column states whether it is positive or negative, and whether it is financial, technical, environmental, organizational, commercial, political, etc.

The Influence/impact column sets out the possible effects of stakeholder interference, which may be helpful or disruptive. This influence could affect the cost, time or performance criteria of the project. Clearly, stakeholders with financial muscle must be of particular interest.

The Probability column can only be completed following a cursory risk analysis based on experience and other techniques such as brainstorming, Delphi and historical surveys.

The Action column relates to positive stakeholders and lists the best ways to generate support, such as maintaining good personal relations, invitations to certain meetings and updated information.

The Reaction column sets out the tactics to assuage unfounded fears, kill malicious rumours and minimize physical disruption.

The key to all these procedures is a good communication and intelligence gathering system.

Whilst it is relatively easy to get to list and classify stakeholders using these techniques the management of the different stakeholders is rather more difficult as this is dependent largely on the personalities of the individuals representing the various organizations pressure groups. Large contracts tend to be particularly prone to the often unwelcome attention of political or environmental pressure groups who will have to be managed by specialist of the PR department. The largest group of stakeholders are probably the men or women employed during the various phases of the project from initial design to commission and hand-over. These workers are

Stakeholder	Interest	Influence impact	Probability	Action to maximize support	Reaction to minimize disruption

Figure 7.2 Stakeholder analysis.

often represented at meetings by their trade union delegate of shop steward and it is prudent to employ an experienced negotiator in labour relations. Indeed many companies employ retired union officials in a form of 'poacher-turned-gamekeeper' role, precisely because of their familiarity with the required procedures. Unless there is a good relationship between the project manager and the union representative, friction will never inevitably occur, often when the project reaches a critical stage.

A stakeholder who could be either positive or negative is the local or national newspaper who, depending on their political persuasion, may wish the project to proceed or be strangled at birth.

When working with print media or broadcasting organizations, it is usual on large projects for the client and the contractors who have joint briefing sessions with the media at which the aims, purpose, function, modes of operation as well as the benefits to the environment, society as a whole or country are explained and highlighted. Having set the tone of these meetings, it is necessary to give monthly or quarterly updates. In that way, all parties can be made aware of progress as well as the potential or anticipated difficulties which may be encountered. The important thing is to be both honest and transparent and be prepared to answer both friendly and aggressive questions.

Once the project has started and the programmes have been prepared and issued, there is no better way to gain the interest, sympathy and involvement of stakeholders, than by showing them on a well-drawn network how they fit into the picture and how their contribution will help to move the project forward. As mentioned in an earlier chapter in this book, the use of a critical path network will not only be the focal point of the planning meetings but also acts as an excellent communication document which shows clearly what the effect the delay of any particular party can have on the programme as a whole. The operative word is involvement. People should be made to feel that they have a real stake in the project and feel proud to be associated with it.

Clearly, the relationship between the project manager and the client or owner is of prime importance, and this stakeholder must be kept abreast of developments and progress on a regular basis. As soon as problems arise, he must be informed and meetings must be convened to discuss and resolve any issues which have arisen since the previous regular meeting. Where the cost of the project has increased due to changes required ordered by the client, it is vital that these are transmitted for approval at the earliest possible time so that the party responsible for the increase has the opportunity to revoke the change order.

There may be instances where the client feels that an increase of the costs, even if it was due to their actions or changes, are unjustified. In such circumstances, the success of any subsequent negotiations can well depend on the personal relationship between the project manager and the client's representative. It is not easy to present a client with new, unwelcome figures and still remain on friendly terms at the end. Certainly, there is no doubt that sharing a meal or drinks will help to keep relations amicable.

Regular or impromptu social functions involving the wives or partners can often be very beneficial in maintaining a relationship with clients and other important stakeholders. On the rare occasions when difficult or intransigent stakeholders are encountered, it may be necessary to physically buy them out. For example on a refinery project, the owner of a small farm, whose land was required for the project, refused to sell his land until he received a highly inflated price for the property together with a guarantee that he would be employed by the company until his retirement age. This resolution was not only quicker but probably cheaper than the cost of a lengthy legal battle with tribunals and expensive lawyers.

Groups of stakeholders which are often overlooked are the families of employees, especially if these have to work away from home for long periods. For this reason, regular social gatherings at which family members are invited to meet each other and senior management will help to engender the feeling that they are part of the project family and are involved in the operation and success.

Problems can arise with suppliers or subcontractors where the main contractor is not able to effect adequate control over both quality and delivery. It is prudent therefore for the contract documents to include a clause giving the project manager the right to take necessary steps to ensure that the specified criteria are complied. This may involve 'parachuting' an expert or specialist into the premises to help solve the technical problems and ensure that the delivery and quality criteria are met.

It may be helpful to draw a matrix similar to the one shown in Fig. 15.1 (Chapter 15 Risk analysis) on which a number of stakeholders (designated by identification letters) can be graded for importance and influence.

Depending on their assessments of importance and influence, the letters can be placed in the appropriate box on the grid to give the project manager an immediate picture (and reminder) as to their relevance, priority and importance should there be the need to deal with any issues with which any of these stakeholders are connected.

In a sense, *stakeholder management* is the wrong term for dealing with stakeholders as it could be interpreted as implying a form of manipulation. A better description would be stakeholder involvement, as it is the project manager's job to ensure that this involvement of stakeholders is harnessed to give the maximum benefit to the project and ensure its success by meeting its objectives.

All positive stakeholders should be encouraged to consider themselves as being part of the team, even if they are only involved in a minor way or geographically apart. Once this has been accepted by them, their help and cooperation will be ensured. This sense of involvement can then be reinforced by appropriate expressions of gratitude for their help and praise (subtle or overt) for their contributions.

For example, the initial negative attitude of a local authority or pressure group who quite naturally object to the possible disruption and pollution caused by the project can often be reversed by pointing out the long-term benefits and asking for their help and suggestions in minimizing the feared discomfort as well as inviting them to regular meetings to listen to complaints and resolve actual or perceived problems. In other words, making them part of the team. There is nothing more disruptive to the project than keeping a stakeholder in the dark (see the railway example in Chapter 38, Communication).

Project-success criteria

8

Chapter outline

Key performance indicators 42

One of the topics in the project-management plan is *project-success criteria*. These are the most important attributes and objectives that must be met to enable the project to be termed a success. The most familiar success criteria are completion on time, keeping the project costs within budget, and meeting the performance and quality requirements set out in the specification. However, in some industries, there are additional criteria that are equally or even more important. These can be safety, sustainability, reliability, legacy (long-term performance) and meeting the desired business benefit. It can be argued that all these are enshrined in performance, but there is undoubtedly a difference of emphasis between industries, organizations and public perceptions. With the realization that climate change has a significant impact on the environment and our future lives, sustainability in the form of conservation of energy and natural resources and the control of carbon emissions have all become performance criteria in their own right, especially as these will be subject to stricter and stricter government legislation across the whole industrial spectrum. In this context, sustainability is of course linked with legacy, as future generations will thrive or suffer depending on our success in meeting these important criteria.

It is always possible that during the life of a project, problems arise that demand that certain changes have to be made which may involve compromises and trade-offs to keep the project either on programme or within the cost boundaries. The extent to which these compromises are acceptable or permissible depends on their scope and nature, and requires the approval of the project manager and possibly also the sponsor and client. However, where such an envisaged change will affect one of the project-success criteria, a compromise of the affected success criterion may not be acceptable under any circumstance.

For example, if one of the project-success criteria is that the project finishes by or before a certain date, then there can be no compromise of the date, but the cost may increase or quality may be sacrificed.

Success criteria can of course be subjective and depend often on the point of view of the observer. Judging by the conventional criteria of a well-managed project, that is cost, time and performance, the Sydney Opera House failed in all three, as it was vastly over budget, very late in completion and is considered to be too small for a grand opera. Despite this, most people consider it to be a great piece of architecture and a wonderful landmark for the city of Sydney.

Project Management, Planning and Control. https://doi.org/10.1016/B978-0-12-824339-8.00008-0

While it is not difficult to set the success criteria, they can only be achieved if a number of success factors are met. The most important of these are given in the following:

- Clear objectives and project brief agreed with client
- Good project definition
- Good planning and scheduling methods
- Accurate time control and feedback system
- Rigorous performance monitoring and control systems
- Rigorous change control (variations) procedures
- Adequate resource availability (finance, labour, plant, materials)
- Full top-management and sponsor support
- Competent project management
- Tight financial control
- Comprehensive quality control procedures
- Motivated and well-integrated team
- Competent design
- Good contractual documentation
- Good internal and external communications
- Good client relationship
- Well-designed reporting system to management and client
- Political stability
- Awareness of environmental issues and related legislation

This list is not exhaustive, but even if one of the functions or systems listed is not performed adequately, the project may well end in failure.

Key performance indicators

A *key performance indicator* (KPI) is a major criterion against which a particular part of the project can be measured. KPI can be a milestone that must be met, such as a predetermined design, delivery, installation, production, testing, erection or commissioning stage, a payment date (in or out) or any other important stage in a project. In process plants, KPIs can include the contractual performance obligations such as output or throughput, pressure, temperature or other quality requirements. Even when the project has been commissioned and handed over, KPIs relating to performance over defined time spans (reliability and repeatability) are still part of the contractual requirements. Some KPIs cannot be measured or proven until the project or the operations following project completion have been running for a number of years, but these, which could also include performance and sustainability criteria, should nevertheless be considered and incorporated at both the planning and execution stages.

Organization structures

Chapter outline

To manage a project, a company or authority has to set up a project organization, which can supply the resources for the project and service it during its life cycle. There are three main types of project organizations:

1. Functional
2. Matrix
3. Project or taskforce

Functional organization

This type of organization consists of specialist or functional departments, each with their own departmental manager responsible to one or more directors. Such an organization is ideal for routine operations where there is little variation of the end product. Functional organizations are usually found where items are mass produced, whether they are motor cars or sausages. Each department is expert at its function and the interrelationship between them is well established. In this sense, a functional organization is not a project-type organization at all and is only included because when small, individual one-off projects have to be carried out, they may be given to a particular department to manage. For projects of any reasonable size or complexity, it will be necessary to set up one of the other two types of organizations.

Matrix organization

This is probably the most common type of project organization, as it utilizes an existing functional organization to provide human resources without disrupting the day-to-day operation of the department.

The personnel allocated to a particular project are responsible to a project manager for meeting the three basic project criteria: time, cost and quality. The departmental manager is, however, still responsible for their 'pay and rations' and their compliance

Project Management, Planning and Control. https://doi.org/10.1016/B978-0-12-824339-8.00009-2

with the department's standards and procedures, including technical competence and conformity to company's quality standards. The members of this project team will still be working at their desks in their department, but will be booking their time to the project. Where the project does not warrant a full-time contribution, only those hours actually expended on the project will be allocated to it.

The advantages of a matrix organization are as follows:

1. Resources are employed efficiently, as staff can switch to different projects if any one of the projects is held up.
2. The expertise built up by the department is utilized and the latest state-of-the-art techniques are immediately incorporated.
3. Special facilities do not have to be provided and disrupting staff movements are avoided.
4. The career prospects of team members are left intact.
5. The organization can respond quickly to changes of scope.
6. The project manager does not have to be concerned with staff problems.

The disadvantages are the following:

1. There may be a conflict of priorities between different projects.
2. There may be split loyalties between the project manager and the departmental manager due to the dual reporting requirements.
3. Communications among team members can be affected if the locations of the departments are far apart.
4. Executive management may have to spend more time to ensure a fair balance of power between the project manager and the department manager.

Matrix organizations can sometimes be categorized as strong or weak, depending on the degree of dominance or authority of the project manager or department manager, respectively. This can of course create friction as both sides will try to assert themselves.

However, all the earlier problems can be resolved if senior management ensures (and indeed insists) that there is a good working relationship between the project manager and the department heads. At times, both sides may have to compromise for sake of the interests of the project and the organization as a whole.

Project organization (taskforce)

From a project manager's point of view, this is the ideal type of project organization, as with such a setup he or she has complete control over every aspect of the project. The project team will usually be located in one area, which can be a room for a small project or a complete building for a very large one.

Lines of communication are short and the interaction of the disciplines reduces the risk of errors and misunderstandings. Not only are the planning and technical functions parts of the team but also the project cost control and project accounting staff. This places an enormous burden and responsibility on the project manager, who will have to delegate much of the day-to-day management to special project coordinators

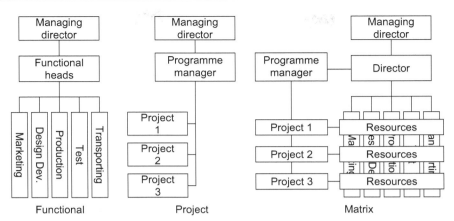

Figure 9.1 Types of organization.

whose prime function is to ensure a good communication flow and timely receipt of reports and feedback information from external sources.

On large projects with budgets often greater than £0.5 billion, the project manager's responsibilities are akin to those of a managing director of a medium-size company. He or she is not only concerned with the technical and commercial aspects of the project but must also deal with the staff, financial and political issues, which are often more difficult to delegate.

There is no doubt that for large projects a taskforce type of project organization is essential, but as with so many areas of business, the key to success lies with the personality of the project manager and his or her ability to inspire the project team to regard themselves as personal stakeholders in the project.

One of the main differences between the two true project organizations (matrix and taskforce) and the functional organization is the method of financial accounting. For the project manager to retain proper cost control during the life of the project, it is vital that a system of project accounting is instituted, whereby all incomes and expenditures, including a previously agreed overhead allocation and profit margin, are booked to the project as if it were a separate self-standing organization. The only possible exceptions are certain corporate financial transactions such as interest payments on loans taken out by the host organization and interest receipts on deposits from a positive cash flow.

Fig. 9.1 shows a diagrammatic representation of the three basic project management organizations, functional, project (or taskforce) and matrix.

Organization roles

10

Every project has a number of key people, who have important roles to play. Each of these role players has specific responsibilities, which, if carried out in their prescribed manner, will ensure a successful project. The following list gives some examples of these organizational roles:

- Project sponsor
- Portfolio manager
- Programme manager
- Project manager
- Project planning manager or project planner
- Cost manager or cost control manager
- Risk manager
- Procurement manager
- Configuration manager
- Quality manager
- Project board or steering committee
- Project support office

In practice, the need for some of these roles depends on the size and complexity of the project and the organizational structure of the company or authority carrying out the project. For example, a small organization carrying out only two or three small projects a year may not require, or indeed be able to afford, many of these roles. Having received the commission from the client, a good project manager supported by a planner, a procurement manager and established well-documented quality control and risk procedures may well be able to deliver the specified requirements.

The detailed descriptions of the earlier roles are given in the relevant sections dealing with each topic, but the last two roles warrant some further discussion.

A project board or steering committee is sometimes set up for large projects to act as a supervising authority, and sometimes as a champion, during the life cycle of the project. Its job is to ensure that the interests of the sponsoring organization (or client) are protected and that the project is run and delivered to meet the requirements of these organizations. The project board appoints the project manager, and the project manager must report to the project board on a regular basis and obtain from it the authority to proceed after certain predetermined stages have been reached.

National or local government projects following the PRINCE methodology frequently have a project board, as the project may affect a number of different departments who are all stakeholders to a greater or lesser extent. The board is thus usually constituted from senior managers (or directors) of the departments most closely involved in the project. A project board may also be desirable where the client consists

Project Management, Planning and Control. https://doi.org/10.1016/B978-0-12-824339-8.00010-9

of a number of companies who are temporary partners in a special consortium set up to deliver the project. A typical example is a consortium of a number of oil producers who each have a share in a refinery, drilling platform or pipeline either during the construction or operating phase, or both.

Ideally, the board should only be consulted or be required to make decisions on major issues, as unnecessary interference with the normal running of a project will undermine the authority of the project manager. However, there may be instances where fundamental disagreements or differences of interests among functional departments or stakeholders need to be discussed and resolved, and it is in such situations that the project board can play a useful role.

The project support office briefly described in Chapter 2 is only required on large projects, where it is desirable to service all the project administration and project technology by a central department or office, supervised by a service manager. The functions most usually carried out by the project support office are the preparation and updating of the project schedule (programme), collection and processing of time sheets, progress reports, project costs and quality reports, operating configuration management and controlling the dissemination of specifications, drawings, schedules and other data.

The project support office is in fact the secretariat of the project and its size, and its constitution will therefore depend on the size of the project, its technical complexity and its administrative procedures and reporting requirements. The administrative procedures and reporting requirements can be very onerous where the client organization is a consortium of companies, each of which requires its own reporting procedures, cost reports and timetable for submission.

Ideally, the accounting system for the project should be self-contained and independent of the corporate accounting system. Clearly, a monthly cost report will have to be submitted to the organization's accounts department, but project accounting will give speedier and more accurate cost information to the project manager, which will enable him or her to take appropriate action before the costs spiral out of control. The project support office can play a vital role in this accounting function provided that the office staff includes the project accountants. These accountants control not only the direct costs such as labour, materials, equipment and plant but also those indirect costs and expenses related to the project.

Project life cycles

11

Most, if not all, projects go through a life cycle that varies with the size and complexity of the project. For medium to large projects, the life cycle will generally follow the following pattern:

1	Concept	Basic ideas, business case, statement of requirements, scope
2	Feasibility	Tests for technical, commercial and financial viability, technical studies, investment appraisal, discounted cash flow, etc.
3	Evaluation	Application for funds, stating risks, options, time, cost, performance criteria
4	Authorization	Approvals, permits, conditions, project strategy
5	Implementation	Development design, procurement, fabrication, installation, commissioning
6	Completion	Performance tests, handover to client, post-project appraisal
7	Operation	Revenue earning period, production, maintenance
8	Termination	Close-down, decommissioning, disposal

Items 7 and 8 are not usually included in a project life cycle, where the project ends with the issue of an acceptance certificate after the performance tests have been successfully completed. Where these two phases are included, as, for example, with defence projects, the term *extended project life cycle* is often used.

The project life cycle of an IT project may be slightly different as the following list shows:

1	Feasibility	Definition, cost benefits, acceptance criteria, time, cost estimates
2	Evaluation	Definitions of requirements, performance criteria, processes
3	Function	Functional and operational requirements, interfaces, system design
4	Authorization	Approvals, permits, firming up procedures
5	Design and build	Detail design, system integration, screen building, documentation
6	Implementation	Integration and acceptance testing, installation, training
7	Operation	Data-loading, support setup, handover

Project Management, Planning and Control. https://doi.org/10.1016/B978-0-12-824339-8.00011-0

Running through the period of the life cycle are control systems and decision stages at which the position of the project is reviewed. The interfaces of the phases of the life cycle form convenient milestones for progress payments and reporting progress to top management, who can then make the decision to abort the project or provide further funding. In some cases, the interfaces of the phases overlap, as in the case of certain design and construct contracts, where construction starts before the design is finished. This is known as concurrent engineering and is often employed to reduce the overall project programme.

As the word *cycle* implies, the phases may have to be amended in terms of content, cost and duration as new information is fed back to the project manager and sponsor. Projects are essentially dynamic organizations that are not only specifically created to effect change, but are also themselves subject to change.

For some projects, it may be convenient to appoint a different project manager at a change of phase. This is often done where the first four stages are handled by the development or sales department, who then hands the project over to the operations department for the various stages of implementation and completion phases.

When decommissioning and disposal are included, it is known as an extended life cycle, as these two stages could occur many years after commissioning and could well be carried out by a different organization.

While all institutions associated with project management stress the importance of the project cycle, both British Standards Institution and International Standards Organisation preferred describing what operations should be carried out during the various phases, rather than giving the phases specific names.

Different organizations tend to have different descriptions and sequences of the phases, and Fig. 11.1 shows two typical life cycles prepared by two different organizations. The first example, as given by Association for Project Management in their latest Body of Knowledge, is a very simple generic life cycle consisting of only six

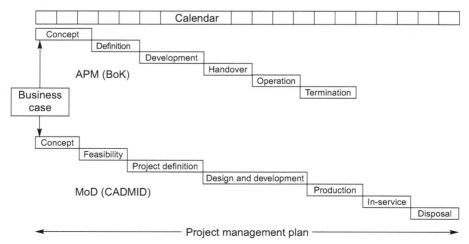

Figure 11.1 Examples of project life cycles.

basic phases. The second life cycle, as formulated by the Ministry of Defence (MoD), clearly shows the phases required for a typical weapons system, where concept, feasibility and project definition are the responsibility of the MoD; design, development and production are carried out by the manufacturer; and in-service and disposal are the phases when the weapon is in the hands of the armed forces.

The diagram also shows a calendar scale over the top. While this is not strictly necessary, it can be seen that if the lengths of the bars representing the phases are drawn proportional to the time taken by the phases, such a presentation can be used as a high-level reporting document, showing which phases are complete or partially complete in relation to the original schedule.

The important point to note is that each organization should develop its own life cycle diagram to meet its particular needs. Where the life cycle covers all the phases from cradle to grave as it were, it is often called a *programme life cycle*, since it spans over the full programme of the deliverable. The term *project life cycle* is then restricted to those phases that constitute a project within the programme (e.g., the design, development and manufacturing periods).

Fig. 11.2 shows how decision points or milestones (sometimes called trigger points or gates) relate to the phases of a life cycle. At each gate, a check should be carried out to ensure that the project is still viable, that it is still on schedule, that costs are still within budget, that sufficient resources are available for the next phase and that the perceived risks can be managed.

Fig. 11.3 shows how the life cycle of the MoD project shown in Fig. 11.1 could be split into the *project life cycle*, that is, the phases under the control of the project team (conception to production), the *product life cycle*, the phases of interest to the sponsor, which now includes the in-service performance and lastly the *extended life cycle*,

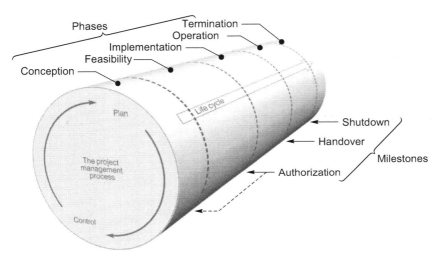

Figure 11.2 Project management life cycle.

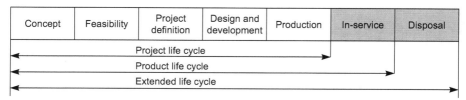

Figure 11.3 Life cycle of MoD project.

which includes disposal. From the point of view of the contractor, the *project life cycle* may only include design, development and production. Therefore, it can be seen that there are no hard-and-fast rules as to where the demarcation points are, as each organization will define its own phases and life cycles to suit its method of working.

Work breakdown structures

Chapter outline

An examination of the project life cycle diagram will show that each phase can be regarded as a project in its own right, although each will be of very different size and complexity. For example, when a company is considering developing a new oil field, the feasibility study phase could be of considerable size although the main project would cover the design, development and production phases. To be able to control such a project, the phases must be broken down further into stages or tasks, which in turn can be broken down further into subtasks or work packages until a satisfactory and acceptable control structure has been achieved.

The choice of tasks incorporated in the *work breakdown structure* (WBS) is best made by the project team drawing on their combined experience or by engaging in a brainstorming session.

Once the main tasks have been decided upon, they can in turn be broken down into subtasks or work packages, which should be coded to fit in with the project cost coding system. This will greatly assist in identifying the whole string of relationships from overall operational areas down to individual tasks. For this reason, the WBS is the logical starting point for subsequent planning networks. Another advantage is that a cost allocation can be given to each task in the WBS and, if required, a risk factor can also be added. This will assist in building up the total project cost and creates a risk register for a subsequent, more rigorous risk assessment.

The object of all this is to be able to control the project by allocating resources (human, material and financial) and giving time constraints to each task. It is always easier to control a series of small entities that make up a whole, than to control the whole enterprise as one operation. What history has proven to be successful for armies — which are divided into divisions, regiments, battalions, companies and platoons — or corporations — which have area organizations, manufacturing units and sales territories — is also true for projects, whether they are large, small, sophisticated or straightforward.

The tasks will clearly vary enormously with the type of project in both size and content, but by representing their relationships diagrammatically, a clear graphical picture can be created. This, when distributed to other members of the project team, becomes a very useful tool for disseminating information as well as a reporting medium to all stakeholders. As the completion of the main tasks are in fact the major project

Project Management, Planning and Control. https://doi.org/10.1016/B978-0-12-824339-8.00012-2

milestones, the WBS is an ideal instrument for reporting progress upwards to senior management, and, for this reason, it is essential that the status of each work package or task is regularly updated.

As the WBS is produced in the very early stages of a project, it will probably not reflect all the tasks that will eventually be required. Indeed, the very act of draughting the WBS often throws up the missing items or work units, which can then be formed into more convenient tasks. As these tasks are decomposed further, they may be given new names such as unit or work package. It is then relatively easy for management to allocate task owners to each task or group of tasks, who have the responsibilities for delivering each task to the normal project criteria of cost, time and quality/performance.

The abbreviation WBS is a generic term for a hierarchy of stages of a project. However, some methodologies like PRINCE call such a hierarchical diagram a *product breakdown* structure (PBS) (Fig. 12.1). The difference is basically what part of speech is being used to describe the stages. If the words used are *nouns*, it is, strictly speaking, a PBS, because we are dealing with products or things. If, on the other hand, we are describing the work that has to be performed on the nouns and use *verbs*, we call it a WBS. Frequently, a diagram starts as a PBS for the first three or four levels and then becomes a WBS as more detail is introduced.

Despite this unfortunate lack of uniformity of nomenclature in the project management fraternity, the principles of subdividing the project into manageable components are the same.

It must be pointed out, however, that the WBS is *not* a programme, although it looks like a precedence diagram. The interrelationships shown by the link lines do not necessarily imply a time-dependence or indeed any sequential operation.

The corresponding WBS shown in Fig. 12.2 uses verbs, and the descriptions of the packages or tasks then become: *assemble* car, *build* power unit, *weld* chassis, *press* body shell, etc.

The degree to which the WBS needs to be broken down before a planning network can be drawn will have to be decided by the project manager, but there is no reason why a whole family of networks cannot be produced to reflect each level of the WBS.

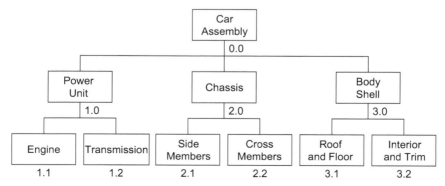

Figure 12.1 Product breakdown structure.

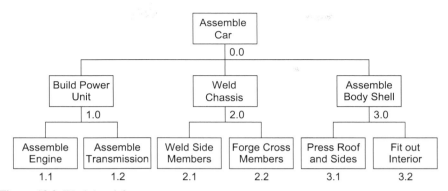

Figure 12.2 Work breakdown structure.

Once the WBS (or PBS) has been drawn, a *bottom-up* cost estimate can be produced starting at the lowest branch of the family tree. In this method, each work package cost is calculated and arranged in such a way that the total cost of the packages on any branch must add up to the cost of the package of the parent package on the branch above. If the parent package has a cost value of its own, this must clearly be added before the next stage of the process. This is shown in Fig. 12.3, which not only explains the bottom-up estimating process, but also shows how the packages can be coded to produce a project cost coding system that can be carried through to network analysis and earned value analysis.

The resulting diagram is now a *cost breakdown structure*.

It can be seen that a WBS is a powerful tool that can show clearly and graphically who is responsible for a task, how much it should cost and how it relates to the other tasks in the project. It was stated earlier that the WBS is not a programme, but once it has been accepted as a correct representation of the project tasks, it will become a good

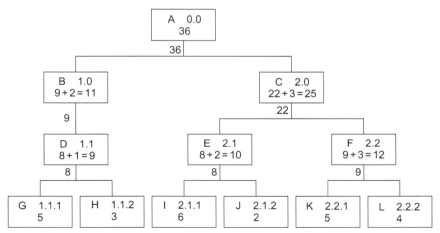

Figure 12.3 Cost breakdown structure.

base for drawing up the network diagram. The interrelationships of the tasks will have to be shown more accurately, and the only additional items of information to be added are the durations.

An alternative to the bottom-up cost allocation is the *top-down* cost allocation. In this method, the cost of the total project (or subproject) has been determined and is allocated to the top package of the WBS (or PBS) diagram. The work packages below are then forced to accept the appropriate costs so that the total cost of each branch cannot exceed the total cost of the package above. Such a top-down approach is shown in Fig. 12.4.

In practice, both methods may have to be used. For example, the estimator of a project may use the bottom-up method on a WBS or PBS diagram to calculate the cost. When this is given to the project manager, he or she may break this total down into the different departments of an organization and allocate a proportion to each, making sure that the sum total does not exceed the estimated cost. Once names have been added to the work packages of a WBS or PBS it becomes an organization breakdown structure (OBS).

It did not take long for this similarity to be appreciated so that another name for such an organization diagram became *organization breakdown*. This is the family tree of the organization in the same way that the WBS is in fact the family tree of the project. It is in fact more akin to a family tree or organization chart (organogram).

Fig. 12.5 shows a typical OBS for a manufacturing project such as the assembly of a prototype motor car. It can be seen that the OBS is not identical in layout to the WBS, as one manager or task owner can be responsible for more than one task.

The OBS shown is typical of a matrix-type project organization where the operations manager is in charge of the actual operating departments for 'pay and rations', but each departmental head (or his or her designated project leader) also has a reporting

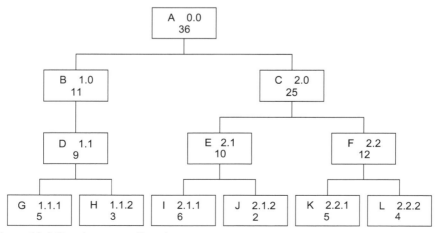

Figure 12.4 Top-down cost allocation.

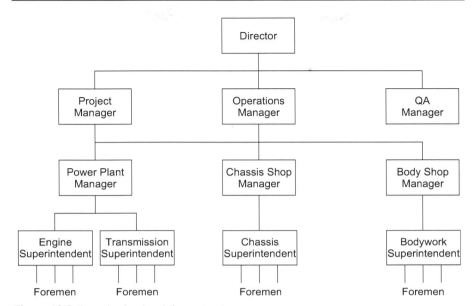

Figure 12.5 Organization breakdown structure.

line to the project manager. If required, the OBS can be expanded into a responsibility matrix to show the responsibility and authority of each member of the organization or project team.

The quality assurance (QA) manager reports directly to the director to ensure independence from the operating and projects departments. He or she will, however, assist all operating departments with producing the quality plans and give ongoing advice on QA requirements and procedures as well as pointing out any shortcomings he or she may discover.

Although the WBS may have been built up by the project team, based on their collective experience or by brainstorming, there is always the risk that a stage or task has been forgotten (Table 12.1).

An early review then opens up an excellent opportunity to refine the WBS and carry out a risk identification for each task, which can be the beginning of a risk register. At a later date, a more rigorous risk analysis can then be carried out. The WBS does in fact give everyone a better understanding of the risk-assessment procedure.

Indeed, a further type of breakdown structure is the *risk breakdown structure*. The main risks are allocated here to the WBS or PBS in either financial- or risk-rating terms, giving a good overview of the project risks.

In another type of risk breakdown structure, the main areas of risk are shown in the first level of the risk breakdown structure, and the possible risk headings are listed in the following. (See Table 16.1, Chapter 16, Risk Management.).

Table 12.1

Project risks			
Organization	Environment	Technical	Financial
Management	Legislation	Technology	Financing
Resources	Political	Contracts	Exchange rates
Planning	Pressure groups	Design	Escalation
Labour	Local customs	Manufacture	Financial stability of
Health and safety	Weather	Construction	(a) Project
Claims	Emissions	Commissioning	(b) Client
Policy	Security	Testing	(c) Suppliers

Responsibility matrix

By combining the WBS with the OBS, it is possible to create a *responsibility matrix*. Using the car assembly example given in Figs. 12.1 and 12.5, the matrix is drawn by writing the WBS work areas vertically and the OBS personnel horizontally, as shown in Fig. 12.6.

	Director	Project Manager	Operations Manager	QA Manager	Power Plant Manager	Chassis Shop Manager	Body Shop Manager	Engine Superintendent	Transmission Superintendent	Chassis Superintendent	Bodywork Superintendent
Car Assembly	A	B	A	B							
Power Unit	A	B	A	B	C						
Chassis	A	B	A	B		C					
Body Shell	A	B	A	B			C				
Engine		B	A	B	C			C			
Transmission		B	A	B	C				C		
Side Members		B	A	B		C				C	
Cross Members		B	A	B		C				C	
Roof and Floor		B	A	B			C				C
Interior and Trim		B	A	B			C				C

Figure 12.6 Responsibility matrix.

By placing a suitable designatory letter into the intersecting boxes, the level of responsibility for any work area can be recorded on the matrix.

A = receiving monthly reports.

B = receiving weekly reports.

C = daily supervision.

The vertical list giving the WBS stages could be replaced by the PBS stages. The horizontal list showing the different departmental managers could instead have the names of the departments or even consultants, contractors, subcontractors and suppliers of materials or services.

Further reading

Haugan, G. T. (2002). *Effective work breakdown structures*. Kogan Page.

Estimating

Chapter outline

Estimating is an essential part of project management, as it becomes the baseline for the subsequent cost control. If the estimate for a project is too low, a company may well lose money in the execution of the work. If the estimate is too high, the company may well lose the contract due to overpricing.

As explained in the section on work breakdown structures, there are two basic methods of estimating: top-down and bottom-up. However, unfortunately only in a few situations, the costs are available in a form for simply slotting into the work package boxes. Therefore, it is necessary to produce realistic estimates for each package, and indeed for the entire project, before a meaningful cost estimate can be carried out. For most of the estimates that require any reasonable degree of accuracy, the method used must be bottom-up. This principle is used in bills of quantities, which literally start at the bottom of the construction process, the ground clearance and foundations, and work up through the building sequence to the final stages such as painting and decorating.

Estimating the cost of a project requires a structured approach, but whatever method is used, the first thing is to decide the level of accuracy required. This depends on the status of the project and the information available. There are four main estimating methods in use, varying from the very approximate to the very accurate:

1. subjective (degree of accuracy $\pm20\%-40\%$)
2. parametric (degree of accuracy $\pm10\%-20\%$)
3. comparative (degree of accuracy $\pm10\%$)
4. analytical (degree of accuracy $\pm5\%$)

Subjective

At the proposal stage, the contractor may well be able to give only a 'ballpark figure' to give a client or sponsor an approximate 'feel' of the possible costs. The estimating method used in this case would either be *subjective or approximate parametric*. In either case, the degree of accuracy would largely depend on the experience of the estimator. When using the subjective method, the estimator relies on his or her experience of similar projects to give a cost indication based largely on 'hunch'. Geographical and political factors as well as the more obvious labour and material content must be taken into account. Such an approximate method of estimating is often given the disparaging name of 'guesstimating'.

Parametric

The *parametric* method would be used at the budget preparation stage, but relies on good historical data-based past jobs or experience. By using well-known empirical formulae or ratios, in which costs can be related to specific characteristics of known sections or areas of the project, it is possible to produce a good estimate on which firm decisions can be based. Clearly, such estimates need to be qualified to enable external factors to be separately assessed. For example, an architect will be able to give a parametric estimate of a new house once he or she is given the cube (height × length × depth) of the proposed building and the standard of construction or finish. The estimate will be in £/cubic metre of structure. Similarly, office blocks are often estimated in £/square metre of floor space. The qualifications would be the location, ground conditions, costs of the land, etc. Another example of a parametric estimate is when a structural steel fabricator gives the price of fabrication in £/tonnes of steel, depending on whether the steel sections are heavy beams and columns or light latticework. In both cases, the estimate may or may not include the cost of the steel itself.

Comparative (by analogy)

As an alternative to the parametric method, the *comparative* method of estimating can be used for the preparation of the budget. When a new project is very similar to another project recently completed, a quick comparison can be made of the salient features. This method is based on the costs of a simplified schedule of major components that were used on previous similar jobs. It may even be possible to use the costs of a similar-sized complete project of which one has had direct (and preferably recent) experience. Due allowance must clearly be made for the inevitable minor differences, inflation and other possible cost escalations. An example of such a comparative estimate is the installation of a new computer system in a building when an almost identical (and proven) system was installed 6 months earlier in another building. It must be stressed that such an estimate does not require a detailed breakdown.

Analytical

Once the project has been sanctioned, a working budget estimate will be necessary against which the cost of the project will be controlled. This will normally require an *analytical* estimate or bill of quantities. This type of estimate may also be required where a contractor has to submit a fixed-price tender, because once the contract is signed there can be no price adjustment except by inflation factors or client-authorized variations.

As the name implies, this is the most accurate estimating method, but it requires the project to be broken down into sections, subsections and finally individual components. Each component must then be given a cost value (and preferably also a cost code) including both the material and labour contents. The values, which are sometimes referred to as 'norms', are usually extracted from a database, or company archives, and must be individually updated or factored to reflect the current political and environmental situation.

Examples of analytical estimates are the norms used by the petrochemical industry where a value exists for the installation of piping depending on pipe diameter, wall thickness, material composition, height from ground level and whether flanged or welded. The norm is given as a cost/linear metre, which is then multiplied by the meterage including an allowance for waste. Contingencies, overheads and profits are then added to the total sum.

Quantity surveyors will cost a building or structure by measuring the architect's drawings and applying a cost to every square metre of wall or roof, every door and window and such systems as heating, plumbing, electrics. Such estimates are known as bills of quantities and together with a schedule of rates for costing variations form the basis of most building and civil engineering contracts. The accuracy of such estimates is better than $\pm 5\%$, depending on the qualifications accompanying the estimate. The rates used in bills of quantities (when produced by a contractor) are usually inclusive of labour, materials, plant, overheads and anticipated profit, but when produced by an independent quantity surveyor the last two items may have to be added by the contractor.

Unfortunately, such composite rates are not ideal for planning purposes as the time factor only relates to the labour content. To overcome this problem, the UK Building Research Station in 1970 developed a new type of bill of quantities called 'operational bills' in which the labour was shown separately from the other components, thus making it compatible with critical path-planning techniques. However, these new methods were never really accepted by industry and especially not by the quantity surveying profession.

A number of estimating books have been published to assist the estimator, which provide the materials and labour costs in great detail for nearly every operation or trade used in the construction business. These costs are separately given for labour based on the number of man-hours required and the materials cost as per the appropriate unit of measurements such as the metre length, square metre or cubic metre. Most of these books also give composite rates including materials, labour, overhead and profit. As rates for materials and labour change every year due to inflation or other factors, these books will have to be republished yearly to reflect the current rates. It is important, however, to remember that these books are only guides and require the given rates to be factorized depending on site conditions, geographical location and any other factor the estimator may consider to be significant.

The percentage variation at all stages should always be covered by an adequate contingency allowance that must be added to the final estimate to cover for possible, probable and unknown risks, which could be technical, political, environmental, administrative, etc. depending on the results of a more formal risk analysis. The further addition of overheads and profit gives the price (i.e. what the customer is being asked to pay).

It must be emphasized that such detailed estimating is not restricted to the construction (building or civil engineering) industry. Every project, given sufficient time, can be broken down into its labour, material, plant and overhead content and costing can be done very accurately.

Sometimes, an estimate produced by the estimator is drastically changed by senior management to reflect market conditions, the volume of work currently in the company or the strength of the perceived competition. However, from a control point of view, such changes to the final price should be ignored, which are in any case normally restricted to the overhead and profit portion and are outside the control of the project manager. When such a price adjustment is downward, every effort should be made to recover these 'losses' by practising value management throughout the period of the project.

Computer systems and software preparation, which are considerably more difficult to estimate than construction work due to their fundamentally innovative and untried processes, can be estimated using the following points:

1. Function-point analysis, where the numbers of software functions, such as inputs, outputs, files, interfaces, etc. are counted, weighted and adjusted for complexity and importance. Each function is then given a cost value and aggregated to find the overall cost.
2. Lines of code to be used in the programme. A cost value can be ascribed to each line.
3. Plain man-hour estimates based on the experience of previous or similar work, taking into account such new factors as inflation, the new environment and the client organization.

While it is important to produce the best possible estimate at every stage, the degree of accuracy will vary with the phase of the project, as shown in Fig. 13.1. As the project develops and additional or more accurate information becomes available, it is inevitable that the estimate becomes more accurate. This is sometimes known as rolling-wave estimating, and while these revised costs should be used for the next estimating stage, once the actual final budget stage has been reached and the price has been accepted by the client, any further cost refinements will only be useful for updating the monthly cost estimate, which may affect the profit or loss without changing the price or control budget as used in earned value methods.

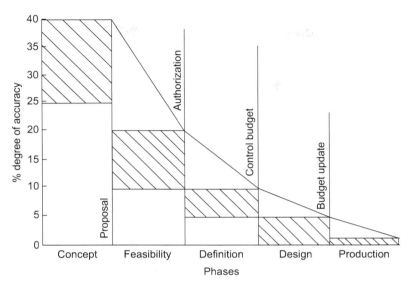

Figure 13.1 Phase/accuracy curve.

When estimating the man-hours related to the activities in a network programme, it may be difficult to persuade certain people to commit themselves to giving a firm man-hour estimate. In such cases, just in order to elicit a realistic response, it may be beneficial to employ the 'three times estimate' approach, $t = (a + 4m + b)/6$, as described in Chapter 20. In this formula, t is the expected or most likely time, a is the most optimistic time, b is the most pessimistic time and m is the most probable time.

In most cases m, the most probable time, is sufficient for the estimate, as the numerical difference between this and the result obtained by rigorously applying the formula is very small in most cases.

Further reading

Stutzke, R. (2005). *Software project estimation*. Addison-Wesley.
Taylor, J. C. (2005). *Project cost estimating tools techniques & perspectives*. St: Lucie Press.
Thacker, N. (2012). *Winning your bid*. Gower.

Project management plan

14

Chapter outline

As soon as the project manager has received his or her brief or project instructions, he or she must produce a document that distils what is generally a vast amount of information into a concise, informative and well-organized form that can be distributed to all members of the project team and indeed all the stakeholders in the project. This document is called a project management plan (PMP). It is also sometimes just called a project plan, or in some organizations a coordination procedure.

The PMP is one of the key documents required by the project manager and his or her team. It lists the phases and encapsulates all the main parameters, standards and requirements of the project in terms of time, cost and quality/performance by setting out the 'Why', 'What', 'When', 'Who', 'Where' and 'How' of the project. In some organizations, the PMP also includes the 'How much', that is, the cost of the project. There may, however, be good commercial reasons for restricting this information to key members of the project team.

The contents of a PMP vary depending on the type of project. Whilst it can run to several volumes for a large petrochemical project, it need not be more than a slim binder for a small, unsophisticated project.

There are, however, a number of areas and aspects that should always feature in such a document. These are set out very clearly in Table 1 of BS 6079-1-2002. With the permission of the British Standards Institution the main headings of the Project Management Plan (PMP), which have been augmented and re-arranged, are given below.

Project Management, Planning and Control. https://doi.org/10.1016/B978-0-12-824339-8.00014-6

General

1. Foreword
2. Contents, distribution and amendment record
3. Introduction
   3.1 Project diary
   3.2 Project history

The Why

4. Project aims and objectives
   4.1 Business case

The What

5. General description
   5.1 Scope
   5.2 Project requirement
   5.3 Project security and privacy
   5.4 Project management philosophy
   5.5 Management reporting system

The When

6. Programme management
   6.1 Programme method
   6.2 Program software
   6.3 Project life cycle
   6.4 Key dates
   6.5 Milestones and milestone slip chart
   6.6 Bar chart and network if available

The Who

7. Project organization
8. Project resource management
9. Project team organization
   9.1 Project staff directory
   9.2 Organizational chart

9.3 Terms of reference
   (a) For staff
   (b) For the project manager
   (c) For the committees and working group

The Where

10. Delivery requirements
   10.1 Site requirements and conditions
   10.2 Shipping requirements
   10.3 Major restrictions

The How

11. Project approvals required and authorization limits
12. Project harmonization
13. Project implementation strategy
   13.1 Implementation plans
   13.2 System integration
   13.3 Completed project work
14. Acceptance procedure
15. Procurement strategy
   15.1 Cultural and environmental restraints
   15.2 Political restraints
16. Contract management
17. Communications management
18. Configuration management
   18.1 Configuration control requirements
   18.2 Configuration management system
19. Financial management
20. Risk management
   20.1 Major perceived risks
21. Technical management
22. Tests and evaluations
   22.1 Warranties and guarantees
23. Reliability management (see also BS 5760: Part 1)
   23.1 Availability, reliability and maintainability
   23.2 Quality management
24. Health and safety management
25. Environmental issues
26. Integrated logistic support management
27. Close-out procedure

The numbering of the main headings should be standardized for all projects in the organization. In this way, the reader will quickly learn to associate a clause number with a subject. This will not only enable him or her to find the required information quickly but will also help the project manager when he or she has to write the PMP. The numbering system will in effect serve as a convenient checklist. If a particular item or heading is not required, it is best to simply enter 'not applicable' (or NA), leaving the standardized numbering system intact.

In addition to giving all the essential information about the project between two covers, for quick reference, the PMP serves another very useful function. In many organizations, the scope and technical and contractual terms of the project are agreed on in the initial stages by the proposals or sales department. It is only when the project becomes a reality that the project manager is appointed. By having to assimilate all these data and write such a PMP (usually within 2 weeks of the handover meeting), the project manager will inevitably obtain a thorough understanding of the project requirements as he or she digests the often voluminous documentation agreed on with the client or sponsor.

Clearly, not every project requires the exact breakdown given in this list, and each organization can augment or expand this list to suit the project. If there are any subsequent changes, it is essential that the PMP is amended as soon as the changes become apparent so that every member of the project team is immediately aware of the latest revision. These changes must be numbered on the amendment record in the beginning of the PMP and annotated on the relevant page and clause with the same amendment number or letter.

The contents of the project management plan are neatly summarized in the first verse of the little poem from *The Elephant's Child* by Rudyard Kipling:

I keep six honest serving-men

(They taught me all I knew);

Their names are What and Why and When.

And How and Where and Who.

Methods and procedures

Methods and procedures are the very framework of project management and are necessary throughout the life cycle of the project. All the relevant procedures and processes are set out in the project management plan, where they are customized to suit the particular project.

Methods and procedures should be standardized within an organization to ensure that project managers do not employ their own pet methods or 'reinvent the wheel'.

All the organization's standard methods and procedures as well as some of the major systems and processes should be enshrined in a company project manual. This should then be read and signed by every project manager who will then be familiar with the company systems, and thus avoid wasteful and costly duplication. The main contents of such a manual are the methods and procedures covering:

- Company policy and mission statement
- Company organization and organization chart
- Accountability and responsibilities
- Estimating
- Risk analysis
- Cost control
- Planning and network analysis
- Earned value management
- Resource management
- Change management (change control)
- Configuration management
- Procurement (bid preparation, purchasing, expediting, inspection, shipping)
- Contract management and documentation
- Quality management and control
- Value management and value engineering
- Issue management
- Design standards
- Information management and document distribution
- Communication
- Health and safety
- Conflict management
- Close-out requirements and reviews

It will be seen that this list is very comprehensive, but in every case a large proportion of the documentation required can be standardized. There are always situations where a particular method or procedure has to be tailored to suit the circumstances or where a system has to be simplified, but the standards set out in the manual form a baseline that acts as a guide for any necessary modification.

Certain UK government departments, a number of local authorities and other public bodies have adopted a project management methodology called PRINCE 2 (an acronym for projects in a controlled environment). This was developed by the Central Computer and Telecommunications Agency for IT and government contracts, but has not found favour in the construction industry due to a number of differences in approach to reporting procedures, management responsibilities and assessing durations with respect to resources.

Risk management

Chapter outline

Every day we take risks. If we cross the street, we risk being run over. If we go down the stairs, we risk missing a step and tumbling down. Taking risks is such a common occurrence that we tend to ignore it. Indeed, life would be unbearable if we are constantly worried whether or not we should carry out a certain task or take an action, because the risk is, or is not, acceptable.

With projects, however, this luxury of ignoring the risks cannot be permitted. By their very nature, because projects are inherently unique and often incorporate new techniques and procedures, they are risk prone and the risk has to be considered right from the start. It then has to be subjected to a disciplined regular review and investigative procedure known as risk management.

Before applying risk management procedures, many organizations produce a *risk management plan*. This is a document produced at the start of the project that sets out the strategic requirements for risk assessment and the whole risk management procedure. In certain situations, the risk management plan should be produced at the estimating or contract-tender stage to ensure that adequate provisions are made in the cost build-up of the tender document.

The project management plan (PMP) should include a résumé of the risk management plan, which will, first of all, define the scope and areas to which risk management applies, particularly the risk types to be investigated. It will also specify which techniques will be used for risk identification and assessment, whether SWOT (strengths, weaknesses, opportunities and threats) analysis is required and which risks (if any) require a more rigorous quantitative analysis such as Monte Carlo simulation methods.

The risk management plan will set out the type, content, and frequency of reports, the roles of risk owners, and the definition of the impact and probability criteria in qualitative and/or quantitative terms covering cost, time and quality/performance.

Project Management, Planning and Control. https://doi.org/10.1016/B978-0-12-824339-8.00015-8

The main contents of a risk management plan are as follows:

- *General introduction.* Explaining the need for the risk management process;
- *Project description.* Only required if it is a stand-alone document and not part of the PMP;
- *Types of risks.* Political, technical, financial, environmental, security, safety, programme, etc.;
- *Risk processes.* Qualitative and/or quantitative methods, maximum number of risks to be listed;
- *Tools and techniques.* Risk identification methods, size of P—I matrix, computer analysis, etc.;
- *Risk reports.* Updating periods of risk register, exception reports, change reports, etc.; and
- *Attachments.* Important project requirements, dangers, exceptional problems, etc.

The risk management plan of an organization should follow a standard pattern in order to increase its familiarity (rather like standard conditions of contract), but each project will require a bespoke version to cover its specific requirements and anticipated risks.

Risk management consists of the following five stages, which, if followed religiously, will enable one to obtain a better understanding of the project risks that could jeopardize the cost, time, quality and safety criteria of the project. The first three stages are often referred to as *qualitative analysis* and are by far the most important stages of the process.

Stage 1: risk awareness

This is the stage at which the project team begins to appreciate that there are risks to be considered. The risks may be pointed out by an outsider, or the team may be able to draw on their own collective experience. The important point is that once this attitude of mind is set, that is, that the project, or certain facets of it, is at risk, it leads very quickly to risk identification.

Stage 2: risk identification

This is essentially a team effort at which the scope of the project, as set out in the specification, contract and WBS (if drawn), is examined and each aspect is investigated for a possible risk.

To get the investigation going, the team may have a brainstorming session and use a prompt list (based on specific aspects such as legal or technical problems) or a checklist compiled from risk issues from similar previous projects. It may also be possible to obtain an expert opinion or carry out interviews with outside parties. The end product is a long list of activities that may be affected by one or a number of adverse situations or unexpected occurrences. The risks that generally have to be considered may be conveniently split into four main areas as shown in Table 15.1. Any applicable risk

Table 15.1

Project risks			
Organization	Environment	Technical	Financial
Management	Legislation	Technology	Financing
Resources	Political	Contracts	Exchange rates
Planning	Pressure groups	Design	Escalation
Labour	Local customs	Manufacture	Financial stability of
Health and safety	Weather	Construction	(a) Project
Claims	Emissions	Commissioning	(b) Client
Policy	Security	Testing	(c) Suppliers

in each area can then be examined by a further screening process as shown by the samples given in the following:

Technical	New technology or materials, test failures
Environmental	Unforeseen weather conditions, traffic restrictions
Operational	New systems and procedures, training needs
Cultural	Established customs and beliefs, religious holidays
Financial	Freeze on capital, bankruptcy of stakeholder, currency fluctuations
Legal	Local laws, lack of clarity of contract
Commercial	Change in market conditions or customers
Resource	Shortage of staff, operatives, plant or materials
Economic	Slow-down of economy, change in commodity prices
Political	Change of central or local government or government policies
Security	Safety, theft, vandalism, civil disturbance

Some risks could be categorized in more than one area or section, such as civil unrest, which could be a political as well as a security problem.

The following gives the advantages and disadvantages:

Brainstorming	
Advantages	Wide range of possible risks suggested for consideration, involves a number of stakeholders
Disadvantages	Time consuming, requires firm control by the facilitator

Continued

Prompt List	
Advantages	Gives benefit of past problems, saves time by focusing on real possibilities
Disadvantages	Restricts suggestions to past experience, past problems may not be applicable
Checklist	
Advantages	Similar to prompt list: company standard
Disadvantages	Similar to prompt list
Work Breakdown Structure	
Advantages	Focused on specific project risks
	Quick and economical
Disadvantages	May limit scope of possible risks
Delphi Technique	
Advantages	Offers wide experience of experts
	Can be wide ranging
Disadvantages	Time consuming if experts are far away
	Expensive if experts have to be paid
	Advice may not be specific enough
Asking Experts	
Advantages	Similar to Delphi
Disadvantages	Similar to Delphi

At this stage, it may be possible to identify who is the best person to manage each risk. This person becomes the *risk owner*.

To reduce the number of risks being seriously considered from what could well be a very long list, some form of screening will be necessary. Only those risks that pass certain criteria need to be examined more closely, which leads to the next stage of risk management:

Stage 3: risk assessment

This is the qualitative stage at which the two main attributes of a risk, *probability* and *impact*, are examined.

The *probability* of a risk becoming a reality has to be assessed using experience and/or statistical data, such as historical weather charts or close-out reports from previous projects. Each risk can then be given a probability rating of high, medium or low.

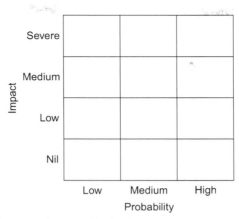

Figure 15.1 Probability versus impact table. Such a table could be used for each risk worthy of further assessment, and to assess, for example, all major risks to a project or programme.

In a similar way, by taking into account all the available statistical data, past project histories and expert opinion, the *impact* or effect on the project can be rated as severe, medium or low.

A simple matrix can now be drawn up which identifies whether a risk should be taken any further. Such a matrix is shown in Fig. 15.1.

Each risk can now be given a *risk number* so that it is possible to draw up a simple chart that lists all the risks considered so far. This chart will show the risk number, a short description, the risk category, the probability rating, the impact rating (in terms of high, medium or low) and the *risk owner* who is charged with monitoring and managing the risk during the life of the project.

Fig. 15.2 shows the layout of such a chart.

A *quantitative analysis* can now follow, known as risk evaluation.

		Risk Summary Chart		
Risk no.	Description	Probability rating	Impact rating	Risk owner

Figure 15.2 Risk summary chart.

Stage 4: risk evaluation

It is now possible to give comparative values, often on a scale 1−10, to the probability and impact of each risk and by drawing up a matrix of the risks, an order of importance or priority can be established. By multiplying the *impact rating* by the *probability rating*, the *exposure rating* is obtained. This is a convenient indicator that may be used to reduce the list to only the top dozen that require serious attention, but an eye should nevertheless be kept on even the minor ones, some of which may suddenly become serious if unforeseen circumstances arise.

An example of such a matrix is shown in Fig. 15.3. Clearly, the higher the value, the greater the risk and the more attention it must receive to manage it.

Another way to quantify both the impact and probability is to number the ratings as shown in Fig. 15.4 from one for very low to five for very high. By multiplying the appropriate numbers in the boxes, a numerical (or quantitative) exposure rating is obtained, which gives a measure of seriousness and hence importance for further investigation.

For example, if the impact is rated 3 (i.e. medium) and the probability 5 (very high), the exposure rating is $3 \times 5 = 15$.

Further sophistication in evaluating risks is possible by using some of the computer software developed specifically to determine the probability of occurrence. These programs use sampling techniques such as 'Monte Carlo simulations' that carry out hundreds of iterative sampling calculations to obtain a probability distribution of the outcome.

One application of the Monte Carlo simulation is determining the probability to meet a specific milestone (e.g. the completion date) by giving three time estimates to every activity. The program will then carry out a great number of iterations resulting in a frequency/time histogram and a cumulative 'S' curve from which the probability of meeting the milestone can be read off (see Fig. 15.5).

Exposure table							
			Probability				
	Rating		Very low	Low	Medium	High	Very high
		Value	0.1	0.2	0.5	0.7	0.9
Impact	Very high	0.8					
	High	0.5					
	Medium	0.2					
	Low	0.1					
	Very low	0.05					

Figure 15.3 Exposure table.

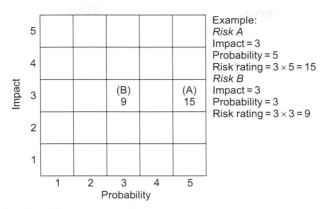

Figure 15.4 5 × 5 matrix.

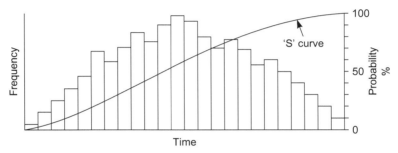

Figure 15.5 Frequency/time histogram.

At the same time, a *Tornado* diagram can be produced, which shows the sensitivity of each activity as far as it affects the project completion (see Fig. 15.6).

Other techniques such as sensitivity diagrams, influence diagrams and decision trees have all been developed in an attempt to make risk analysis more accurate or more reliable. It must be remembered, however, that any answer is only as good as the initial assumptions and input data, and the project manager must give serious consideration as to the cost-effectiveness of these methods for his/her particular project.

Stage 5: risk management

Having listed and evaluated the risks and established a table of priorities, the next stage is to decide how to manage the risks; in other words, what to do about them and who should be responsible for managing them. For this purpose, it is advisable to appoint a *risk owner* for every risk that has to be monitored and controlled. A risk owner may, of course, be responsible for a number, or even all, of the risks. There are a number of

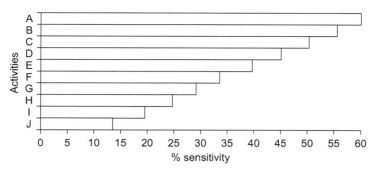

Figure 15.6 Tornado diagram.

options available to the project manager when faced with a set of risks. These are as follows:

* Avoidance
* Reduction
* Sharing
* Transfer
* Deference
* Mitigation
* Contingency
* Insurance
* Acceptance

These options are perhaps most easily explained by a simple example.

The owner of a semi-detached house decides to replace part of his roof with solar panels to save on his water-heating bill. The risks involved in carrying out this work are as follows:

Risk 1	The installer may fall off the roof
Risk 2	The roof may leak after completion
Risk 3	The panels may break after installation
Risk 4	Birds may befoul the panels
Risk 5	The electronic controls may not work
Risk 6	The heat recovered may not be sufficient to heat the water on a cold day
Risk 7	It may not be possible to recover the cost if the house is sold within 2–3 years
Risk 8	The cost of the work will probably never pay for itself
Risk 9	The cost may escalate due to unforeseen structural problems

These risks can all be managed by applying one or several of the following options:

Risk 1	Transfer	Employ a builder who is covered by insurance
Risk 2	Transfer	Insist on a 2-year guarantee for the work (at least two season cycles)
Risk 3	Insurance	Add the panel replacement to the house insurance policy
Risk 4	Mitigation	Provide access for cleaning (this may increase the cost)
Risk 5	Reduction	Ensure a control unit is used which has been proven for a number of years
Risk 6	Contingency	Provide for an electric immersion heater for cold spells
Risk 7	Deference	Wait for 3 years before selling the house
Risk 8	Acceptance	This is a risk one must accept if the work goes ahead
Risk 8	Avoidance	Do not go ahead with the work
Risk 9	Sharing	Persuade the neighbour in the adjoining house to install a similar system at the same time

Monitoring

To keep control of the risks, a *risk register* should be produced that lists all the risks and their method of management. Such a list is shown in Fig. 15.7. When risk owners are appointed, these will be identified on the register. The risks must be constantly monitored and the register must be reassessed at preset periods, and if necessary amended to reflect the latest position. Clearly, as the project proceeds, the risks reduce in number so that the contingency sums allocated to cover the risk of the completed activities can be allocated to other sections of the budget. These must be recorded in the register under the heading of *risk closure*. However, sometimes new rules emerge that must be taken into account.

The summary of the risk management procedure is then as follows:

1. Risk awareness
2. Risk identification (checklists, prompt lists, brainstorming)
3. Risk owner identification
4. Qualitative assessment
5. Quantification of probability
6. Quantification of impact (severity)
7. Exposure rating
8. Mitigation
9. Contingency provision
10. Risk register
11. Software usage (if any)
12. Monitoring and reporting

| Project:.. | Prepared by: .... | | | | Reference:..................... | | |
| Key: H, high: M, medium: L, low | .... | | | | Date:............................. | | |

Type of risk	Description of risk	Probability			Impact			Risk reduction strategy	Contingency plans	Risk owner
		H	M	L	Perf.	Cost	Time			

Figure 15.7 Risk register (risk log).

To aid the process of risk management, a number of software tools have been developed. The most commonly used ones are *@Risk, Predict, Pandora* and *Plantrac Marshal*, but new ones will be undoubtedly developed in the future.

Example of effective risk management

One of the most striking and beneficial advantages of risk analysis is associated with the temporary jetties system known as Mulberry Harbour, which was towed across the Channel to support Allied landings in Normandy on D-Day, 6 June 1944.

The two jetties (A) at Gold beach for the British army and (B) at Omaha for the American army consisted of floating roadways, pontoons and caissons, protected by breakwaters, which were prefabricated at numerous sites across Britain and towed across the Channel where the caissons were sunk onto the seabed. All site construction of the roadways was carried out by British and American military engineers. The harbours were completed 12 days later on 18 June, and on 19 June an enormous storm blew up, which by 22 June destroyed the American and badly damaged the British harbour.

Fortunately, someone carried out a risk analysis that identified the possibility of such a disaster. As a result, the provision had been made for a large quantity of spare units to be ready at British ports to be towed out as replacements. By using these spare sections and cannibalizing the wrecked American harbour, the British harbour could be repaired and until it was abandoned following the capture of Cherbourg 8 months later, more than 7000 tons of vital military supplies per day were delivered to the allied armies in France.

Positive risk or opportunity

Although most risks are generally regarded as negative or undesirable, and indeed most mitigation strategies have been devised to reduce the impact or probability of negative risk, paradoxically, there is also such a thing as positive risk or opportunistic risk. This is basically the risk that any entrepreneur or investor takes when he/she invests in a new enterprise. A simple case of 'Nothing ventured, nothing gained'. A case may also arise where a perceived negative risk becomes a positive risk or opportunity. For example, in an attempt to reduce the risk of skidding, a car manufacturer may invent an anti-skid device that can be marketed independently at a profit. If there had been no risk, there would have been no need for the antidote.

Local authorities tend to use the term 'positive risk taking' when referring to the proactive approach of providing services and facilities to certain sections of the community (usually challenged by a disability). However, in addition to discussing and agreeing on their risks with clients, the process is in effect a comprehensive programme of ascertaining, prioritizing and mitigating the perceived risks before they occur.

Further reading

Barkley. (2004). *Project risk management.* McGraw-Hill.
Chapman, C. B., & Ward, S. C. (2003). *Project risk management* (2nd ed.). Wiley.
Garlick, A. (2007). *Estimating risk.* Gower.
Hillson, D. (2009). *Managing risk projects.* Gower.
Hopkinson, M. (2010). *The project risk maturity model.* Gower.
Hulett, D. (2009). *Practical schedule risk analysis.* Gower.
Ward, S., & Chapman, C. (2011). *How to manage project opportunity and risk.* Wiley.

Quality management

16

Chapter outline

Quality is remembered long after the price is forgotten.

Gucci.

Quality (or performance) forms the third corner of the time—cost—quality triangle, which is the basis of project management.

A project may be completed on time and within the set budget, but if it does not meet the specified quality or performance criteria it will at best attract criticism and

Project Management, Planning and Control. https://doi.org/10.1016/B978-0-12-824339-8.00016-X

at worst be considered a failure. Striking a balance between meeting the three essential criteria of time, cost and quality is one of the most onerous tasks for a project manager, and in practice usually one will be paramount. When quality is synonymous with safety, as with aircraft or nuclear design, there is no question of which point of the project management triangle is the most important. However, even if the choice is not so obvious, a failure in quality can be expensive and dangerous and can destroy an organization's reputation far quicker than it took to build up.

Quality management is therefore an essential part of project management, and as with any other attribute, it does not just happen without a systematic approach. To ensure a quality product, it has to be defined, planned, designed, specified, manufactured, constructed (or erected) and commissioned to an agreed set of standards that involve every department of the organization from top management to dispatch.

It is not possible to build quality into a product. If a product meets the specified performance criteria for a minimum specified time, it can be said to be a quality product. Whether the cost of achieving these criteria is high or low is immaterial, but to ensure that the criteria are met will almost certainly require additional expenditure. If these costs are then added to the normal production costs, a quality assured product will normally cost more than an equivalent one that has not gone through a quality control process.

Quality is an attitude of mind, and for it to be most effective, every level of an organization should be involved and committed to achieving the required performance standards by setting and operating procedures and systems which ensure this. It should permeate right through an organization from the board of directors down to the operatives on the shop floor.

Ideally, everyone should be responsible for ensuring that his or her work meets the quality standards set down by the management. To ensure that these standards are met, quality assurance requires checks and audits to be carried out on a regular basis.

However, producing a product that has not undergone a series of quality checks and tests and therefore not met customers' expectations could be much more expensive, as there will be more returns of faulty goods and fewer returns of customers. In other words, quality assurance is good business. It is far better to get it right first time, every time, than to have a second attempt or carry out a repair.

To enable this consistency of performance to be obtained (and guaranteed), the quality assurance, control, review and audit procedures have to be carried out in an organized manner and the following functions and actions need to be implemented:

1. The quality standards have been defined.
2. The quality requirements have been disseminated.
3. The correct equipment has been set up.
4. The staff and operatives have been trained.
5. The materials have been tested and checked for conformity.
6. Adequate control points have been set up.
7. The designated components have been checked at predetermined stages and intervals.
8. A feedback and rectification process has been set up.
9. Regular quality audits and reviews are carried out.
10. All these steps, which make up quality control, are enshrined in the quality manual together with the quality policy, quality plan and quality programme.

History

The first quality standards were produced in the United States for the military as MIL-STD and were subsequently used by NATO.

In the 1970s, the MOD issued the Defence Standard series 05 to 20 (Def. Std.) based on the American MIL-Q-9858A, but it was then superseded by 15 parts of the Allied Quality Assurance Publication.

Defence contractors and other large companies adopted the MOD system until BSI produced the BS 5750 series of Quality Systems in 1979. These were updated in 1987 and then became an international standard (ISO), the ISO 9000:1987 series, which also covers the European standard EU 29,000 series.

To understand the subject, it is vital that the definitions of the various quality functions are understood. These are summarized in the following list and explained more fully in the subsequent sections.

Quality management definitions

Quality

The totality of features and characteristics of a product, service or facility that bear on its ability to satisfy a given need.

Quality policy

The overall quality intentions and direction of an organization as regards quality, as formally expressed by top management.

Quality management

That aspect of overall quality functions that determines and implements the quality policy.

Quality assurance (QA)

All planned and systematic actions necessary to provide confidence that an item, service or facility will meet the defined requirements.

Quality systems (quality management systems or QMS)

These include the organizational structures, responsibilities, procedures, processes and resources for implementing the quality management.

Quality control (QC)

Those quality assurance actions that provide a means of control and measure the characteristics of an item, process or facility to established requirements.

Quality manual

These are a set of documents that communicates the organization's quality policy, procedures and requirements.

Quality programme

A contract (project)-specific document that defines the quality requirements, responsibilities, procedures and actions to be applied at various stages of the contract.

Quality plan

A contract (project)-specific document defining the actions and processes to be undertaken together with the hold points for reviews and inspections. It also defines the control document, applicable standards, inspection methods and inspection authority. This authority may be internal and/or may include the client's inspectors or an independent/statutory inspection authority.

Quality audit

A periodic check to ensure that the quality procedures set out in the quality plan have been carried out.

Quality reviews

Periodic reviews of the quality standards, procedures and processes to ensure their applicability to current requirements.

Total quality management (TQM)

The company-wide approach to quality beyond the prescriptive requirements of a quality management standard such as ISO 9001.

Explanation of the definitions

Quality policy

The quality policy has to be set by top management and issued to the whole organization so that everyone is aware what the aims of management regarding quality are. The quality policy might be to produce a component that lasts a specific period of time under normal use, withstands a set number of reversing cycles before cracking, withstands a defined load or pressure or, on the opposite scale, lasts only a limited number of years so that a later model can be produced to replace it. A firm of house-builders might have a quality policy to build all their houses to the highest standards in only the most desirable locations, or the top management of a car manufacturer might dictate to their design engineers to design a car using components that will not fail for at least 5 years. There is clearly a cost and marketing implication in any quality policy that must be taken into account.

Quality management

Quality management can be divided into two main areas: quality assurance (QA) and quality control (QC). All the quality functions, such as the procedures, methods, techniques, programmes, plans, controls, reviews and audits, make up the science of quality management.

It also includes all the necessary documentation and its distribution, the implementation of the procedures and the training and appointment of quality managers, testers, checkers, auditors and other staff involved in quality management.

Quality assurance

QA is the process that ensures that adequate quality systems, processes and procedures are in place. It is the term given to a set of documents that provide evidence of how and when the different quality procedures and systems are actually being implemented. These documents give proof that quality systems are in place and adequate controls have been set up to ensure compliance with the quality policy. To satisfy him- or herself that the quality of a product he or she needs is to the required standard, the buyer may well ask all tenderers or suppliers to produce their quality assurance documents with their quotations or tenders.

Guidelines for quality management and quality assurance standards are published by various national and international institutions including the British Standards Institution, which publishes the following quality standards:

ISO 9000	Quality management and quality assurance standards
ISO 9001	Quality systems — model for quality assurance in design and development, production, installation and servicing
ISO 9002	Quality systems — model for quality assurance in production and installation
ISO 9003	Quality systems — model for quality assurance in inspection and testing
ISO 9004	Guide to quality management and quality systems elements
ISO 10006	Guidelines for quality in project management
ISO 10007	Guidelines for configuration management

Quality systems

Quality systems or quality management systems, as they are often called, are the structured procedures that in fact enable quality control to be realized. The systems required include the levels of responsibility for quality control, such as hierarchical diagrams (family trees) showing who is responsible to whom and for which part of the quality spectrum they are accountable for, as well as the procedures for recruiting and training staff and operatives. Other systems cover the different quality procedures and processes that may be common to all, as well as for all the components or specifics for particular ones.

Quality systems also include the procurement, installation, operation and maintenance of equipment for carrying out quality checks. These cover such equipment as measuring tools, testing bays, non-destructive testing equipment for radiography (X-rays), magnetic particle scans, ultrasonic inspection and all the other different techniques being developed for testing purposes.

Documentation plays an important role in ensuring that the tests and checks have been carried out as planned, and the results accurately recorded and forwarded to the specified authority. Suitable action plans for recovering from deviations of set criteria will also form parts of the quality systems. The sequence of generating the quality-related documentation is shown in Fig. 16.1.

Quality control

The means to control and measure characteristics of a component, and the methods employed for monitoring and measuring a process or facility are parts of quality control. Control covers the actions to be taken by different staff and operators employed in the quality environment and the availability of the necessary tools to enable this work to be done. Again, the provision of the right documentation to the operatives and their correct, accurate and timely completion has to be controlled. This control covers the design, material specification, manufacture, assembly and distribution stages. The performance criteria are often set by the feedback from market research and customer requirements, and confirmed by the top management.

Quality manual

The quality manual is in fact the 'bible' of quality management. It is primarily a communication document which, between its covers, contains the organization's quality policy, various quality procedures, quality systems to be used and the list of personnel involved in implementing the quality policy. The manual will also contain the various test certificates required for certain operatives such as welders, the types of tests to be carried out on different materials and components, and the sourcing trails required for specified materials.

Quality programme

This is a document written specifically for the project in hand and contains all the requirements for that project. Various levels and stages for quality checks or tests will be

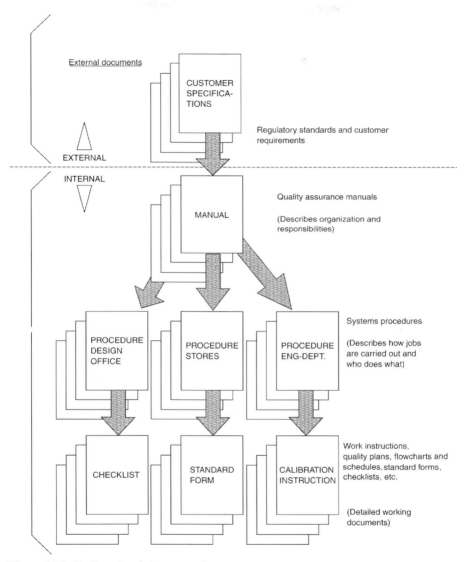

Figure 16.1 Quality-related documentation.

listed together with the names of the staff and operatives required at each stage, along with the reporting procedures and the names or organizations authorized to approve or reject the results, and instruct what remedial action (if required) has to be taken, especially when concessions for noncompliance are requested.

Quality plan

This is also a contract-specific document that can greatly vary in content and size from company to company. As a general rule, it defines in great detail:

1. The processes to be employed;
2. The hold points of each production process;
3. The tests to be carried out for different materials and components, including:
    3.1 Dimensional checks and weight checks
    3.2 Material tests (physical and chemical)
    3.3 Non-destructive tests (r/adiography, ultrasonic, magnetic particle, etc.)
    3.4 Pressure tests
    3.5 Leak tests
    3.6 Electrical tests (voltage, current, resistance, continuity, etc.)
    3.7 Qualification and capability tests for operatives
4. The control documents including reports and concession requests;
5. The standards to be applied for the different components;
6. The method of inspection;
7. The percentage of items or processes (such as welds) to be checked;
8. The inspection authorities, whether internal, external or statutory and
9. The acceptance criteria for the tests and checks.

Most organizations have their own standards and test procedures but may additionally be required to comply with the client's quality standards. A sample quality plan of components of a boiler is shown in Fig. 16.2.

The quality plan is part of the project management plan and because of its size is usually attached as an appendix.

Quality audit

To ensure that various procedures are implemented correctly, regular quality audits must be carried out across the whole spectrum of the quality systems. These audits vary in scope and depth and can be carried out by internal members of staff or external authorities. When an organization is officially registered as being in compliance with a specific quality standard, such as ISO 9000, an annual audit by an independent inspection authority may be carried out to ensure that the standards are still being met.

Quality reviews

As manufacturing or distribution processes change, a periodic review must be carried out to ensure that the quality procedures are still relevant and applicable in light of the changed conditions. Statutory standards may also have been updated, and these reviews check that the latest versions have been incorporated.

As part of the reviewing process, existing, proposed or new suppliers and contractors have to confirm their compliance with the quality assurance procedures. Fig. 16.3 shows the type of letter that should be sent periodically to all vendors.

CLIENT : X Y Z	QUALITY PLAN	Sheet : 3 of 3
CONTRACT No : 2-31-07797	FOR	Rev : 0 Issued by :
CODE : DS. 1113 :	BOILERS	Approved by:

LEGEND
D = Domestic Inspection
C = Certificate Required
T = Internal Record
R = Random Insp. (i.e. R10 = 10%)
E = External Surveillance = Mandatory Inspection

SURVEY AUTHORITY
B = British Engine
P = P.S.A.

Description of Part		REFERENCE DOCUMENT	BR. 600/108M BS. 1113	BS. 1113	F.W.P.P.	F.W.P.P.	F.W.P.P.	BS. 1113	BS. 1113	BS. 1113	F.W.P.P.				
		OPERATION	Material Identification	Inspection of Set Up	Inspection of Marking Off/ Drilling and Machining	Tube Manipulation	VisualInspection of Welding	Radiographic Examination	Magnetic Particle Examination	Hydraulic Test	Release for Despatch				
SUPERHEATER AND PANELS			1	2	3	4	5	6	7	8	9				
Headers and End Caps	D		C	I	I										
	E		B												
Header Circ. Welds	D						I	C							
	E							B							
Stubs/Panel Tubes	D		C		I10										
	E														
Stubs to Header/ Panel to Hdr.Welds	D			I10			I	C10							
	E							B							
Tubes	D		C	I10			I10								
	E														
Tube Butt Welds	D						I	C10							
	E							B							
Attachments	D		C	I5											
	E														
Attachment Welds	D						I								
	E														
Complete Superheater	D									C	C				
	E									B					
Completed Panels	D									C	C				
	E									B					

Figure 16.2 Quality assurance approval.

Quality assurance approval

                        1121/DAR/QA
            Our ref: Your ref:

DATE

F.A.O: <u>Quality Assurance Manager</u>

Dear Sirs,

<u>QUALITY ASSURANCE APPROVAL</u>

In order to meet the increasing Quality Assurance demands of our Clients, we
are revising our Approved Vendor Lists. Should you wish to either remain on,
or be added to these lists it will be necessary for you to complete the
attached document and return it to us without delay.

It is of the utmost importance that the document is fully completed and gives
all relevant information asked for regarding your existing Q.A. Approvals
including the following details:-

            <u>Sub-Contract Quality Assurance - Form A</u>
1)   The level approved at.

2)   The organisation or body who have awarded the approval.

3)   Certificate Number

4)   The date approved.

5)   The period of validity of the approval.

6)   Commodity and materials approved.

The completed form should be returned to:-

                        (state address here)

        Marked for the attention of Mr. John Brown - Procurement Manager

Figure 16.3 Confirmation of compliance request.

All the earlier procedures are sometimes described as the 'tools' of quality management to which the following techniques can be added:

- Failure mode analysis (cause and effect analysis)
- Pareto analysis
- Trend analysis

Failure mode analysis (cause and effect analysis)

This technique involves selecting certain (usually critical) items and identifying all the
possible modes of failure that could occur during its life cycle. The probability, causes
and impact of such a failure are then assessed and the necessary controls and

rectification processes are put in place. Clearly, as with risk analysis, the earlier in the project this process is carried out, the more opportunity there is to anticipate a problem and, if necessary, change the design to 'engineer' it out.

The following example illustrates how this technique can be applied to find the main causes affecting the operability of a domestic vacuum cleaner.

The first step is to list all the main causes that are generally experienced when using such a machine. These causes (or quality shortcomings), which may require a brainstorming session to generate them, are as follows:

- Electrical
- Physical (weight and size)
- Mechanical (brush wear)
- Suction (dust collection)

The second step is drawing a *cause and effect* diagram as shown in Fig. 16.4, which is also known as an *Ishikawa* or *fishbone* diagram, from which it is possible to see clearly how these causes affect the operation of the vacuum cleaner.

The third step requires all the sub-causes (or reasons) of a main cause to be written against the tributary lines (or fish bones) of each cause. For example, the sub-causes of electrical failure are the lead being too short, thus pulling the plug out of the socket or hauling the cleaner by the lead and causing a break in the cable.

The last step involves an assessment of the number of times over a measured period each cause has resulted in a failure. However, it is highly advantageous to concentrate on those causes that are responsible for the most complaints, and when this has been completed and assessed by applying the next technique, Pareto analysis, appropriate action can be taken to resolve any problem or rectify any error.

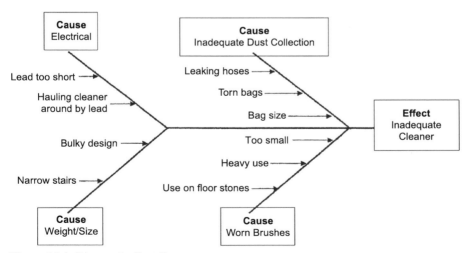

Figure 16.4 Cause and effect diagram.

Pareto analysis

In the 19th century, Vilfredo Pareto discovered that in Italy 90% of income was earned by 10% of the population. Further study showed that this distribution was also true for many other situations from political power to industrial problems. He therefore formulated Pareto's law, which states that 'In any series of elements to be controlled, a small fraction in terms of the number of elements always account for a large fraction in terms of effect'.

In the case of the vacuum cleaner, this is clearly shown in the Pareto chart in Fig. 16.5, which plots the impact Y in terms of percentage of problems encountered against the number of causes X identified. The survey of faults shows that of the four main causes examined, inadequate dust collection is responsible for 76% (nearly 80%) of the failures or complaints. This is why Pareto's law is sometimes called the 80/20 rule.

The percentage figure can be calculated by tabulating the causes and the number of times they resulted in a failure over a given period, say 1 year, and then converting these into a percentage of the total number of failures. This is shown in Fig. 16.6. Clearly, such ratios are only approximate and can vary widely, but in general only a relatively small number of causes are responsible for the most serious effects. Anyone who is involved in club committee activities will know that there are always a few keen members who have the greatest influence.

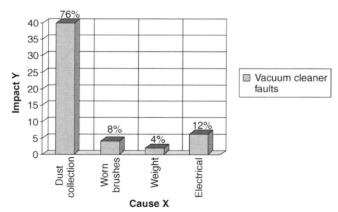

Figure 16.5 Pareto chart.

Cause	No. of failures	% of failures
Electrical	6	12
Physical	2	4
Mechanical	4	8
Suction	38	76
Total	50	100

Figure 16.6 Failure table.

Trend analysis

Part of the quality control process is to issue regular reports that log non-conformance, accepted or non-accepted variances, delays, cost overruns or other problems. If these reports are reviewed on a regular basis, trends may be discerned which, if considered to be adverse, can then be addressed by taking appropriate corrective action. At the same time, the opportunity can be taken to check whether there has been a deficiency in the procedures or documents and whether other components could be affected. Most importantly, the cause and source of a failure have to be identified, which may require a review of all the suppliers and subcontractors involved.

Further reading

BSI. (2008). *BS EN ISO 90001.2008 Quality management systems requirements*. BSI.
Rose, K. (2005). *Project quality management*. J. Ross.

Change management

17

Chapter outline

There are very few projects that do not change in some way during their life cycle. Equally, there are very few changes that do not affect in some way either (or all) the time, cost or quality aspects of the project. For this reason, it is important that all changes are recorded, evaluated and managed to ensure that the effects are appreciated by the originator of the change, and the party carrying out the change is suitably reimbursed when the change is a genuine extra to the original specification or brief.

In cases where a formal contract exists between the client and the contractor, an equally formal procedure of dealing with changes (or variations) is essential to ensure that:

1. No unnecessary changes are introduced.
2. The changes are only issued by an authorized person.
3. The changes are evaluated in terms of cost, time and performance.
4. The originator is made aware of these implications before the change is put into operation (In practice, this may not always be possible if the extra work has to be carried out urgently for safety or security reasons. In such a case, the evaluation and report of the effect must be produced as soon as possible).
5. The contractor is compensated for the extra costs and given extra time to complete the contract.

Unfortunately, clients do not always appreciate what effect even a minor change can have on a contract. For example, a client might think that eliminating an item of equipment such as a small pump a few weeks into the contract would reduce the cost. He or she might well find, however, that the changes in the design documentation, datasheets, drawings, bid requests, etc. will actually cost more than the capital value of the pump so that the overall cost of the project will increase. Therefore, the watchwords must be: *is the change really necessary?*

Project Management, Planning and Control. https://doi.org/10.1016/B978-0-12-824339-8.00017-1

In practice, as soon as a change or variation is requested either verbally or by a change order, it must be confirmed back to the originator with a statement to the effect that the cost and time implications will be advised as soon as possible.

A change of contract-scope notice must then be issued to all departments who may be affected to enable them to assess the cost, time and quality implications of the change.

A copy of such a document is shown in Fig. 17.1, which should contain the following information:

- Project or contract no.
- Change of scope no.

FOSTER WHEELER POWER PRODUCTS LTD. HAMSTEAD ROAD LONDON NW1 7QN	**ADVICE OF CHANCE OF CONTRACT SCOPE** DEPARTMENT: ENGINEERING No 82

To: Contract Management Department	Distribution:
☑ Please note that the scope of the subject contract has been altered due to the change(s) detailed below. To: Contract Management Department ☐ The following is a statement of the manhours and expenses incurred due to Contract Variation Notice reference ____ dated 17 Dec. 1982 ICI BILLINGHAM 2-32-07059	Project Manager Estimating Department Management Services Deaprtmental Manager Manager Engineering (see note 4 below) File N. Smith J. Harris By internal mail

BRIEF DESCRIPTION OF CHANGE AND EFFECT ON DEPARTMENTAL WORK

The provision of an 'Air to Igniters' control valve.
Scope of work includes purchasing and adding to drawings.
The clients preferred-specified vendor for control valves is Fisher Controls.

Manhour requirements are as follows:-
Dept 1104-63 manhours (1104 Split)
Dept 1102- 8 manhours Req. 60
Drg. 2
Dept 1105-38 manhours MH. 1
63

CHANGE NOTIFIED BY		MANHOURS AND COSTS INCURRED IN		
☐ Minutes of Meeting with client	Date of Meeting : Subject of Meeting : Minute Number :	Department No.	1104, 1102, 1105	
		☐ Increase	☐ Decrease	
☐ Client's telex	Date of Telex : Reference : Signed by :	Engineering	69	Manhours
		Design/drghtg	37	Manhours
☒ Client's letter	Date of Letter : 10.12.1992 Reference : SGP 3641 Signed by : B. Francis	Tech clerks	3	Manhours
		TOTAL	109	Manhours
☐ Client's request by telephone	Date of Call : Name of Contact :	COSTS	£T.B.A.	Manhours
☐ Client's Variation Order	V.O.Ref. : Date of V.O. :	Remarks		

NOTES	Initiated by N. Smith	Date 17.12.82
1. The 'change notified by' section need not be completed if form is used to advise manhours and costs only. 2. This form to be completed IMMEDIATELY ON RECIEPT of definite instructions.	Checked by MWN	Date 22.12.82
3. Manhours MUST BE REALISTIC. Make FULL ALLOWANCE for all additional and re-cycle work. Take into account 'chain reaction' affect throughout department. 4. Submit copy of this form to Manager Engineering if manhours involved exceed 250.	Approved by	Date

Figure 17.1 Change of contract-scope form.

- Issue date
- Name of originator of change
- Method of transmission (letter, fax, telephone, e-mail, etc.)
- Description of change
- Date of receipt of change order or instruction

When all the affected departments have inserted their cost and time estimates, the form is sent to the originator for permission to proceed, or for the advice of the implications if the work has had to be started before the form could be completed. The method of handling variations will probably have been set out in the contract documentation, but it is important to follow the agreed procedures, especially if there are time limitations for submitting the claims at a later stage.

As soon as a change has been agreed, the cost and time variations must be added to the budget and programme, respectively, to give the revised target values against which costs and progress will be monitored. However, while all variations have to be recorded and processed in the same way, the *project budget* can only be changed (increased or decreased) when the variation has been requested by the client. When the change was generated internally, for example, by one of the design departments due to a discovered error, omission or necessary improvement, it is not possible to increase the budget (and hence the price) unless the client has agreed to this. The extra cost must still be clearly recorded and monitored but will only appear as an increase (or decrease) in the *actual* cost column of the cost report. The result will be a reduction or increase in the profit, depending on whether the change required more or fewer resources.

The accurate and timely recording and managing of changes could decide the fate of a project as making profit or losing money.

Change management must not be confused with the *management of change*, which is the art of changing the culture or systems of an organization and managing human reactions. Such a change can have far-reaching repercussions on the lives and attitudes of all the members of the organization, from the board level to the operatives on the shop floor. The way such changes are handled and the psychological approaches used to minimize stress and resistance are outside the scope of this book.

Document-control

Invariably, a change to even the smallest part of a project requires the amendment of one or more documents. These may be programmes, specifications, drawings, instructions and, of course, financial records. The amendment of each document is in itself a change, and it is vital that the latest version of the document is issued to *all* the original recipients. In order to ensure that this takes place, a document- or version-control procedure, must be a part of the project-management plan.

In practice, a document-control procedure can either be a single page of A4 or several pages of a spreadsheet as a part of the computerized project-management system. The format should, however, feature the following columns:

- Document number
- Document title
- Originator of document
- Original issue date
- Issue code (general or restricted)
- Name of originator (or department) of revision
- Revision (or version) number
- Date of revision (version)

The sheet should also include a list of recipients.

A separate sheet records the date the revised document is sent to each recipient and the date of acknowledgement of receipt.

Where changes have been made to one or more pages of a multipage document, such as a project-management plan, it is only necessary to issue the revised pages under a page revision number. This requires a discrete version-control sheet for this document with each clause listed and its revision and date of issue recorded.

Issue management

An issue is a threat that could affect the objectives or operations of a project. Unlike a risk, which is an uncertain event, an issue is a reality that may have already occurred or is about to occur.

The difference between an issue and a problem is that an issue is a change or potential change of external circumstances outside the control of the project manager. A problem, on the other hand, is a day-to-day adverse event the project manager has to deal with as a matter of routine. An issue, therefore, has to be brought to the attention of a higher authority such as the project board, steering group or programme manager, as it may well require additional resources, either human, financial or physical (material), which require sanctioning by senior management.

As with other types of changes, a register of issues, often called an issue log, must be set up and kept up to date. This not only records the type, source and date of the issue to be addressed, but also shows to whom it was circulated, the effect in terms of cost, time and performance it has on the project, and how it was resolved. If the issue is large or complex, it may, as with particular risks, be necessary to appoint an issue owner (or resolution owner) who will be responsible for implementing the agreed actions to resolve the matter.

The way an issue is resolved depends clearly on its size and complexity. Again, as with change requests, other departments that may be affected have to be advised and consulted, and in serious situations, special meetings may have to be convened to discuss the matter in depth and sufficient detail to enable proposals for a realistic solution to be tabled. Care must be taken not to escalate inevitable problems that arise during the course of a project to the status of an issue, as this would take up valuable management time of the members of the steering group or senior management. As is so often the case with project management, judgement and experience are the key attributes for handling the threats and vicissitudes of a project.

Baseline

The three baselines in project management are associated with the three project-management criteria, that is cost, time and performance. The baselines are set by the project manager at the beginning of the project and are incorporated into the project-management plan. All subsequent data gathered by the project reporting system which affect the cost or duration of the project are compared with the baseline, which can only be changed by agreed variations issued by the client. This is normally shown by a step change in the budget curve (usually straight line) of the earned value set of control curves.

There are instances where the originator of the change to the baseline is the project manager instead of the client. This occurs when the delay or cost overrun is due to an error by the project-management team or the contractor. The baseline will be changed, but the cost cannot be passed on to the client. Indeed, if the change delays the completion date, the party responsible may be liable for liquidated damages or other forms of compensation payment.

Where the change is originated by the client, the project manager has to issue a change of contract-scope notice as shown in Fig. 17.1.

Baseline review

The Association for Project Management published a book titled *Guide to Conducting Integrated Baseline Reviews* which sets out the procedures necessary for effective baseline management.

As set out in this book, the integrated baseline review (IBR) is a formal procedure carried out at prearranged intervals to assess the status of the project and take remedial action where necessary. The process is divided into four main stages: pre-IBR, preparation, execution and post-review.

Pre-IBR

This is when the need, policy, timing, objectives and acceptance criteria of the IBR are discussed and agreed.

Preparation

This stage includes assessing the competencies of and recruiting the IBR team, producing an IBR handbook and collating the required documents.

Execution

In this stage, every facet of the project such as cost, programme and performance is examined and discussed. After collating the discussions and feedback, the results are presented and issued as a final IBR report.

Post-review

This stage deals with the preparation of an action plan in which all the risks and problems have been identified and logged. A root-cause analysis should be carried out to identify areas which can be improved and remedial actions for this should be assigned to the appropriate members of the IBR team.

Further reading

APM. (2011). *Directing change*. APM.
Balogun, J., et al. (2008). *Exploring strategic change*. Prentice-Hall.
Ludovino, E. M. (2016). *Change management*. EM Press Ltd.

Configuration management

18

Although in the confined project management context, configuration management is often assumed to be synonymous with version control of documentation or software, it is of course much more far-reaching in the total project environment. Developed originally for the aerospace industry, it was created to ensure that changes and modifications to physical components, software, systems and documentation are recorded and identified in such a way that replacements, spares and assembly documentation conforms to the version in service. It was also developed to ensure that the design standards and characteristics were reflected accurately in the finished product.

It can be seen that when the projects involve complex systems as in the aerospace, defence or petrochemical industry, configuration management is of the utmost importance as the very nature of these industries involves development work and numerous modifications, not only from the original concept or design but also during the whole life cycle of the product.

Keeping track of all these changes to specifications, drawings, support documentation and manufacturing processes is the essence of configuration management, which can be split into the following five main stages:

1. *Configuration management and planning.* This covers the necessary standards, procedures, support facilities, resources and training, and sets out the scope, definitions, reviews, milestones and audit dates.
2. *Configuration identification.* This encompasses the logistics and systems and procedures. It also defines the criteria for selection in each of the project phases.
3. *Configuration-change management.* This deals with the proposed changes and their investigation before acceptance. At this stage, changes are compared with the configuration baseline including defining the stages when formal departure points have reached.
4. *Configuration-status accounting.* This records and logs the accepted (registered) changes and notifications as well as providing traceability of all baselines.
5. *Configuration audit.* This ensures that all the previous stages have been correctly applied and incorporated in the organization. The output of this stage is the audit report.

In all these stages, resources and facilities must always be considered, and arrangements must be made to feedback the comments to the management stage.

Essentially, the process of identification, evaluation and implementation of changes requires accurate monitoring and recording, and subsequent dissemination of documentation to the interested parties. This is controlled by a master record index (MRI). An example of such an MRI for controlling documents is shown in Fig. 18.1.

For large, complex and especially multinational projects, where the design and manufacture are carried out in different countries, great effort is required to ensure

Project Management, Planning and Control. https://doi.org/10.1016/B978-0-12-824339-8.00018-3

Master Record Index

Document Title	Reference Number	Documents		Responsibility	Distribution
		Issue	Date		
Business Case	Rqmt SR 123	Draft A	14/6/86	Mr Sponsor	PM, Line Mgmt
		Draft B	24/7/86		
		Issue 1	30/7/86		
		Issue 2	30/9/86		
Project Mgmt Plan	PMP/MLS/34	Draft A	28/7/86	Ms MLS PM	All Stakeholders
		Issue 1	30/9/86		
WBS	WBS/PD1	Draft A	30/7/86	Mr MLS Deputy PM	IPMT (Project Team)
		Issue 1	2/8/86		
Risk Mgmt Plan, etc.	RMP/MLS/1				

Figure 18.1 Master record index.

that the product configuration is adequately monitored and controlled. To this end, a *configuration-control committee* is appointed to head up special *interface-control groups* and *configuration-control boards* that investigate and, when accepted, approve all proposed changes.

Basic network principles

19

Chapter outline

It is true to say that whenever a process requires a large number of separate but integrated operations, a critical path network can be used to advantage. This does not mean, of course, that other methods are not successful or that the critical path method (CPM) is a substitute for these methods — indeed, in many cases network analysis can be used in conjunction with traditional techniques — but if correctly applied CPM will give a clearer picture of the complete programme than other systems evolved to date.

Every time we do anything, we string together, knowingly or unknowingly, a series of activities that make up the operation we are performing. Again, if we so desire, we can break down each individual activity into further components until we end up with the movement of an electron around a nucleus. Clearly, it is ludicrous to go to such a limit, but we can call a halt to this successive breakdown at any stage to suit our requirements. The degree of the breakdown depends on the operation we are performing or intend to perform.

In the United Kingdom, it was the construction industry that first realized the potential of network analysis, and most of the large, if not all, construction, civil engineering and building firms now use CPM regularly for their larger contracts. However, a contract does not have to be large before CPM can be usefully employed. If any process can be split into 20 or more operations or 'activities', a network will show their

Project Management, Planning and Control. https://doi.org/10.1016/B978-0-12-824339-8.00019-5

interrelationship in a clear and logical manner so that it may be possible to plan and rearrange these interrelationships to produce either a shorter or a cheaper project, or both.

Network analysis

Network analysis, as the name implies, consists of two basic operations:

1. Drawing the network and estimating the individual activity times
2. Analysing these times in order to find the critical activities and the amount of float in the non-critical ones

The network

Basically, the network is a flow diagram showing the sequence of operations of a process. Each individual operation is known as an activity and each meeting point or transfer stage between one activity and another is an event or node. If the activities are represented by straight lines and the events by circles, it is very simple to draw their relationships graphically, and the resulting diagram is known as the network. In order to show whether an activity has to be performed before or after its neighbour, arrowheads are placed on the straight lines, but it must be explained that the length or orientation of these lines is quite arbitrary. This format of network is called activity on arrow (AoA), as the activity description is written over the arrow.

It can be seen, therefore, that each activity has two nodes or events; one at the beginning and one at the end (Fig. 19.1). Thus, events 1 and 2 in the figure show the start and finish of activity A. The arrowhead indicates that 1 comes before 2, that is the operation flows towards 2.

We can now describe the activity in two ways:

1. By its activity title (in this case, A)
2. By its starting and finishing event nodes 1−2

For analysis purposes (except when using a computer), the second method must be used.

Basic rules

Before proceeding further it may be prudent at this stage to list some very simple but basic rules for network presentation, which must be rigidly adhered to the following:

Figure 19.1

1. When the starting node of an activity is also the finishing node of one or more other activities, it means that *all* the activities with this finishing node must be completed before the activity starting from that node can be commenced. For example, in Fig. 19.2, 1–3 (A) and 2–3 (B) must be completed before 3–4 (C) can be started.

2. Each activity must have a different set of starting and finishing node numbers. This poses a problem when two activities start and finish at the same event node, and means that the example shown in Fig. 19.3 is incorrect. In order to apply this rule, therefore, an artificial or 'dummy' activity is introduced into the network (Fig. 19.4). This 'dummy' has a duration of zero time and thus does not affect the logic or overall time of the project. It can be seen that activity A still starts at 1 and takes 7 units of time before being completed at event 3. Activity B also still takes 7 units of time before being completed at 3, but it starts at node 2. The activity between 1 and 2 is a timeless dummy.

3. When two chains of activities are interrelated, this can be shown by joining the two chains either by a linking activity or a 'dummy' (Fig. 19.5). The dummy's function is to show that

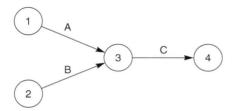

Figure 19.2

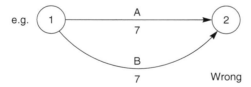

Figure 19.3

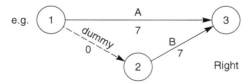

Figure 19.4

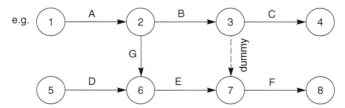

Figure 19.5

all the activities preceding it, that is 1−2 (A) and 2−3 (B) shown in Fig. 20.5, must be completed before activity 7−8 (F) can be started. Needless to say, activities 5−6 (D), 6−7 (E) and 2−6 (G) must also be completed before 7−8 (F) can be started.

4. Each activity (except the last) must run into another activity. Failure to do so creates a loose end or 'dangle' (Fig. 19.6). Dangles create premature 'ends' of a part of a project so that the relationship between this end and the actual final completion node cannot be seen. Hence the loose ends must be joined to the final node (in this case, node 6 in Fig. 19.7) to enable the analysis to be completed.

5. No chain of activities must be permitted to form a loop, that is such a sequence that the last activity in the chain has an influence on the first. Clearly, such a loop makes any logic sense-less as, if one considers activities 2−3 (B), 3−4 (C), 4−5 (E) and 5−2 (F) in Fig. 19.8, one finds that B, C and E must precede F, yet F must be completed before B can start. Such a situation cannot occur in nature and defies analysis.

Apart from strictly following the basic rules 1 to 5 set out earlier, the following points are worth remembering to obtain the maximum benefit from network techniques.

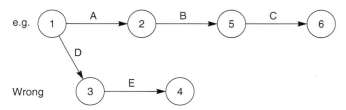

Figure 19.6

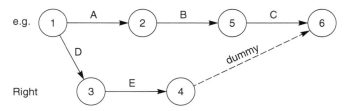

Figure 19.7

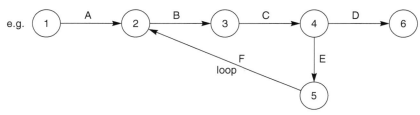

Figure 19.8

1. Maximize the number of activities that can be carried out in parallel. This obviously (resources permitting) cuts down the overall programme time.
2. Beware of imposing unnecessary restraints on any activity. If a restraint is convenient rather than imperative, it should best be omitted. The use of resource restraints is a trap to be particularly avoided since additional resources can often be mustered — even if at additional cost. The exception is when using 'critical chain' project management methods.
3. Start activities as *early* as possible and connect them to the rest of the network as *late* as possible (Figs. 19.9 and 19.10). This avoids unnecessary restraints and gives maximum float.
4. Resist the temptation to use a conveniently close node point as a 'staging post' for a dummy activity used as a restraint. Such a break in a restraint could impose an additional unnecessary restraint on the succeeding activity. In Fig. 19.11, the intent is to restrain activity E by B and D and activity G by D. However, because the dummy from B uses node 6 as a staging post, activity G is also restrained by B. The correct network is shown in Fig. 19.12. It must be remembered that the restraint on G may have to be added at a later stage so that the effect of B in Fig. 19.11 may well be overlooked.
5. When drawing ladder networks beware of the danger of trying to economize on dummy activities as described later (Figs. 19.24 and 19.25).

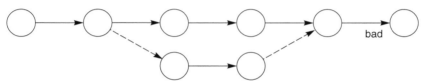

Figure 19.9

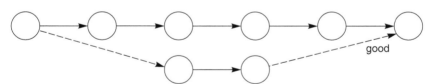

Figure 19.10

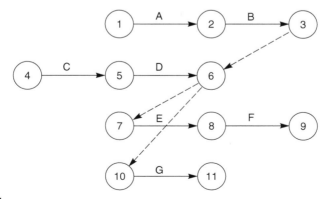

Figure 19.11

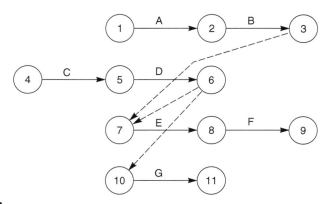

Figure 19.12

Durations

Having drawn the network in accordance with the logical sequence of the particular project requirements, the next step is to ascertain the duration or time of each activity. These can be estimated in the light of experience, in the same manner that any programme times are usually ascertained, for example using standard industry or company norms, but it must be remembered that the shorter the durations, the more accurate they usually are.

The durations are then written against each activity in any convenient time unit, but this must, of course, be the same for every activity. For example, referring to Fig. 19.13, if activities 1–2 (A), 2–5 (B) and 5–6 (C) took 3, 2 and 7 days, respectively, one would show this by merely writing these times under the activity.

Numbering

The next stage of network preparation is numbering the events or nodes. Depending on the method of analysis, the following systems shown in Fig. 19.14 can be used.

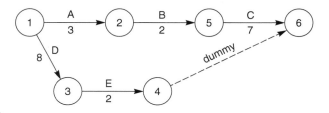

Figure 19.13

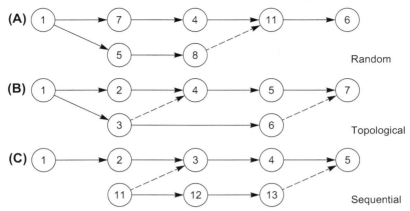

Figure 19.14

Random

This method, as the name implies, follows no pattern and merely requires each node number to be different. All computers (if used) can, of course, accept this numbering system, but there is always the danger that a number may be repeated.

Topological

This method was necessary for the original early computer programs, which demanded that the starting node of an activity be smaller than the finishing node of that activity. If this law is applied throughout the network, the node numbers will increase in value as the project moves towards the final activity. It had some value for beginners using network analysis since loops are automatically avoided. However, it is very time consuming and requires constant back-checking to ensure that no activity has been missed. The real drawback is that if an activity is added or changed, the whole network has to be renumbered from that point onwards. Clearly, this is an unacceptable restriction in practice and, with the development of modern computer programs, can now be consigned to the history books of project planning.

Sequential

This is a random system from an analysis point of view, but the numbers are chosen in blocks so that certain types of activities can be identified by the nodes. The system therefore clarifies activities and facilitates recognition. The method is quick and easy to use, and should always be used whatever method of analysis is employed. Sequential numbering is usually employed when the network is banded (see Chapter 20). It is useful in such circumstances to start the node numbers in each band with the same prefix number, that is the nodes in band 1 would be numbered 101, 102, 103, etc. while the nodes in band 2 are numbered 201, 202, 203, etc. Fig. 20.1 would lend itself to this type of numbering.

Coordinates

This method of activity identification can only be used if the network is drawn on a gridded background. In practice, thin lines are first drawn on the *back* of the translucent sheet of drawing paper to form a grid. This grid is then given coordinates or map references with letters for the vertical coordinate and numbers for the horizontal (Fig. 19.15).

The reason for drawing the lines on the back of the paper is, of course, to leave the grid intact when the activities are changed or erased. A fully drawn grid may be confusing to some people, so it may be preferable to draw a grid showing the intersections only (Fig. 19.16).

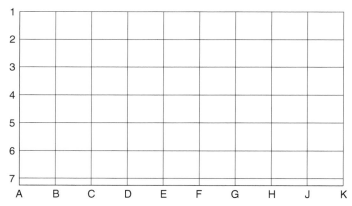

Figure 19.15 Grid.

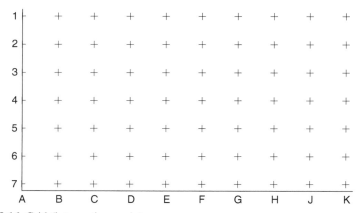

Figure 19.16 Grid (intersections only).

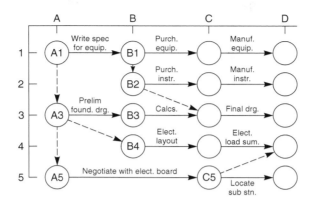

Figure 19.17

When activities are drawn, they are confined in length to the distance between two intersections. The node is drawn on the actual intersection so that the coordinates of the intersection become the node number. The number can be written in or the node left blank, as the analyst prefers.

As an alternative to writing the grid letters on the nodes, it may be advantageous to write the letters *between* the nodes as in Fig. 23.3. This is described in more detail in Chapter 23.

Fig. 19.17 shows a section of a network drawn on a gridded background representing the early stages of a design project. As can be seen, there is no need to fill in the nodes, although, for clarity, activities A1−B1, B1−B2, A3−B3, A3−B4 and A5−C5 have had the node numbers added. The node numbers for 'electrical layout' would be B4−C4, and the map reference principle helps to find the activity on the network when discussing the programme on the telephone or quoting it in e-mail.

There is no need to restrict an activity to the distance between two adjacent intersections of coordinates. For example, A5−C5 takes up two spaces. Similarly, any space can also be used as a dummy, and there is no restriction on the length or direction of dummies. It is, however, preferable to restrict activities to horizontal lines for ease of writing and subsequent identification.

When required additional activities can always be inserted in an emergency by using suffix letters. For example, if activity 'preliminary foundation drawings' A3−B3 had to be preceded by, say, 'obtain loads', the network could be redrawn as shown in Fig. 19.18.

Quickly identifying or finding activities on a network can be of great benefit, and the earlier method has considerable advantages over other numbering systems. The use of coordinates is particularly useful in minimizing the risk of duplicating node numbers in a large network. Since each node is, as it were, pre-numbered by its coordinates, the possibility of double numbering is virtually eliminated.

Unfortunately, in the earlier computer programs, if the planner entered any number twice, the results could be disastrous, since the machine will, in many instances, interpret the error as a logical sequence. The following example shows how this is possible.

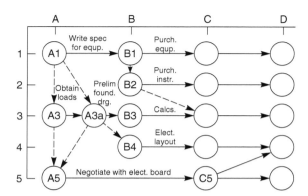

Figure 19.18

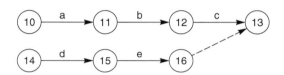

Figure 19.19

The intended sequence is shown in Fig. 19.19. If the planner by mistake enters a number 11 instead of 15 for the last event of activity D, the sequence will in fact be as shown in Fig. 19.20, but the computer will interpret the error as in Fig. 19.21. Clearly, this will give a wrong analysis. If this little network had been drawn on a grid with coordinates as node numbers, it would have appeared as in Fig. 19.22. As the planner knows that all activities on line B must start with a B, the chance of the error occurring is considerably reduced.

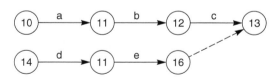

Figure 19.20

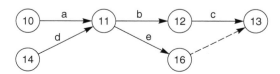

Figure 19.21

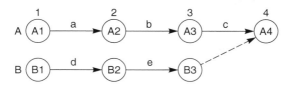

Figure 19.22

Hammocks

When a number of activities are in series, they can be summarized into one activity encompassing them all. Such a summary activity is called a *hammock*. It is assumed that only the first activity is dependent on another activity outside the hammock and only the last activity affects another activity outside the hammock.

On bar charts, hammocks are frequently shown as summary bars above the constituent activities and can therefore simplify the reporting document for a higher management who are generally not concerned with too much detail. For example, in Fig. 19.22, activities A1–A4 could be written as one hammock activity since only A1 and A4 are affected by work outside this activity string.

Ladders

When a string of activities repeats itself, the set of strings can be represented by a configuration known as a ladder. For a string consisting of, say, four activities relating to two stages of excavation, the configuration is shown in Fig. 19.23.

This pattern indicates that, for example, hand trim of Stage II can only be done if,

1. Hand trim of Stage I is complete
2. Machine excavation of Stage II is complete

This, of course, is what it should be.

However, if the work were to be divided into three stages, the ladder could, on the face of it, be drawn as shown in Fig. 19.24.

Again, in Stage II all the operations are shown logically in the correct sequence, but closer examination of Stage III operations will throw up a number of logic errors that the inexperienced planner may miss.

What we are trying to show in the network is that Stage III hand trim cannot be performed until Stage III machine excavation and Stage II hand trim are complete.

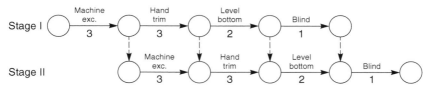

Figure 19.23

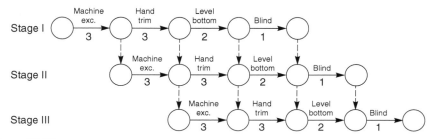

Figure 19.24

However, what the diagram says is that, in addition to these restraints, Stage III hand trim cannot be performed until Stage I level bottom is also complete.

Clearly, this is an unnecessary restraint and cannot be tolerated. Therefore, the correct way of drawing a ladder when more than two stages are involved is shown in Fig. 19.25.

We must, in fact, introduce a dummy activity in Stage II (and any intermediate stages), between the starting and completion node of every activity except the last. In this way, the Stage III activities will not be restrained by Stage I activities, except by those of the same type.

An examination of Fig. 19.26 shows a new dummy between the activities in Stage II, that is:

This concept led to the development of a new type of network presentation called the 'Lester' diagram, which is described in more detail in Chapter 23. This has considerable advantages over the conventional arrow diagram and the precedence diagram, also described later.

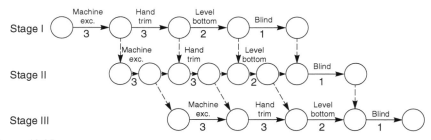

Figure 19.25

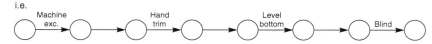

i.e.

Figure 19.26

Precedence or activity on node diagrams

Some planners prefer to show the interrelationship of activities by using the node as the activity box and interlinking them by lines. The durations are written in the activity box or node, and are therefore called activity on node (AoN) diagrams. This has the advantage that separate dummy activities are eliminated. In a sense, each connecting line is, of course, a dummy because it is timeless. The network produced in this manner is also called variously a precedence diagram or a circle and link diagram. Precedence diagrams have a number of advantages over arrow diagrams in that

1. No dummies are necessary.
2. They may be easier to understand for people familiar with flowsheets.
3. Activities are identified by one number instead of two so that a new activity can be inserted between two existing activities without changing the identifying node numbers of the existing activities.
4. Overlapping activities can be shown very easily without the need for the extradummies shown in Fig. 19.25.

Analysis and float calculation (see Chapter 21) is identical to the methods employed for arrow diagrams and if the box is large enough, the earliest and latest start and finishing times can also be written.

A typical precedence network is shown in Fig. 19.27 where the letters in the box represent the description or activity numbers. Durations are shown above-centre, and the earliest and latest starting and finish times are given in the corners of the box, as explained in the key diagram. The top line of the activity box gives the earliest start (ES), duration (D) and earliest finish (EF).

Therefore,

$$EF = ES + D.$$

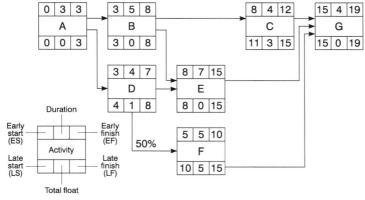

Figure 19.27 AoN diagram.

The bottom line gives the latest start and the latest finish. Therefore,

   LS = LF − D.

The centre box is used to show the total float.

ES is of course the *highest* EF of the previous activities leading into it, that is the ES of activity E is 8, taken from the EF of activity B.

LF is the *lowest* LS of the previous activity *working backwards*, that is the LF of A is 3, taken from the LS of activity B.

The ES of activity F is 5 because it can start after activity D is 50% complete, that is ES of activity D is 3.

Duration of activity D is 4.

Therefore, 50% of duration is 2.

Therefore, ES of activity F is 3 + 2 = 5.

Sometimes, it is advantageous to add a percentage line on the bottom of the activity box to show the stage of completion before the next activity can start (Fig. 19.28). Each vertical line represents 10% completion. In addition to showing when the next activity starts, the percentage line can also be used to indicate the percentage completion of the activity as a statement of progress once work has started as shown in Fig. 19.29.

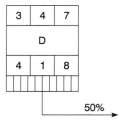

Figure 19.28

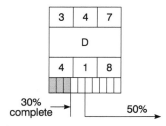

Figure 19.29 Progress indication.

There are four other advantages of the precedence diagram over the arrow diagram.

1. The risk of making logic errors is virtually eliminated. This is because each activity is separated by a link so that the unintended dependency on another activity is just not possible.
2. This is made clear by referring to Fig. 19.30, which is the precedence representation of Fig. 19.25.
3. As can be seen, there is no way for an activity like 'Level bottom' in Stage I to affect activity 'Hand trim' in Stage III, as is the case in Fig. 19.30.
4. In a precedence diagram, all the important information of an activity is shown in a neat box.

A close inspection of the precedence diagram (Fig. 19.31) shows that in order to calculate the total float, it is necessary to carry out the forward and backward pass. Once this has been done, the total float of any activity is simply the difference between the latest finishing time (LF) obtained from the backward pass and the earliest finishing time (EF) obtained from the forward pass.

On the other hand, the free float can be calculated from the forward pass alone, because it is simply the difference of the earliest start (ES) of a subsequent activity and the earliest finishing time (EF) of the activity in question.

This is clearly shown in Fig. 19.31.

Despite the earlier-mentioned advantages, which are especially appreciated by people familiar with flow diagrams as used in manufacturing industries, many prefer the arrow diagram because it resembles more closely to a bar chart. Although the arrows

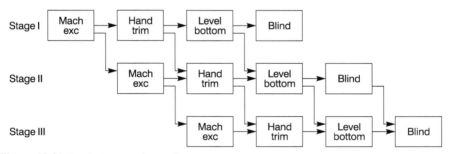

Figure 19.30 Logic to precedence diagram.

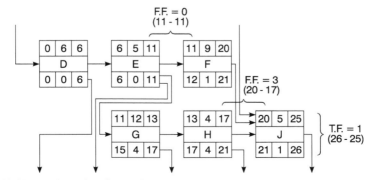

Figure 19.31 Total and free float calculation.

are not drawn to scale, they do represent a forward-moving operation and, by thickening up the actual line in approximately the same proportion as the reported progress, a 'feel' for the state of the job is immediately apparent.

One major practical disadvantage of precedence diagrams is the size of the box. The box has to be large enough to show the activity title, duration and earliest and latest times so that the space taken up on a sheet of paper reduces the network size. By contrast, an arrow diagram is very economical, as the arrow is a natural line over which a title can be written, and the node need be no larger than a few millimetres in diameter − if the coordinate method is used.

The difference (or similarity) between an arrow diagram and a precedence network can be seen most easily by comparing the two methods in the following example. Fig. 19.32 shows a project programme in AoA format and Fig. 19.33 the same programme as a precedence diagram, or AoN format. The difference in the area of paper required by the two methods is obvious (see also Chapter 33).

Fig. 19.33 shows the precedence version of Fig. 19.32.

In practice, the only information necessary when drafting the original network is the activity title, duration and, of course, the interrelationships of the activities. A precedence diagram can therefore be modified by drawing ellipses just big enough to contain the activity title and duration, leaving the computer (if used) to supply the other information at a later stage. The important thing is to establish an acceptable logic before the end date and the activity floats are computed. For explaining the principles of network diagrams in textbooks (and in examinations), letters are often used as activity titles, but in practice when building up a network, the real descriptions have to be used.

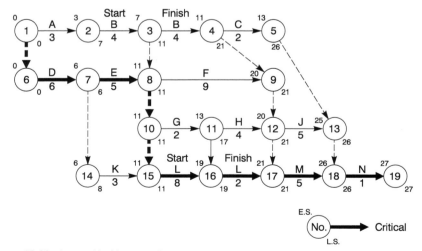

Figure 19.32 Arrow (AoA) network.

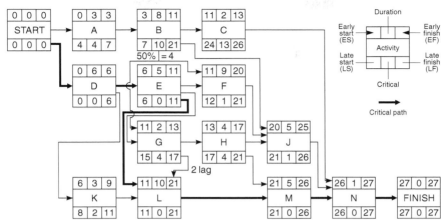

Figure 19.33 Precedence (AoN) network.

An example of such a diagram is shown in Fig. 19.34. Care must be taken not to cross the nodes with the links and to insert the arrowheads to ensure the correct relationship.

One problem of a precedence diagram is that when large networks are being developed by a project team, the drafting of the boxes takes up a lot of time and paper space, and the insertion of links (or dummy activities) becomes a nightmare, because it is confusing to cross the boxes, which are in fact nodes. Therefore, it is necessary to restrict the links to run horizontally or vertically between the boxes, which can lead to congestion of the lines, making the tracing of links very difficult.

When a large precedence network is drawn by a computer, the problem becomes even greater because the link lines can sometimes be so close together that they will appear as one thick black line. This makes it impossible to determine the beginning or end of a link, thus nullifying the whole purpose of a network, that is to show the interrelationship and dependencies of the activities (see Fig. 19.35).

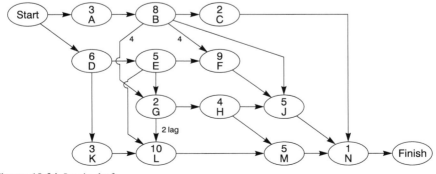

Figure 19.34 Logic draft.

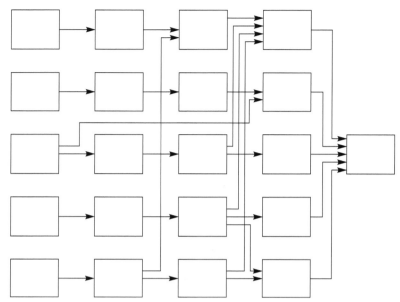

Figure 19.35 Computer-generated AoN diagram.

For small networks with fewer dependencies, precedence diagrams are no problem, but for networks with 200–400 activities per page, it is a different matter. The planner must not feel restricted by the drafting limitations to develop an acceptable logic, and the tendency by some irresponsible software companies to advocate eliminating the manual drafting of a network altogether must be condemned. This manual process is after all the key operation for developing the project network and the distillation of the various ideas and inputs of the team. In other words, it is the thinking part of network analysis. The number crunching can then be left to the computer.

Once the network has been numbered and the times or durations added, it must be analysed. This means that the earliest start and completion dates must be ascertained and the floats or 'spare times' calculated. There are three main types of analysis:

1. Arithmetical
2. Graphical
3. Computer

Since these three different methods (although obviously giving the same answers) require very different approaches, a separate chapter is devoted to each technique (see Chapters 21, 22 and 24).

Constraints

By far the most common logical constraint of a network is as given in the examples on the previous pages, that is 'finish to start' (F–S) where activity B can only start when

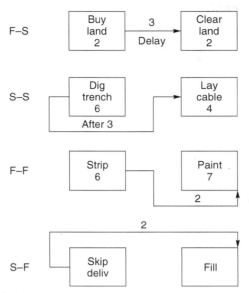

Figure 19.36 Dependencies.

activity A is complete. However, it is possible to configure other restraints. These are start to start (S—S), finish to finish (F—F) and start to finish (S—F). Fig. 19.36 shows these less usual constraints, which are sometimes used when a lag occurs between the activities. Analysing a network manually with such restraints can be very confusing, and should there be a lag or delay between any two activities, it is better to show this delay as just another activity. In fact, all these three less usual constraints can be redrawn in the more conventional finish to start (F—S) mode as shown in Fig. 19.37.

When an activity can start before the previous one has been completed, that is, when there is an overlap, it is known as *lead*. When an activity cannot start until part of the previous activity has been completed, it is called a *lag*.

Bar (Gantt) charts

The bar chart was originally first invented by a Polish engineer, Karol Adamiecki, but it was an American production engineer, Henry Laurence Gantt, who developed it at the beginning of the 20th century, and the chart is consequently known now as the Gantt chart. The Gantt chart was used in a number of planning applications during World War I and was the main planning tool for production and construction engineers until the invention of the critical path network. Each activity on the Gantt chart is represented by a straight horizontal line. The length of the line is proportional to its duration, and the starting and finishing times specified by the planner are plotted against the calendar scale provided at the top (or bottom) of the paper. The original set of lines is called the baselines, and progress can be monitored against them. By performing this graphical representation for each activity, an overall picture of the

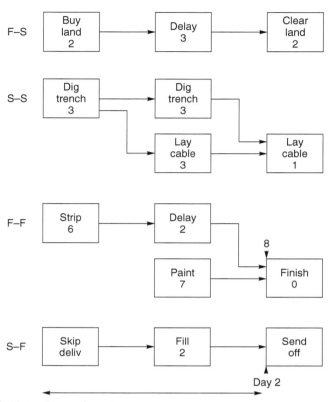

Figure 19.37 Alternative configurations.

required work on a project can be seen at a glance. Progress of an activity can be recorded either by drawing a second line beneath the baseline or colouring up the baseline to show the progress on a daily, weekly or monthly basis.

If the schedule is first drafted manually on a network, it is very quick to produce a Gantt chart as all the thinking has been done at the network stage. However, when using a computer for scheduling, although the Gantt chart is generated automatically as the data is being inputted, the operation takes longer because the sequences and dependencies have to be considered and firmed up as the activities are being typed in. It is for this reason that the first draft of a schedule should be on a network, as the dependencies and basic logic as well as the durations can be very easily and quickly modified to produce the optimum (earliest) completion date.

A Gantt chart is most useful as a method for showing and allocating resources. When the quantity of a common resource, such as labourers, is recorded against each period of an activity, the vertical addition of the resources for any time period gives the total resources for this period. This can then be shown graphically as a vertical bar or column. By carrying out this operation for all the time periods, a series of vertical bars show what resources are required for the project, which is known as a

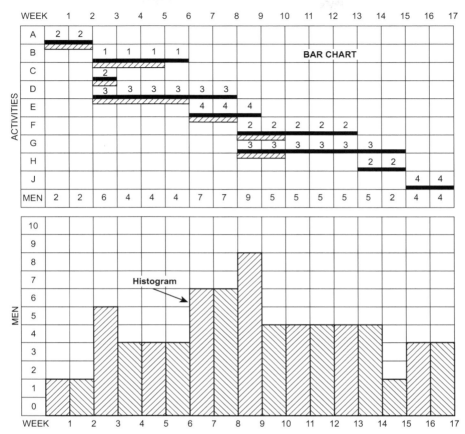

Figure 19.38 Gantt (bar) chart and histogram. Note: The horizontal scale on all the graphs must be in day numbers or week numbers. On no account must they be calendar dates. Numbers must be written on the vertical lines, not in the spaces between the lines.

resource histogram. Fig. 19.38 shows a simple bar chart where progress is represented by the hatched lines below the activity lines. The resources (men) have been added for each activity and after vertical summation, converted into a histogram. This is discussed further in Chapter 30.

Time-scale networks and linked bar charts

When preparing presentation or tender documents, or when the likelihood of the programme being changed is small, the main features of a network and bar chart can be combined in the form of a time-scale network, or a linked bar chart. A time-scale network has the length of the arrows drawn to a suitable scale in proportion to the duration of the activities. The whole network can in fact be drawn on a gridded background where each square of the grid represents a period of time such as a day, week or month. Free float is easily ascertainable by inspection, but total float must be calculated in the conventional manner.

By drawing the activities to scale and starting each activity at the earliest date, a type of bar chart is produced that differs from the conventional bar chart in that some of the activity bars are on the same horizontal line. The disadvantage of such a presentation is that a part of the network has to be redrawn 'downstream' from any activity that changes its duration. It can be seen that if one of the early activities changes in either duration or starting point, the whole network has to be modified.

However, a time-scale network (especially if restricted to a few major activities) is a clear and concise communication document for reporting up. It loses its value in communicating down because changes increase with detail and constant revision would be too time consuming.

A further development of the Gantt chart, called a linked bar chart, is very similar to a normal bar chart, that is each activity is on a separate line and the activities are listed vertically at the edge of the paper. However, by drawing vertical lines connecting the end of one activity with the start of another, one can show the dependencies as a Gantt chart format. Unfortunately, these links, like the original bar charts, only show the implied relationship, as opposed to a critical path network, which shows the logically absolute relationship.

Chapter 24 describes the graphical analysis of networks, and it can be seen that if the ends of the activities were connected by the dummies a linked bar chart would result. This would, however, be based on the logic of the original AoA or AoN network.

Fig. 19.39 shows a small time-scale network, and Fig. 19.40 shows the same programme drawn as a linked bar chart.

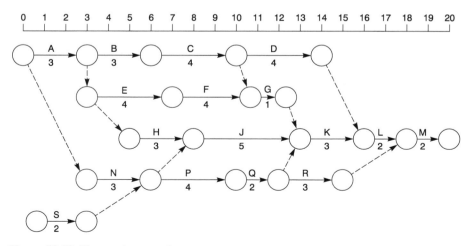

Figure 19.39 Time-scale network.

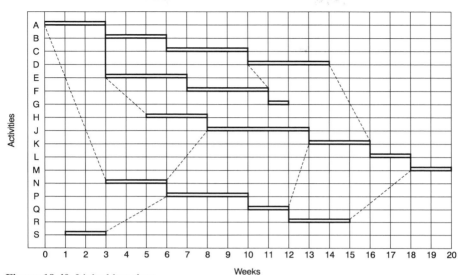

Figure 19.40 Linked bar chart.

Further reading

Gordon, J., & Lockyer, K. (2005). *Project management and project network techniques* (7th ed.). Prentice-Hall.

Planning blocks and subdivision of blocks

Chapter outline

Before any meaningful programme can be produced, it is essential that careful thought is given to the number and size of the networks required. Not only it is desirable to limit the size of the network, but each 'block' of networks should be considered in relation to the following aspects:

1. The geographical location of the various portions or blocks of the project;
2. The size and complexity of each block;
3. The systems in each block;
4. The process or work being carried out in the block when the plant is complete;
5. The engineering disciplines required during the design and construction stage;
6. The erection procedures;
7. The stages at which individual blocks or systems have to be completed, that is the construction programme;
8. The site organization envisaged; and
9. Any design or procurement priorities.

For convenience, a block can be defined as a *geographical process area within a project*, which can be easily identified, usually because it serves a specific function. The importance of choosing the correct blocks, that is drawing the demarcation lines in the most advantageous way, cannot be overemphasized. This decision has an effect

Project Management, Planning and Control. https://doi.org/10.1016/B978-0-12-824339-8.00020-1

not only on the number and size of planning networks but also on the organization of the design teams and, in the case of large projects, on the organizational structure of the site-management setup.

Because of its importance, a guide is given in the following which indicates the type of block distribution that may be sensibly selected for various projects. The list is obviously limited, but it should not be too difficult to abstract some firm guidelines to suit the project under consideration.

Pharmaceutical factory

Block A	Administration block (offices and laboratories)
Block B	Incoming goods area, raw material store
Block C	Manufacturing area 1 (pills)
Block D	Manufacturing area 2 (capsules)
Block E	Manufacturing area 3 (creams)
Block F	Boiler house and water treatment
Block G	Air-conditioning plant room and electrical distribution control room
Block H	Finished goods store and dispatch

For planning purposes, general site services such as roads, sewers, fencing and guard houses can be incorporated into Block A or, if extensive, can form a block of their own.

New housing estate

Block A	Low-rise housing area – North
Block B	Low-rise housing area – East
Block C	Low-rise housing area – South
Block D	Low-rise housing area – West
Block E	High-rise – block 1
Block F	High-rise – block 2
Block G	Shopping precinct
Block H	Electricity substation

Obviously, the number of housing areas or high-rise blocks can vary with the size of the development. Roads and sewers and statutory services are part of their respective housing blocks unless they are constructed earlier as a separate contract, in which case they would form their own block or blocks.

Portland cement factory

Block A	Quarry crushing plant and conveyor
Block B	Clay pit and transport of clay
Block C	Raw-meal mill and silos
Block D	Nodulizer plant and precipitators
Block E	Preheater and rotary kiln
Block F	Cooler and dust extraction
Block G	Fuel storage and pulverization
Block H	Clinker storage and grinding
Block I	Cement storage and bagging
Block J	Administration, offices, maintenance workshops and lorry park

Here again, the road and sewage system could form a block on its own incorporating the lorry park.

Oil terminal

Block A	Crude reception and storage
Block B	Stabilization and desalting
Block C	Stabilized crude storage
Block D	Natural gas liquid (NGL) separation plant
Block E	NGL storage
Block F	Boiler and water treatment
Block G	Effluent and ballast treatment
Block H	Jetty loading
Block J	Administration block and laboratory
Block K	Jetty 1

Continued

Block L	Jetty 2
Block M	Control room 1
Block N	Control room 2
Block P	Control room 3

Here, roads, sewers and underground services are divided into various operational blocks.

Multi-storey block of offices

Block A	Basement and piling work
Block B	Ground floor
Block C	Plant room and boilers
Block D	Office floors 1—4
Block E	Office floors 5—8
Block F	Lift well and service shafts
Block G	Roof and penthouse
Block H	Substation
Block J	Computer room
Block K	External painting, access road and underground services

Clearly, in the construction of a multi-storey building, whether for offices or flats, the method of construction has a great bearing on the programme. There is obviously quite a different sequence required for a block with a central core — especially if sliding formwork is used — than with a more conventional design using reinforced concrete or structural steel columns and beams. The degree of precasting will also have a great influence on the split of the network.

Colliery surface reconstruction

Block A	Headgear and airlocks
Block B	Winding house and winder
Block C	Mine car layout and heapstead building

Block D	Fan house and duct
Block E	Picking belt and screen building
Block F	Wagon loading and bunkering
Block G	Electricity substation, switch room and lamp room
Block H	Administration area and amenities
Block J	Baths and canteen (welfare block)

Roads, sewers and underground services could be part of Block J or be a separate block.

Bitumen refinery

Block A	Crude line and tankage
Block B	Process unit
Block C	Effluent treatment and oil/water separator
Block D	Finished product tankage
Block E	Road loading facility, transport garage and lorry park
Block F	Rail loading facility and sidings
Block G	Boiler house and water treatment
Block H	Fired heater area
Block J	Administration building, laboratory and workshop
Block K	Substation
Block L	Control room

Depending on size, the process unit may have to be subdivided into more blocks, but it may be possible to combine K and L. Again, roads and sewers may be separate or part of each block.

Typical manufacturing unit

Block A	Incoming goods ramps and store
Block B	Batching unit
Block C	Production area 1

Continued

Block D	Production area 2
Block E	Production area 3
Block F	Finishing area
Block G	Packing area
Block H	Finished goods store and dispatch
Block J	Boiler room and water treatment
Block K	Electrical switch room
Block L	Administration block and canteen

Additional blocks can, of course, be added where complexity or geographical location dictates this.

It must be emphasized that these typical block breakdowns can, at best, be a rough guide, but they do indicate the splits that are possible. When establishing the boundaries of a block, the main points given at the beginning of this chapter must be considered.

The interrelationship and interdependence between blocks during the construction stage is, in most cases, remarkably small. The physical connections are usually only a number of pipes, conveyors, cables, underground services and roads. None of these offer any serious interface problems and should not, therefore, be permitted to unduly influence the choice of blocks. Construction restraints must, of course, be taken into account, but they too must not be allowed to affect the basic block breakdown.

This very important point is only too frequently misunderstood. On a refinery site, for example, a delay in the process unit has hardly any effect on the effluent treatment plant except, of course, right at the end of the job.

In a similar way, the interrelationship at the design stage is often overemphasized. Design networks are usually confined to work in the various engineering departments and need not include such activities as planning and financial approvals or acceptance of codes and standards. These should preferably be obtained in advance by project management. Once the main flow sheets, plot plans and piping and instrument diagrams have been drafted (i.e. they need not even have been completed), design work can proceed in each block with a considerable degree of independence. For example, the tank farm may be designed quite independently of the process unit or the NGL plant, etc. and the boiler house has little effect on the administration building or the jetties and loading station.

In the case of a single building being divided into blocks, the roof can be designed and detailed independently of the other floors or the basement, provided, of course, that the interface operations such as columns, walls, stairwell, lift shaft and service ducts have been located and more or less finalized. In short, therefore, the choice of blocks is made as early as possible, taking into account all or most of the factors mentioned before, with particular attention being given to design and construction requirements.

This split into blocks or work areas, of course, takes place in practice in any design office or site, whether the programme is geared to it or not. One is, in fact, only formalizing an already well-proven and established procedure. Depending on size, most work areas in the design office are serviced by squads or teams, even if they only consist of one person in each discipline who looks after that particular area. The fact that on a small project the person may look after more than one area does not change the principle; it merely means that the team is half an operator instead of one.

Natural breakdown into work areas is even more obvious on site. Most disciplines on a site are broken down into gangs, with a ganger or foreman in charge, and, depending again on size and complexity, one or more gangs are allocated to a particular area or block. On very large sites, a number of blocks are usually combined into a complete administrative centre with its own team of supervisors, inspectors, planners, subcontract administrators and site engineers, headed by an area manager.

No difficulty should, therefore, be experienced in obtaining the cooperation of an experienced site manager when the type, size and number of blocks are proposed. Indeed, this early discussion serves as an excellent opportunity to involve the site team in the whole planning process, the details of which are added later. By that time, the site team is at least aware of the principles and a potential communication gap, so frequently a problem among construction people, has been bridged.

Subdivision of blocks

One major point that requires stressing covers the composition of a string of activities. It has already been mentioned that the site should be divided into blocks that are compatible with the design networks. However, each block could in itself be a very large area and a complex operational unit. It is necessary, therefore, to subdivide each block into logical units. There are various ways for doing this. The subdivision could be by the following:

1. Similar items of equipment
2. Trades and disciplines
3. Geographical proximity
4. Operational systems
5. Stages of completion

Each subdivision has its own merits and justifies further examination.

Similar items of equipment

Here, the network shows a series of strings that collect together similar items of equipment, such as pumps, tanks, vessels, boilers and roads. This is shown in Fig. 20.1.

Advantages:

1. Equipment items are quickly found.
2. Interface with design network is easily established.

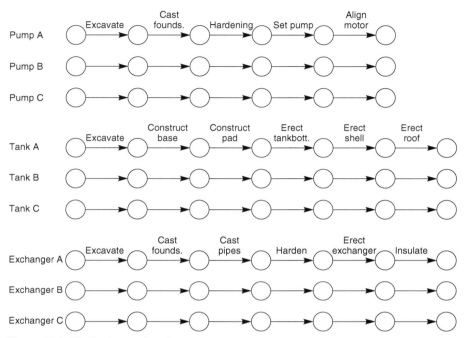

Figure 20.1 Similar items of equipment.

Trades and disciplines

This network groups the work according to type. It is shown in Fig. 20.2.

Advantages:

1. Suitable when it is desirable to clear a trade off the site as soon as completed and
2. Eases resource loading of individual trades.

Geographical proximity

It may be considered useful to group together activities that are geographically close to each other without further segregation into types or trades. This is shown in Fig. 20.3.

Advantages:

1. Makes a specific area self-contained and eases control.
2. Coincides frequently with natural subdivision on site for construction management.

Operational systems

Here, the network consists of all the activities associated with a particular system such as the boiler plant, the crude oil loading and the quarry crushing and screening. A typical system network is shown in Fig. 20.4.

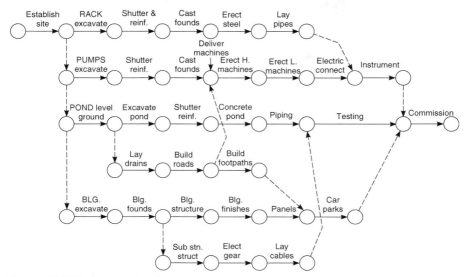

Figure 20.2 Trades and disciplines.

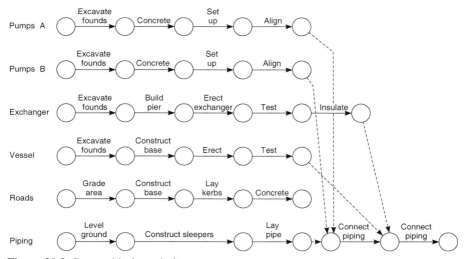

Figure 20.3 Geographical proximity.

Advantages:

1. Easy to establish and monitor the essential interrelationships of a particular system;
2. Particularly useful when commissioning is carried out by system as a complete 'package' can be programmed very easily; and
3. Ideal where stage completion is required.

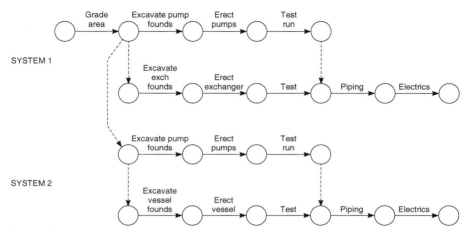

Figure 20.4 Operational system.

Stages of completion

If particular parts of the site have to be completed earlier than others (i.e. if the work has to be handed over to the client in well-defined stages), it is essential that each stage is programmed separately. There will, of course, be interfaces and links with preceding and succeeding stages, but within these boundaries, the network should be self-contained.

Advantages:

1. Attention is drawn to activities requiring early completion;
2. Predictions for completion of each stage are made more quickly;
3. Resources can be deployed more efficiently; and
4. Temporary shut-off and blanking-off operations can be highlighted.

In most cases, a site network is in fact a combination of a number of the earlier sub-divisions. For example, if the boiler plant and water-treatment plant are first required to service an existing operational unit, it would be prudent to draw a network based on operational systems but also incorporating stages of completion. In practice, geographical proximity would almost certainly be equally relevant as the water-treatment plant and boiler plant would be adjacent.

It must be emphasized that the networks shown in Figs. 20.1−20.4 are representative only and do not show the necessary interrelationships or degree of detail normally shown in a practical construction network. The oversimplification of these diagrams may in fact contradict some of the essential requirements discussed in other sections of this book, but it is hoped that the main point, that is the differences between the various types of construction network formats, is highlighted.

Banding

If we study Fig. 20.1, we note that it is very easy to find a particular activity on the network. For example, if we wanted to know how long it would take to excavate the foundations of exchanger B, we would look down the column EXCAVATE until we found the line EXCHANGER B, and the intersection of this column and line shows the required excavation activity. This simple identification process was made possible because Fig. 20.1 was drawn using very crude subdivisions or bands to separate the various operations.

For certain types of work, this splitting of the network into sections can be of immense assistance in finding required activities. By listing various types of equipment or materials vertically on the drawing paper and writing the operations to be performed horizontally, one produces a grid that almost defines the activity. In some instances, the line of operations may be replaced by a line of departments involved. For example, the involvement of the electrical department in designing a piece of equipment can be found by reading across the equipment line until one comes to the electrical department column.

The principle is shown clearly in Fig. 20.5, and it can be seen that the idea can be applied to numerous types of networks. A few examples of banding networks are given in the following, but these are only for guidance since the actual selection of bands depends on the type of work to be performed and the degree of similarity of operation between the different equipment items.

Vertical listing (horizontal line)	Horizontal listing (vertical column)
Equipment	Operations
Equipment	Departments
Material	Operations
Design stages	Departments
Construction stages	Subcontracts
Decision stages	Departments
Approvals	Authorities (clients)
Operations	Department responsibilities
Operations	Broad time periods

It may, of course, be advantageous to reverse the vertical and horizontal bands; when considering, for example, the fifth item on the list, the subcontracts could be listed vertically and the construction stages horizontally. This would most likely be

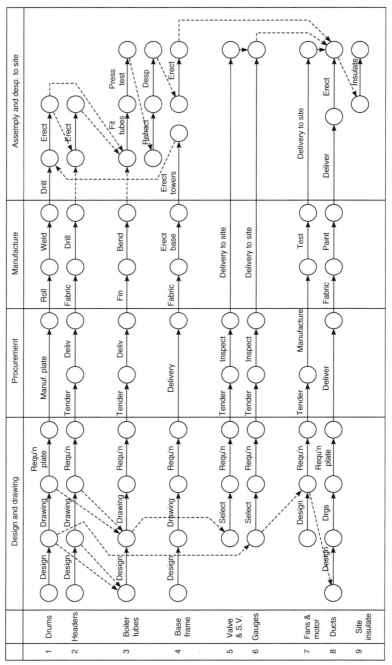

Figure 20.5 Simplified boiler network.

the case when the subcontractors perform similar operations as the actual work stages would then follow logically across the page in the form of normally timed activities. It may indeed be beneficial to draw a small trial network of a few (say, 20—30) activities to establish the best banding configuration.

It can be seen that banding can be combined with the coordinate method of numbering by simply allocating a group of letters of the horizontal coordinates to a particular band.

Banding is particularly beneficial on master networks which cover, by definition, a number of distinct operations or areas, such as design, manufacture, construction and commissioning. Fig. 20.5 is an example of such a network.

Arithmetical analysis and floats

Chapter outline

Arithmetical analysis

This method is the classical technique and can be performed in a number of ways. One of the easiest methods is to add up the various activity durations on the network itself and writing the sum of each stage in a square box at the end of that activity, that is next to the end event (Fig. 21.1). It is essential that each route is examined separately and where the routes meet, the *largest* sum total must be inserted in the box. When the complete network has been summed in this way, the *earliest* starting will have been written against each event.

Now, the reverse process must be carried out. The last event sum is now used as a base from which the activities leading into it are subtracted. The results of these subtractions are entered in the triangular boxes against each event (Fig. 21.2). Similar to the addition process for calculating the earliest starting times, a problem arises when a node is reached where two routes or activities meet. As the *latest* starting times of an activity are required, the *smallest* result is written against the event.

The two diagrams are combined in Fig. 21.3. The difference between the earliest and latest times gives the 'float', and if this difference is zero (i.e. if the numbers in the squares and triangles are the same) the event is on the critical path.

A table can now be prepared setting out the results in a concise manner (Table 21.1).

Project Management, Planning and Control. https://doi.org/10.1016/B978-0-12-824339-8.00021-3

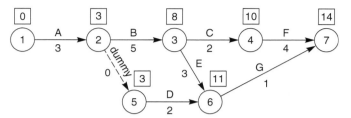

Figure 21.1 Forward pass.

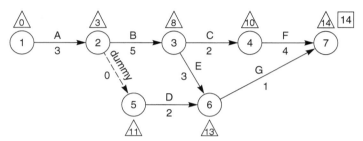

Figure 21.2 Backward pass.

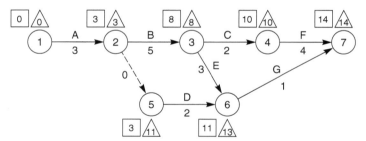

Figure 21.3

Slack

The difference between the latest and earliest times of any event is called 'slack'. As each activity has two events, a beginning event and an end event, it follows that there are two slacks for each activity. Thus, the slack of the beginning event can be expressed as $TL_B - TE_B$ and called beginning slack, and the slack of the end event, appropriately called end slack, is $TL_E - TE_E$. The concept of slack is useful when discussing various types of float, as it simplifies the definitions.

Float

This is the name given to the spare time of an activity, and it is one of the most important by-products of network analysis. The four types of float possible will now be explained.

Table 21.1 Activity summary.

a	b	c	d	e	f	g	h
Title	Activity	Duration, D	Latest time end event	Earliest time end event	Earliest time beginning event	Total float (d − f − c)	Free float (e − f − c)
A	1–2	3	3	3	0	0	0
B	2–3	5	8	8	3	0	0
DUMMY	2–5	0	11	3	3	8	0
C	3–4	2	10	10	8	0	0
E	3–6	3	13	11	8	2	0
F	4–7	4	14	14	10	0	0
D	5–6	2	13	11	3	8	6
G	6–7	1	14	14	11	2	2

Column a: activities by the activity titles. Column b: activities by the event numbers. Column c: activity durations, D. Column d: latest time of the activities' end event, TL$_E$. Column e: earliest time of the activities' end event, TE$_E$. Column f: earliest time of the activities' beginning event, TE$_B$. Column g: total float of the activity. Column h: free float of the activity.

Total float

It can be seen that activities 3—6 in Fig. 21.3 *must* be completed after 13 time units, but can be started after 8 time units. Clearly, as the activity itself takes 3 time units, the activity *could* be completed in $8 + 3 = 11$ time units. Therefore, there is a leeway of $13 - 11 = 2$ time units on the activity. This leeway is called total float and is defined as the latest time of end event minus earliest time of beginning event minus duration, or $TL_E - TE_B - D$.

Fig. 21.3 shows that total float is, in fact, the same as beginning slack. Also, free float is the same as total float minus end slack. The proof is given at the end of this chapter.

Total float has an important role to play in network analysis. By definition, it is the time between the anticipated start (or finish) of an activity and the latest permissible start (or finish).

The float can be either positive or negative. A positive float means that the operation or activity will be completed earlier than necessary, and a negative float indicates that the activity will be late. A prediction of the status of any particular activity is, therefore, a very useful and important piece of information for a manager. However, this information is of little use if not transmitted to management as soon as it becomes available, and every day of delay reduces the manager's ability to rectify the slippage or replan the mode of operation.

The reason for calling this type of float 'total float' is because it is the total of all the 'free floats' in a string of activities when working back from where this string meets the critical path to the activity in question.

It is very easy to calculate the total floats and free floats in a precedence or Lester diagram. For any activity, the total float is the difference between the *latest finish* and *earliest finish* (or *latest start* and *earliest start*). The free float is the difference between the *earliest finish* of the activity in question and the *earliest start* of the following activity. Fig. 21.15 makes this clear.

Calculation of float

By far, the quickest way to calculate the float of a particular activity is to do it manually. In practice, one does not need to know the float of *all* activities at the same time. A list of floats is, therefore, unnecessary. The important point is that the float of a particular activity of immediate interest is obtainable quickly and accurately.

Consider the string of activities in a simple construction process. This is shown in Fig. 21.5 in activity on arrow (AoA) format and in Fig. 21.6 in the simplified activity on node (AoN) format.

It can be seen that the total duration of the sequence is 34 days. By drafting the network in the method shown, and by using the day numbers at the end of *each* activity, including dummies, an accurate prediction is obtained immediately and the float of any particular activity can be seen almost by inspection. It will be

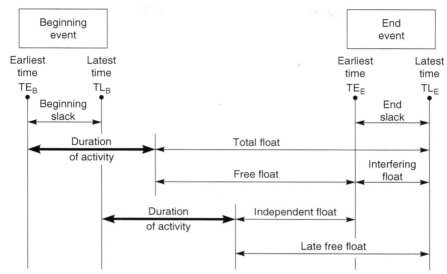

Figure 21.4 Floats.

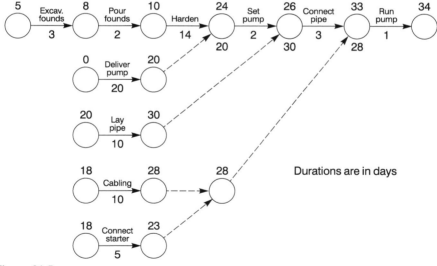

Figure 21.5

noted that each activity has two dates or day numbers — one at the beginning and one at the end (Fig. 21.7). Therefore, where two (or more) activities meet at a node, all the end-day numbers are inserted (Fig. 21.8). The highest number is now used to calculate the overall project duration, that is $30 + 3 = 33$, and the difference between the highest and the other number immediately gives the float of

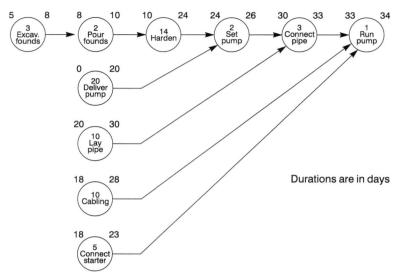

Figure 21.6

Figure 21.7

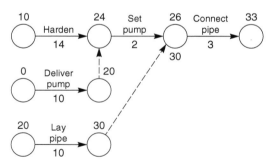

Figure 21.8

the other activity and *all* the activities in that string up to the previous node at which more than one activity meets. In other words, 'set pumps' (Fig. 21.5) has a float of $30 - 26 = 4$ days, and have all the activities preceding it except 'deliver pump', which has an additional $24 - 20 = 4$ days.

If, for example, the electrical engineer needs to know for how long he can delay the cabling because of an emergency situation on another part of the site, without delaying the project, he can find the answer right away. The float is $33 - 28 = 5$ days. If the labour he needs for the emergency can be drawn from the gang erecting the starters, he can gain another $28 - 23 = 5$ days. This gives him a total of 10 days' grace to start the starter installation without affecting the total project time.

A few practice runs with small networks will soon emphasize the simplicity and speed of this method. We have in fact only dealt in this exposition with small − indeed, tiny − networks. How about large ones? It would appear that this is where the computer is essential, but, in fact, a well-drawn network can be analysed manually just as easily, whether it is large or small. Provided the very simple base rules are adhered to, a very fast-forward pass can be inserted. The float of any *string* can then be seen by inspection, that is by simply subtracting the lower node number from the higher node number, which forms the termination point of the string in question. This point can be best illustrated by the example given in Fig. 21.9. For simplicity, the activities have been given letters instead of names, as the importance lies in understanding the principle, and the use of letters helps to identify the string of activities. In this example, there are 50 activities. Normally, a practical network should have between 200 and 300 activities maximum (i.e. 4−6 times the number of activities shown), but this does not pose any greater problem. All the times (day numbers) were inserted, and the floats of activities in strings A, B, C, E, F, G and H were calculated in 5 minutes. A 300-activity network would, therefore, take 30 minutes.

It can in fact be stated that any practical network can be 'timed', that is the forward pass can be inserted and the important float reported in 45 minutes. It is, furthermore,

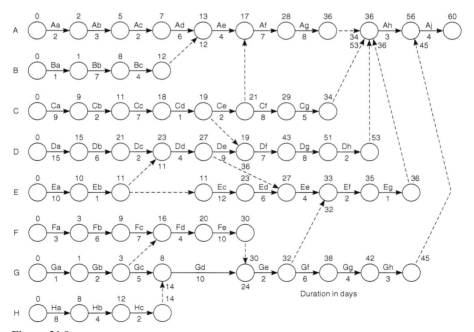

Figure 21.9

very easy to find the critical path. Clearly, it runs along the strings of activities with the highest node times. This is most easily calculated by working back from the end. Therefore, the path runs through Aj, Ah, dummy, Dh, Dg, Df, De, Dd, Dc, Db and Da.

An interesting little problem arises when calculating the float of activity Ce, as there are two strings emanating from the end node of that activity. By conventional backward pass methods — and indeed this is how a computer carries out the calculation — one would insert the backward pass in the nodes starting from the end (see Fig. 21.10). When arriving at Ce, one finds that the latest possible time is 40 when calculating back along string Cg−Cf, while it is 38 when calculating back along string Ag−Af. Clearly, the actual float is the difference between the earliest date and the earliest of the two latest dates, that is day 38 instead of day 40. The float of Ce is therefore $38 - 21 = 17$ days.

As described earlier, the calculation is tedious and time consuming. A far quicker method is available by using the technique shown in Fig. 21.9, that is, one simply inserts the various forward passes on each string and then looks at the end node of the activity in question — in our case, activity Ce. It can be seen by following the two strings emanating from Ce that string Af−Ag joins Ah at day 36. String Cf−Cg, on the other hand, joins Ah at day 34. The float is, therefore, the *smallest* difference between the *highest* day number and one of the 2-day numbers just mentioned. Therefore, the float of activity Ce is $53 - 36 = 17$ days. Cf and Cg, of course, have a float of $53 - 34 = 19$ days.

The time required to inspect and calculate the float by the second method is literally only a few minutes. All one has to do is to run through the paths emanating from the end node of the selected activity and note the *highest* day number where the strings *meet the critical path*. The difference between the day number of the critical string and the highest number on the tributary strings (emanating from the activity in question) is the float.

Suppose we now wish to find the float of activity Gb:
Follow string Fd−Fe.
Follow string Gc−Gd−Ge.
Follow string Gf−Gg−Gh.
Follow string Ef−Eg−Ah.
Fe and Gd meet at Ge; therefore, they can be ignored.

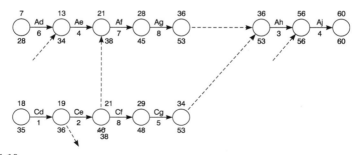

Figure 21.10

String Gf—Gh meets Aj at day 45

String Ef—Eg meets Ah at day 36

Therefore, the float is either 56 − 45 = 11

or 53 − 36 = 17

Clearly, the correct float is 11 as it is the smaller of the two. The time taken to inspect and calculate the float was exactly 21 seconds.

All the floats calculated earlier were total floats. Free float can only occur on activities entering a node when more than one enters that node. It can be calculated very easily by subtracting the total float of the incoming activity from the total float of the outgoing activity, as shown in Fig. 21.11. It should be noted that one of the activities entering the node *must* have zero *free* float.

When more than one activity leaves a node, the value of the free float to be subtracted is the *lowest* of the outgoing activity floats, as shown in Fig. 21.12.

Free float

Some activities, for example, 5—6 (Fig. 21.13), as well as having total float have an additional leeway. It will be noted that the activities 3—6 and 5—6 both

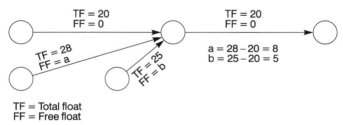

TF = 20
FF = 0

TF = 28
FF = a

TF = 25
FF = b

TF = 20
FF = 0

a = 28 − 20 = 8
b = 25 − 20 = 5

TF = Total float
FF = Free float

Figure 21.11

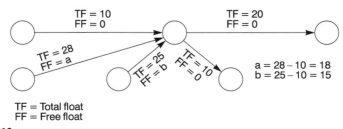

TF = 10
FF = 0

TF = 28
FF = a

TF = 25
FF = b

TF = 10
FF = 0

TF = 20
FF = 0

a = 28 − 10 = 18
b = 25 − 10 = 15

TF = Total float
FF = Free float

Figure 21.12

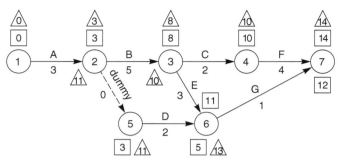

Figure 21.13

affect the activity 6–7. However, one of these two activities will delay 6–7 by the same time unit by which it itself may be delayed. The remaining activity, on the other hand, may be delayed for a period without affecting 6–7. This leeway is called free float, and can only occur in one or more activities where several of them meet at one event, that is, if x activities meet at a node, it is possible that $x - 1$ of these have a free float. This free float may be defined as earliest time of end event minus earliest time of beginning event minus duration, or $TE_E - TE_B - D$.

The concept of free float

Students often find it difficult to understand the concept of free float. The mathematical definitions are unhelpful, and the graphical representation of Fig. 21.4 can be confusing. The easiest way to understand the difference between total and free float is to inspect the *end* node of the activity in question. As stated earlier, the free float can only occur where two or more activities enter a node. If the *earliest* end times (i.e. the forward pass) for each individual activity are placed against the node, the free float is simply the difference between the highest number of the earliest time on the node and the number of the earliest time of the activity in question.

In the example given in Fig. 21.13, the earliest times are placed in squares; so following the same convention it can be seen from the figure (which is a redrawing of Fig. 21.1 with *all* the earliest and latest node times added) that:

Fig. 21.14 shows the equivalent precedence (AoN) diagram from which the free float can be easily calculated by subtracting the early *finish* time of the *preceding* node from the early *start* time of the *succeeding* node.

Free float of activity D = $11 - 5 = 6$.

Free float of activity G = $14 - 12 = 2$.

Activity E, because it is not on the critical path, has a total float of $13 - 11 = 2$ but has no free float.

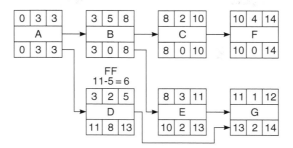

Figure 21.14

The check of the free float by the formal definition is as follows:

Free float = TE$_E$ − TE$_B$ − D

For activity D = 11 − 3 − 2 = 6

For activity G = 14 − 11 − 1 = 2

The check of the total float by the formal definitions is given as follows:

Total float = TL$_E$ − TE$_B$ − D

For activity E = 13 − 8 − 3 = 2

D = 13 − 3 − 2 = 8

G = 14 − 11 − 2 = 2

It was stated earlier that the total float is the same as beginning slack. This can be shown by rewriting the definition of total float = TL$_E$ − TE$_B$ − D as total float = TL$_E$ − D − TE$_B$ but TL$_E$ − D = TL$_B$. Therefore,

Total float = TL$_B$ − TE$_B$

= Beginning slack

To show that free float = total float − end slack, consider the following definitions:

Free float = TL$_E$ − TE$_B$ − D   (21.1)

Total float = TL$_E$ − TE$_B$ − D   (21.2)

End slack = TL$_E$ − TE$_E$ (21.3)

Subtracting Eq. (21.3) from Eq. (21.2):

$$TL_E - TE_B - D - TL_E - TE_E$$

$$TL_E - TE_B - D - TL_E + TE_E$$

$$TE_E - TE_B - D$$

Free float

Therefore,

Eq. (21.1) = Eq. (21.2) − Eq. (21.3) or free float = total float − end slack.

If a computer is not available, free float on an arrow diagram can be ascertained by inspection, as it can only occur when more than one activity meets a node. This was described in detail earlier with Figs. 21.5 and 21.6. If the network is in the precedence format, the calculation of free float is even easier. All one has to do is to subtract the early finish time in the preceding node from the early start time of the succeeding node. This is clearly shown in Fig. 21.15, which is the precedence equivalent to Fig. 21.5.

One of the phenomena of a computer printout is the comparatively large number of activities with free float. Closer examination shows that the majority of these are in fact dummy activities. The reason for this is, of course, obvious, as, by definition, a free float can only exist when more than one activity enters a node. As dummies nearly

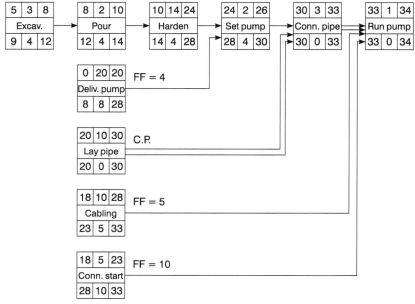

Figure 21.15

always enter a node with another (real) activity, they all tend to have a free float. Unfortunately, no computer program exists that automatically transfers this free float to the preceding real activity, so the benefit of the free float is not immediately apparent and is consequently not taken advantage of.

Interfering float

The difference between the total and free float is known as interfering float. Using the previous notation, this can be expressed as follows:

$$(TL_E - TE_B - D) - (TE_E - TE_B - D) = TL_E - TE_B - D - TE_E + TE_B + D$$

$$= TL_E - TE_E$$

that is, as the latest time of the end event minus the earliest time of the end event. It is, therefore, the same as the end slack.

Independent float

The difference between the free float and the beginning slack is known as independent float:

Since free float $= TE_E - TE_B - D$

Beginning slack $= TL_B - TE_B$

Independent float $= TE_E - TE_B - D - (TL_B - TE_B)$

$$= TE_E - TE_B - D$$

This problem does not exist with precedence diagrams as there are no dummy activities in this network format.

Thus, independent float is given by the earliest time of the end event minus the latest time of the beginning event minus the duration.

In practice, neither the interfering float nor the independent float finds much application, and for this reason they will not be referred to in later chapters. The use of computers for network analysis enables these values to be produced without difficulty or extra cost, but they only tend to confuse the user and are therefore best ignored.

Summarizing all the earlier definitions, Fig. 21.4 and the following expressions can be of assistance.

Notation

D_B = duration of activity.
TE_B = earliest time of beginning event.
TE_E = earliest time of end event.

TL_B = latest time of beginning event.

TL_E = latest time of end event.

Definitions

Beginning slack = $TL_B - TE_B$

End slack = $TL_E - TE_E$

Total float = $TL_E - TE_B - D$

Free float = $TE_E - TE_B - D$

Interfering float = $TL_E - TL_B$ (end slack)

Independent float = $TE_E - TL_B - D$

Last free float = $TL_E - TL_B - D$

Because float is such an important part of network analysis and because it is frequently quoted — or misquoted — by computer protagonists as another reason why computers *must* be used, a special discussion of the subject may be helpful to those readers not too familiar with its use in practice.

Of the three types of float shown on a printout, that is, the total float, free float and independent float, only the first — the total float — is in general use. Where resource smoothing is required, knowledge of free float can be useful, as the activities with free float can be moved backwards or forwards in time without affecting any other activity. Independent float, on the other hand, is really quite a useless piece of information and should be suppressed (when possible) from any computer printout. Of the many managers, site engineers and planners interviewed, none has been able to find a practical application of independent float.

Critical path

Some activities have zero total float, that is, no leeway is permissible for their execution, and hence any delays incurred on the activities will be reflected on the overall project duration. These activities are therefore called critical activities, and every network has a chain of such critical activities running from the beginning event of the first activity to the end event of the last activity, without a break. This chain is called the critical path.

A project network has more than one critical path frequently, that is, two or more chains of activities all have to be carried out within the stipulated duration to avoid a delay to the completion date. In addition, a number of activity chains may have only one or two units of float, therefore, for all intents and purposes, they are also critical. It can be seen, therefore, that it is important to keep an eye on all activity chains that are either critical or near-critical, as a small change in the duration of one chain could quickly alter the priorities of another.

One disadvantage of the arithmetical method of analysis using the table or matrix shown in Table 21.1 is that all the floats must be calculated before the critical path can be ascertained. This method (which has only been included for historical interest) is very laborious and is therefore not recommended.

In any case, this drawback is eliminated when the method of analysis shown in Fig. 21.9 is employed.

Critical chain project management

This planning technique differs from conventional CPA in that resource restraints and dependencies are considered as well as operation logic. This requires buffers to be introduced to allow for possible resource issues. Progress monitoring can then be carried out by measuring the rate at which the buffers are used up instead of just recording the individual task performance.

In practice, this method is probably most appropriate where resources are not easily assessable at the planning stage. Contingencies in the form of buffers must therefore be incorporated with the result that the overall duration will probably be longer, and as a consequence, the planned completion date will more likely be met.

In the construction industry, where the program completion date is often set by the client, that is, the opening of a venue or start of production, the project program will be based on the construction logic of the operational activities, which must then be provided with the necessary resources (plant, material and labour) to carry out the work. It is inconceivable that at the start of a project a contractor in the construction industry would tell his or her client that the project cannot be completed by the specified date because the necessary resources cannot be obtained.

The case for manual analysis 22

Although network analysis is applicable to almost every type of organization, as shown by the examples in Chapter 29, most of the planning functions described in this book have been confined to those related to engineering construction projects. The activities described cover the full spectrum of operations from the initial design stage, through detailing of drawings and manufacture, up to and including construction, in other words, from conception to handover.

In this age of specialization, there is a trend to create specialist groups to do the work previously carried out by the members of more conventional disciplines. One example is teaching, where teaching methods, previously devised and perfected by practising teachers, are now developed by a new group of people called educationalists.

Another example of specialization is planning. In the days of bar charts, planning was carried out by engineers or production staff using well-known techniques to record their ideas on paper and transmit them to other members of the team. Nowadays, however, the specialist planner or scheduler has come to the fore, leaving the engineer time to get on with engineering.

The planner

Planning, in its own right, does not exist. It is always associated with another activity or operation, that is design planning, construction planning, production planning, etc. It is logical, therefore, that a design planner should be or should have been a designer, a construction planner should be familiar with construction methods and techniques and a production planner should be knowledgeable in the process and manufacturing operations of production — whether it be steelwork, motors cars or magazines.

In construction, as long as the specialist planner has graduated from one of the accepted engineering disciplines and is familiar with the problems of a particular project, a realistic network can be probably produced. By calling in specialists to advise in

Project Management, Planning and Control. https://doi.org/10.1016/B978-0-12-824339-8.00022-5

the fields with which the engineer not completely conversant, it can be ensured that the network will be received with confidence by all the interested parties.

The real problem arises when the planner does not have the right background, that is when he or she has not spent a time designing or has not experienced the holdups and frustrations of a construction site. Strangely enough, the less familiar a planner is with the job being planned, the less is the inclination to seek help. This may well be due to the engineer's inability to ask the right questions, or reluctance to discuss technical matters for fear of revealing the personal lack of knowledge. One thing is certain: A network that is not based on sound technical knowledge is not realistic, and an unrealistic network is dangerous and costly, since decisions may well be made for the wrong reasons.

All that has been said so far is a truism that can be applied not only to planning but to any human activity where experts are required to achieve acceptable results. However, in most disciplines, it does not take long for the effects of an inexperienced assistant to be discovered, mainly because the results of the work done can be monitored and assessed within a relatively short time period. In planning, however, the effects of a programme decision may not be felt for months, so it may be very difficult to ascertain the cause of the subsequent problem or failure.

The role of the computer

Unfortunately, the use of computers has enabled inexperienced planners to produce impressive outputs that are frequently utterly useless. There is a great danger in shifting the emphasis from the creation of the network to the analysis and report production of the machine so that many people believe that to carry out an analysis of a network one must have a computer. In fact, the computer is only a sophisticated number cruncher. It does not see the whole picture, including access problems, political or cultural restraints, labour issues and staff idiosyncrasies. The kernel of network analysis is the drafting, checking, refining and redrafting of the network itself, an operation that must be carried out by a team of experienced participants of the job being planned. To understand this statement, it is necessary to go through the stages of network preparation and subsequent updating.

Preparation of the network

The first function of the planner in conjunction with the project manager is to divide the project into manageable blocks. The name is appropriate since, like building blocks, they can be handled individually and shaped to suit the job but are still only a part of the whole structure to be built.

The number and size of each block are extremely important since, if correctly chosen, a block can be regarded as an entity that suits both the design and the construction phases of a project. Ideally, the complexity of each block should be about the same, but this is rarely possible in practice as other criteria such as systems and geographical location have to be considered. If a block is very complex, it can be broken down further, but a more convenient solution may be to produce more than one network for such a block. The aim should be to keep the number of activities down to 200–300 so that they can be analysed manually if necessary.

As the planner sketches his or her logic roughly, and in pencil on the back of an old drawing, the construction specialists are asked to comment on the type and sequence of the activities. In practice, these sessions — if properly run — generate an enthusiasm that is a delight to experience. Often consecutive activities can be combined to simplify the network, thus easing the subsequent analysis. Gradually, the job is 'built', difficulties are encountered and overcome, and even specialists who have never been involved in network planning before are carried away by this visual unfolding of the programme.

The next stage is to ask each specialist to suggest the duration of the activities in his discipline. These are entered onto the network without question. Now comes the moment of truth. Can the job be built on time? With all the participants present, the planner adds up the durations and produces the forward pass. Almost invariably, the total time is longer than the deadlines permit. This is when the real value of network analysis emerges. Logics are re-examined, durations are reduced and new construction methods are evolved to reduce the overall time. When the final network — rough though it may be — is complete, a sense of achievement can be felt pervading the atmosphere.

This procedure, which is vital to the production of a realistic programme, can, of course, only be carried out if the 'blocks' are not too large. If the network has more than 300 activities, it may well pay the planner or project manager to re-examine that section of the programme with a view to dividing it into two smaller networks. If necessary, it is always possible to draw a master network, usually quite small, to link the blocks together.

One of the differences between the original program evaluation and revue technique program and the normal Critical Path Method (CPM) programs was the facility to enter three time estimates for every activity. The purpose of the three estimates is to enable the computer to calculate and subsequently use the most probable time, on the assumption that the planner is unwilling or unable to commit to one time estimate. The actual duration used is calculated from the expression known as the β distribution:

$$t_e = \frac{a + 4m + b}{6}$$

where t_e is the expected time, a the optimistic time, b the pessimistic time and m the most likely time.

However, this degree of sophistication is not really necessary, as the planner can insert the most probable time considered. For example, a foreman, upon being pressed, estimated the times for a particular operation to be:
Optimistic = 5 days.
Pessimistic = 10 days.

Probable = 7 days.

The planner will probably insert 7 days or 8 days. The computer, using the earlier distribution, would calculate:

$$t_c = \frac{5 + (4 \times 7) + 10}{6}$$

$$= 7.16 \text{ days}$$

With the much larger variables found in real-life projects, such finesse is a waste of time. A single time estimate by an experienced planner is all that is required.

Typical site problems

Once construction starts, problems begin to arise. Drawings or other data arrive late on-site, materials are delayed, equipment is held up, labour becomes scarce or goes on strike, underground obstructions are found, the weather deteriorates, etc.

Each new problem must be examined in the light of the overall project programme. It will be necessary to repeat the initial planning meeting to revise the network, to reflect on these problems and to possibly help reduce their effects. It is at these meetings that ingenious innovations and solutions are suggested and tested.

For example, Fig. 22.1 shows the sequence of a section of a pipe rack. Supposing the delivery of pipe will be delayed by 4 weeks, completion now looks like week 14. However, someone suggests that the pump bases can be cast early with starter bars bent down to bond the plinths at a later date. The new sequence appears in Fig. 22.2. Completion time is now only week 11, a saving of 3 weeks.

This type of approach is the very heart of successful networking and keeps the whole programme alive. It is also very rapid. The very act of discussing problems in the company of interested and knowledgeable colleagues generates an enthusiasm that carries the project forward. With good management, this momentum is passed right down the line to the people who are actually doing the work at the sharp end.

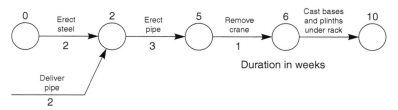

Figure 22.1

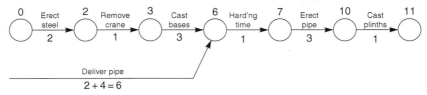

Figure 22.2

The National Economic Development Office report

Perhaps the best evidence that networks are most effective when kept simple is given by the National Economic Development Office report, which is still applicable today even though it was produced way back in the early 1980s.

The relevant paragraphs are reproduced in the following by permission of Her Majesty's Stationery Office.

1. Even if it is true that UK clients build more complex plants, it should still be possible to plan for and accommodate the extra time and resources this would entail. By and large, the UK projects were more generously planned but, nonetheless, the important finding of the case studies is that, besides taking longer, the UK projects tended also to encounter more overrun against planned time. There was no correlation across the case studies between the sophistication with which programming was done and the end result in terms of successful completion on time. On the German power station, the construction load represented by the size and height of the power station was considerable, but the estimated construction time was short and was achieved. This contrasts with the UK power stations, where a great deal of effort and sophistication went into programming, but schedules were overrun. In most of the case studies, the plans made at the beginning of the project were thought realistic at that stage, but they varied in their degree of sophistication and in the importance attached to them.
2. One of the British refineries provided the one UK example where the plan was recognized from the start by both client and contractor to be unrealistic. Nonetheless, the contractor claimed that he believed planning to be very important, particularly in the circumstances of the United Kingdom, and the project was accompanied by a wealth of data collection. This contrasts with the Dutch refinery project where planning was clearly effective but where there was no evidence of very sophisticated techniques. There is some evidence in the case studies to suggest that UK clients and contractors put more effort into planning, but there is no doubt that the discipline of the plan was more easily maintained on the foreign projects. Complicated networks are useful in developing an initial programme, but subsequently, though they may show how badly one has done, they do not indicate how to recover the situation. Networks need, therefore, to be developed to permit simple rapid updates, pointing where action must be taken. Meanwhile, the evidence from the foreign case studies suggests that simple techniques, such as bar charts, can be successful.
3. The attitudes to planning on UK1 and the Dutch plant were very different, and this may have contributed to the delay of UK1 although it is impossible to quantify the effect. The Dutch contractor considered planning to be very important and had two site planning engineers

attached to the home office during the design stage. The programme for UK1 on the other hand was considered quite unrealistic by both the client and the contractor, not only after the event but while the project was under way, but neither of them considered this important in itself.

In the case of UK1, it was not until the original completion date arrived that the construction was rescheduled to take 5 months longer. At this point, the construction was only 80% complete and in the event there was another 8 month's work to do. Engineering had been 3 months behind schedule for some time. A wealth of progress information was being collected but no new schedule appears to have been made earlier.

Progress control and planning was clearly a great deal more effective on the Dutch project; the contractor did not believe in particularly sophisticated control techniques, however.

Clearly, modern computer programs are more sophisticated and user-friendly and have far greater functionality, but it is precisely because these programs are so attractive that there is a risk of underestimating, or even ignoring, the fundamental and relatively simple planning process described earlier in this chapter.

Lester diagram

23

Chapter outline

With the development of the *network grid*, the drafting of an arrow diagram enables the activities to be easily organized into disciplines or work areas and eliminates the need to enter reference numbers into the nodes. Instead, the grid reference numbers (or letters) can be fed into the computer. The grid system also makes it possible to produce acceptable arrow diagrams on a computer that can be used 'in the field' without converting them into the conventional bar chart. An example of a computer-generated arrow diagram is shown in Fig. 23.1. It can be noticed that the link lines never cross the nodes.

A grid system can, however, pose a problem when it becomes necessary to insert an activity between two existing ones. In practice, resourceful planners can overcome the problem by combining the new activity with one of the existing activities.

If, for example, two adjoining activities were 'Cast Column, 4 days' and 'Cast Beam, 2 days', and it was necessary to insert 'Strike Formwork, 2 days' between the two activities, the planner would simply restate the first activity as 'Cast Column and Strike Formwork, 6 days' (Fig. 23.2).

While this overcomes the drafting problem, it may not be acceptable from a cost-control point of view, especially if the network is geared to an earned value analysis (EVA) system (see Chapter 32). Furthermore, the fact that the grid numbers were on the nodes meant that when it was necessary to move a string along one or more grid spaces, the relationship between the grid number and the activity changed. This could complicate the EVA analysis. To overcome this, the grid number was placed *between* the nodes (Fig. 23.3).

It can be argued that a precedence network lends itself admirably to a grid system because the grid number is always and permanently related to the activity and is therefore ideal for EVA. However, the problem of the congested link lines (especially the vertical ones) remains.

Now, however, the perfect solution has been found. It is in fact a combination of the arrow diagram and precedence diagram, and like the marriage of Henry VII, which ended the Wars of the Roses, this marriage should end the war of the networks!

The new diagram, which could be called the 'Lester' diagram, is simply an arrow diagram where each activity is separated by a short link in the same way as in a precedence network (Fig. 23.4).

In this way, it is possible to eliminate or at least reduce logic errors, show total float and free float as easily as in a precedence network, but it has the advantages of an arrow

Project Management, Planning and Control. https://doi.org/10.1016/B978-0-12-824339-8.00023-7

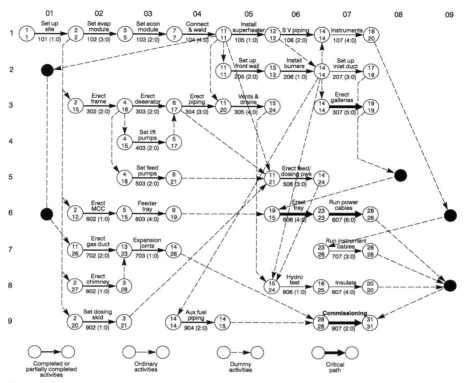

Figure 23.1 Activity on arrow network drawn on grid.

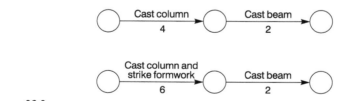

Figure 23.2

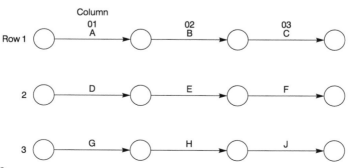

Figure 23.3

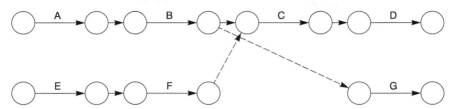

Figure 23.4 Lester diagram principle.

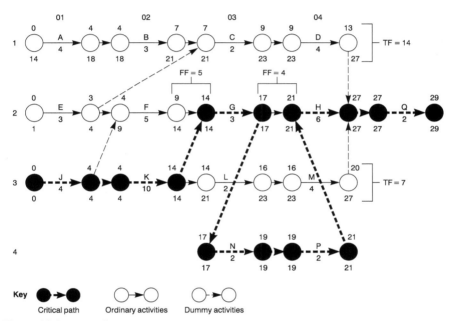

Figure 23.5 Lester diagram.

diagram in speed of drafting, clarity of link presentation and the ability to insert new activities in a grid system without altering the grid number–activity relationship. Fig. 23.5 shows all these features.

If a line or box is drawn around any activity, the similarity between the Lester diagram and precedence diagram becomes immediately apparent (see Fig. 23.6).

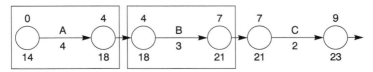

Figure 23.6

'It can be seen from Fig. 23.4 that the danger of the logic errors highlighted on Fig. 19.24 (page 118) is eliminated in the Lester diagram. As all dummies must emanate from the end node of an activity and all dummies leading to an activity must enter the beginning node, unwanted restraints cannot occur.

For example, in Fig. 23.4, activity G is only dependent on activity B. If there were only one node connecting activities B and C (as in a normal arrow or precedence diagram) activity G would also be dependent on activity F, as the dummy from activity F enters the end node of activity B, which, in the arrow diagram, is also the beginning node of activity C.

Although all the examples in subsequent chapters use arrow diagrams, or precedence diagrams, 'Lester' diagrams could be substituted in most cases. The choice of technique is largely one of personal preference and familiarity. Provided the user is satisfied with one system and is able to extract the maximum benefit, there is little point in changing to another.

Basic advantages

The advantages of a Lester diagram are as follows:

1. Faster to draw than precedence diagram — about the same speed as an arrow diagram;
2. As in a precedence diagram:
   a. Total float is vertical difference;
   b. Free float is horizontal difference;
3. Room under arrow for duration and total float value;
4. Logic lines can cross the activity arrows;
5. Requires less space on paper when drafting the network;
6. Good for examinations due to speedy drafting and elimination of node boxes;
7. Can be updated for progress by 'redding' up activity arrows as arrow diagram;
8. Uses same procedures for computer inputting as precedence networks;
9. Output from computer similar to precedence network;
10. Can be used on a grid;
11. Less chance of error when calculating backward pass due to all lines emanating from one node point instead of one of the four sides of a rectangular node;
12. Shows activity as flow lines rather than points in time;
13. Looks like an arrow diagram, but is in fact more like a precedence diagram;
14. No risk of individual link lines being merged into a thick black line when printed out and
15. No possibility of creating the type of logic error often associated with ladders.
16. The risk of creating an unwanted restraint is greatly reduced.

Graphical and computer analysis

24

Chapter outline

Graphical analysis

It is often desirable to present the programme of a project in the form of a bar chart, and when the critical path and floats are found by either arithmetical or computer methods, the bar chart has to be drawn as an additional task. (Most computer programs can actually print a bar chart, but these often run to several sheets.)

As explained in Chapter 30, bar charts, while they are not as effective as networks for the actual planning function, are still one of the best methods for allocating and smoothing resources. If resource listing and subsequent smoothing is an essential requirement, graphical analysis can give the best of both worlds. Naturally, any network, however analysed, can be converted very easily into a bar chart, but if the network is analysed graphically the bar chart can be 'had for free', as it were.

Modern computer programs will of course almost automatically produce the bar charts (or Gantt charts) from the inputs. Indeed, the input screen itself often generates the bar chart as the data are entered. However, when a computer is not available or the planner is not conversant with the particular computer program, the graphical method becomes a useful alternative.

The following list gives some of the advantages over other methods, but before the system is used on large jobs, planners are strongly advised to test it for themselves on smaller contracts so that they can appreciate the shortcut methods and thus save even more planning time.

1. The analysis is extremely rapid, much quicker than the arithmetical method. This is especially the case when, after some practice, the critical path can be found by inspection.
2. As the network is analysed, the bar chart is generated automatically and no further labour needs to be expended to do this at a later stage.

Project Management, Planning and Control. https://doi.org/10.1016/B978-0-12-824339-8.00024-9

3. The critical path is produced *before* the floats are known. (This is in contrast to the other methods, where the floats have to be calculated first before the critical path can be seen.) The advantage of this is that users can see at once whether the project time is within the specified limits, permitting them to make adjustments to the critical activities without bothering about the uncritical ones.
4. Since the results are shown in bar chart form, they are more readily understood by persons familiar with this form of programme. The bar chart will show the periods of heavy resource loading more vividly compared to a printout, and highlights the periods of comparative inactivity. Smoothing is therefore much more easily accomplished.
5. By marking the various trades or operational types in different colours, a rapid approximate resource-requirement schedule can be built up. The resources in any one time period can be ascertained by simply adding up vertically, and any smoothing can be done by utilizing the float periods shown on the chart.
6. The method can be employed for single or multi-start projects. For multi-project work, two or more bar charts can (provided they are drawn to the same time and calendar scale) be superimposed on transparent paper and the amount of resource overlap can be seen very quickly.

Limitations

The limitations of the graphical method are basically the size of the bar chart paper and therefore, the number of activities. Most programmes are drawn on either A1 or A0 size paper and the number of different activities must be compressed into the 840 mm width of this sheet. (It may, of course, be possible to divide the network into two, but then the interlinking activities must be carefully transferred.) Normally, the division between bars is about 6 mm, which means that a maximum of 120 activities can be analysed. However, bearing in mind that in a normal network, 30% of the activities are dummies, a network of 180−200 activities could be analysed graphically on one sheet.

Briefly, the mode of operation is as follows:

1. Draw the network in arrow diagram or precedence format and write in the activity titles (Fig. 24.1 or 24.2). Although a forward pass has been carried out on both these diagrams, this is not necessary when using the graphical method of analysis.

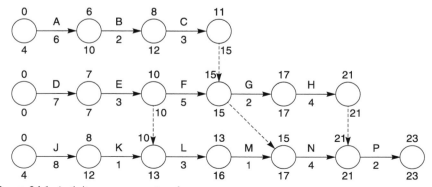

Figure 24.1 Activity on arrow network.

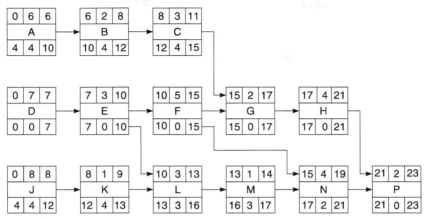

Figure 24.2 Activity on node network.

2. Insert the durations.
3. List the activities on the left-hand vertical edge of a sheet of graph paper (Fig. 21.11) showing:
   a. activity title
   b. duration (in days, weeks, etc.)
   c. node no (only required when using these for bar chart generation)
4. Draw a time scale along the bottom horizontal edge of the graph paper.
5. Draw a horizontal line from day 0 of the first activity, which is proportional to the duration (using the time scale selected), for example 6 days would mean a line six divisions long (Fig. 24.3). To ease identification an activity letter or no. can be written above the bar.

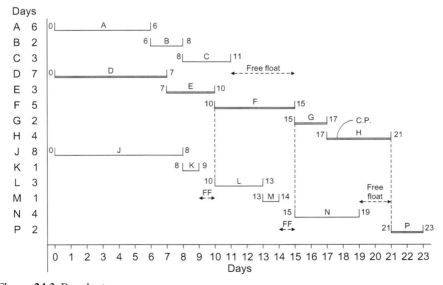

Figure 24.3 Bar chart.

6. Repeat this operation with the next activity on the table starting on day 0.
7. When using activity on arrow networks, mark dummy activities by writing the end time of the dummy next to the start time of the dummy, for example 4 → 7 would be shown as 4,7 (Fig. 21.13).
8. All subsequent activities must be drawn with their start time (start day no.) directly below the end time (end day no.) of the previous activity having the same time value (day no.).
9. If more than one activity has the same end time (day no.), draw the new activity line from the activity end time (day no.) furthest to the *right*.
10. Proceed in this manner until the end of the network.
11. The critical path can now be traced back by following the line (or lines), which runs back to the start without a horizontal break.
12. The break between consecutive activities on the bar chart is the *free float* of the preceding activity.
13. The summation of the free floats in one string, before that string meets the critical path, is the *total float* of the activity from which the summation starts, for example in Fig. 21.11 the total float of activity K is $1 + 1 + 2 = 4$ days, the total float of activity M is $1 + 2 + 3$ days, and the total float of activity N is 2 days.

The advantage of using the start and end times (day nos.) of the activities to generate the bar chart is that there is no need to carry out a forward pass. The correct relationship is given automatically by the disposition of the bars. This method is therefore equally suitable for arrow and precedence diagrams.

An alternative method can, however, be used by substituting the day numbers by the node numbers. Clearly, this method, which is sometimes quicker to draw, can only be used with arrow diagrams, because precedence diagrams do not have node numbers. When using this method, the node numbers are listed next to the activity titles (Fig. 24.5) and the bars are drawn from the starting node of the first activity with a length equal to the duration. The next bar starts vertically below the end node with the same node number as the starting node of the activity being drawn.

As with the day no. method, if more than one activity has the same end node number, the one furthest to the right must be used as a starting time. Fig. 24.4 shows the same network with the node numbers inserted, and Fig. 24.5 shows the bar chart generated using the node numbers.

Fig. 24.6 shows a typical arrow diagram, and Fig. 24.7 shows a bar chart generated using the starting and finishing node numbers. Note that these node numbers have been listed on the left-hand edge together with the durations to ease plotting.

Time for analysis

Probably the most time-consuming operation in bar chart preparation is the listing of the activity titles, and for this there is no shortcut. The same time, in fact, must be expended typing the titles straight into the computer. However, to arrive at a quick answer, it is only necessary at the initial stage to insert the node numbers, and once this listing has been done (together with the activity times) the analysis is very rapid. It is possible to determine the critical path for a 200-activity network (after the listing has been carried out) in less than an hour. The backward pass for ascertaining floats takes about the same time.

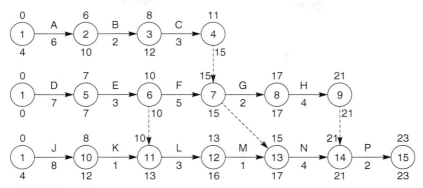

Figure 24.4 Arrow diagram.

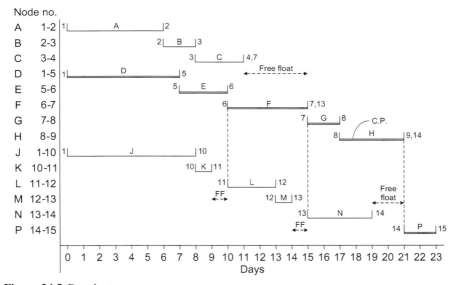

Figure 24.5 Bar chart.

Computer analysis

Most manufacturers of computer hardware, and many suppliers of computer software, have written programs for analysing critical path networks using computers. While the various commercially available programs differ in detail, they all follow a basic pattern, and give, by and large, a similar range of outputs. In certain circumstances, a contractor may be obliged by contractual commitments to provide a computerized output report for the client. Indeed, when a client organization has standardized on a particular project management system for controlling the overall project, the contractor may well be required to use the same proprietary system, so that the contractor's reports can be integrated into the overall project control system on a regular basis.

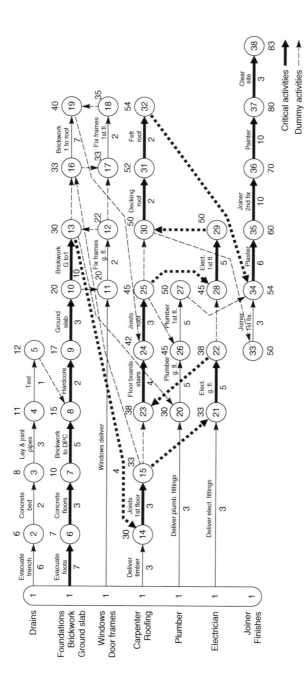

Figure 24.6 Arrow diagram of house.

Activity	S	F	Time	Dummies	Floats
Excavate trench	1	2	6		3
Concrete bed	2	3	2		3
Lay and joint pipes	3	4	3		3
Test	4	5	1		3
Excavate foots	1	6	7	8	0
Concrete foots	6	7	3		0
Brickwork to DPC	7	8	7		0
Hardcore	8	9	2		0
Ground slab	9	10	3	11,	0
Brickwork g - 1	10	13	10	16, 20, 17, 14,	0
Windows delivered	1	11	4	13, 14, 16, 17, 20,	24
Fix frames g. fl.	11	12	2		8
Deliver timber	1	14	3	16, 21, 23, 17,	27
Joists 1st fl.	14	15	3	24,	0
Brickwork 1 - r	16	19	7		0
Fix frames 1st fl.	17	18	2	19, 24,	7
Deliver plumb. ftgs	1	20	5		39
Plumber g. fl.	20	26	5		14
Deliver elec. ftgs	1	21	3	23, 28	30
Electric g. fl.	21	22	5		0
Floor boards stair	23	24	4	26, 28, 30, 33	0
Joists roof	24	25	3		0
Plumber 1st fl.	26	27	5	34,	4
Electr. 1st fl.	28	29	5	30, 33	0
Decking roof	30	31	2		0
Felt roof	21	32	2	34	0
Joiner 1st fix	33	34	3		1
Plasterer	34	35	6		0
Joiner 2nd fix	35	36	10		0
Painter	36	37	10		0
Clear site	37	38	3		0

Figure 24.7 Bar chart of house.

History

The development of network analysis techniques more or less coincided with that of the digital computer. The early network analysis programs were, therefore, limited by the storage and processing capacity of the computer as well as the input and output facilities.

The techniques employed mainly involved producing punched cards (one card for each activity) and feeding them into the machine via a card reader. These procedures were time consuming and tedious, and, because the punching of the cards was carried out by an operator who usually understood little of the program or its purpose, mistakes occurred, which only became apparent after the printout was produced.

Even then, the error was not immediately apparent — only the effect. It then often took hours to scan through the reams of printout sheets before the actual mistake could be located and rectified. To add to the frustration of the planner, the new printout may still have given ridiculous answers because a second error was made on another card. In this way, it often required several runs before a satisfactory output could be issued.

In an endeavour to eliminate punching errors, attempts were made to use two separate operators, who punched their own set of input cards. The cards were then automatically compared and, if not identical, were thrown out, indicating an error. Needless to say, such a practice cost twice as much in manpower.

Because these early computers were large and very expensive, usually requiring their own air-conditioning equipment and a team of operators and maintenance staff, few commercial companies could afford them. Computer bureaux were therefore set up by the computer manufacturers or special processing companies, to whom the input sheets were delivered for punching, processing and printing.

The cost of processing was usually a lump sum fee plus x pence per activity. As the computer could not differentiate between a real activity and a dummy one, planners tended to go to considerable pains to reduce the number of dummies to save cost. The result was often a logic sequence, which may have been cheap in computing cost but was very expensive in application, as frequently important restraints were overlooked or eliminated. In other words, the tail wagged the dog — a painful phenomenon in every sense. It was not surprising, therefore, that many organizations abandoned computerized network analysis or, even worse, discarded the use of network analysis altogether as being unworkable or unreliable.

There is no doubt that manual network analysis is a perfectly feasible alternative to using computers. Indeed, one of the largest petrochemical complexes in Europe was planned entirely using a series of networks, all of which were analysed manually.

The PC

The advent of the personal computer (PC) significantly changed the whole field of computer processing. In place of the punched card or tape, we now have the computer keyboard and video screen, which enable the planner to input the data directly into the computer without filling in input sheets and relying on a punch operator. The information is taken straight from the network and displayed on the video screen as it is 'typed' in. In this way, the data can be checked or modified almost instantaneously.

Provided sufficient information has been entered, trial runs and checks can be carried out at any stage to test the effects and changes envisaged. Modern planning programs (or project management systems, as they are often called) enable the data to be inputted in a random manner to suit the operator, provided, of course, that the relationship between the node numbers (or activity numbers) and duration remains the same.

There are some programs that enable the network to be produced graphically on the screen as the information − especially the logic sequence − is entered. This, as claimed, eliminates the need to draw the network manually. Whether this practice is as beneficial as suggested is very doubtful.

For a start, the number of activities that can be viewed simultaneously on a standard video screen is very limited, and the scroll facility that enables larger networks to be accommodated does not enable an overall view to be obtained at a glance. The greatest drawback of this practice, however, is the removal of the team spirit from the network planning process, which is engendered when a number of specialists sit down with the planner around a conference table to 'hammer out' the basic shape of the network. Most problems have more than one solution, and the discussions and suggestions, both in terms of network logic and durations, are invaluable when drafting the first programs. These meetings are, in effect, a brainstorming session at which the ideas of various participants are discussed, tested and committed to paper. Once this draft network has been produced, the planner can very quickly input it into the computer and call up a few test runs to see whether the overall completion date can, in fact, be achieved. If the result is unsatisfactory, logic and/or duration changes can be discussed with the project team before the new data are processed again by the machine. The speed of the new hardware makes it possible for the computer to be a part of the planning conference, so that (provided the planner/operator is quick enough) the 'what if' scenarios can be tested while the meeting is in progress. A number of interim test runs can be carried out to establish the optimum network configuration before proceeding to the next stage. Even more important, errors and omissions can be corrected and durations of any or all activities can be altered to achieve a desired interim or final completion date.

The relatively low cost of modern PCs has enabled organizations to install planning offices at the head office and sites as well as at satellite offices, associate companies and offices of vital suppliers, contractors and subcontractors. All these PCs can be linked to give simultaneous printouts as well as supply up-to-date information to the head office where the master network is being produced. In other words, the information technology (IT) revolution has made an important impact on the whole planning procedure, irrespective of the type or size of organization.

The advantages of PCs are as follows:

1. The great reduction in the cost of the hardware, making it possible for small companies, or even individuals, to purchase their own computer system.
2. The proliferation of inexpensive, proven software of differing sophistication and complexity, enabling relatively untrained planners to operate the system.

3. The ability to allow the planner to input his or her own program or information via a keyboard and VDU.
4. The possibility to interrogate and verify the information at any stage on the video screen.
5. The speed with which information is processed and printed out either in numerical (tabular) or graphical form.

Programs

During the last few years, a large number of proprietary programs have been produced and marketed. All these programs have the ability to analyse networks and produce the standard output of early and late starts and the three main types of float, that is total, free and independent. Most programs can deal with either arrow diagrams or precedence diagrams, although the actual analysis is only carried out via one type of format.

The main differences between the various programs available at the time of writing are the additional facilities available and the degree of sophistication of the output. Many of the programs can be linked with 'add-on' programs to give a complete project management system covering not only planning but also cost control, material control, site organization, procurement, stock control, EVA, etc. It is not possible to describe the various intricacies of all the available systems within the confines of this chapter, nor is it the intention to compare one system with another. Such comparison can be made in terms of cost, user-friendliness, computing power, output sophistication, functionality or range of add-ons. Should such surveys be required, it is best to consult the internet or some of the specialist computer magazines or periodicals, which carry out such comparisons from time to time.

Most of the programs more commonly available can be found on the internet, but to give a better insight into the versatility of a modern program, one of the more sophisticated systems is described in some detail in Chapter 51. The particular system was chosen because of its ability to be linked with the EVA system described in Chapter 32 of this book. The chosen system, Primavera P6, is capable of fully integrating critical path analysis with earned value analysis and presenting the results on one sheet of A4 paper.

At the time of publication, about 140 project management programs were listed and compared for functionality in Wikipedia. Many of these will probably not exist anymore by the time this book is being read, while no doubt many more will have been created to take their place. It is futile therefore to even attempt to list them. The cost of these systems can vary between $150 and $6000, and the reader is therefore advised to investigate each 'offer' in some depth to ensure value for money. A simple inexpensive system may be adequate for a small organization running small projects or wishing to become familiar with computerized network analysis. Larger companies, whose clients may demand more sophisticated outputs and reports, may require more expensive systems. Indeed, the choice of a particular system may well be dictated by the lead company of a consortium or the client, as described earlier.

It is recommended that the decision to produce any but the most basic printouts, as well as any printouts of reports or summaries, be delayed until the usefulness and degree of detail of a report have been studied and discussed with department managers. There is always a danger with computer outputs that recipients request more reports than they can digest, merely because they know they are available at the press of a button. Too much information or paper becomes self-defeating, as the very bulk frightens the reader to the extent of it not being read at all.

With the proliferation of the PC and the expansion of IT, especially the internet, many of the project management techniques can now be carried out online. The use of e-mail and intranets allows information to be distributed to the many stakeholders of a project almost instantaneously. Where time is important — and it nearly always is — such a fast distribution of data or instructions can be of enormous benefit to the project manager. It does, however, require all information to be carefully checked before dissemination, precisely because so many people receive it at the same time. (See Chapter 52 on BIM.) It is an unfortunate fact that computer errors are more serious for just this reason as well as the naive belief that computers are infallible. Unfortunately, as with all computer systems, RIRO, rubbish in, rubbish out applies.

Milestones and line of balance

Milestones

Important deadlines in a project programme are highlighted by specific points in time called *milestones*. These are timeless activities usually at the beginning or end of a phase or stage and are used for monitoring purposes throughout the life of the project. Needless to say, they should be SMART, which is an acronym for Specific, Measurable, Achievable, Realistic, Timebound. Often milestones are used to act as trigger points for progress payments or deadlines for receipt of vital information, permits or equipment deliveries.

Milestone reports are a succinct way of advising top management of the status of the project and should act as a spur to the project team to meet these important deadlines. This is especially important if they relate to large tranches of progress payments.

Milestones are marked on bar charts or networks by a triangle or diamond and can be turned into a monitoring system in their own right when used in milestone *slip charts*, sometimes also known as *trend charts*.

Fig. 25.1 shows such a slip chart, which was produced at reporting period five of a project. The top scale represents the project calendar, and the vertical scale represents the main reporting periods in terms of time. If both calendars are drawn to the same scale, a line drawn from the top left-hand corner to the bottom right-hand corner will be at 45 degrees to the two axes.

The pre-planned milestones at the start of the project are marked on the top line with a black triangle (▼).

As the project progresses, the predicted or anticipated dates of achievement of the milestones are inserted so that the slippage (if any) can be seen graphically. This should then prompt management action to ensure that the subsequent milestones do not slip! At each reporting stage, the anticipated slippages of milestones as given by the programme are re-marked with an × while those that have not been re-programmed are marked with an O. Milestones that *have* been met will be on the diagonal and will be marked with a triangle (∇).

Project Management, Planning and Control. https://doi.org/10.1016/B978-0-12-824339-8.00025-0

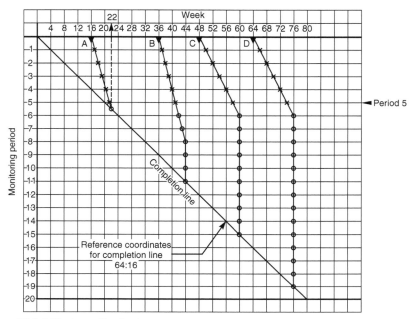

Figure 25.1 Milestone slip chart.

As the programmed slippage of each milestone is marked on the diagram, a pattern emerges, which acts not only as a historical record of the slippages, but can also be used to give a crude prediction of future milestone movements.

A slip chart showing the status at reporting period 11 is shown in Fig. 25.2. It can be seen that milestone A was reached in week 22 instead of the original prediction of week 16. Milestones B, C and D have all slipped, with the latest prediction for B being week 50, for C being week 62, and for D being week 76. It will be noticed that before reporting period 11, the programmed predictions are marked X, and the future predictions, after week 11, are marked O.

If a milestone is not on the critical path, it may well slip on the slip chart without affecting the next milestone. However, if two adjacent milestones on the slip chart *are* on the critical path, any delay on the first one must cause a corresponding slippage on the second. If this is then marked on the slip chart, it will in effect become a prediction, which will then alert the project manager to take action.

Once the milestone symbol meets the diagonal line, the required deadline has been achieved.

Line of balance

Network analysis is essentially a technique for planning one-off projects, whether this is a construction site, a manufacturing operation, a computer software development or

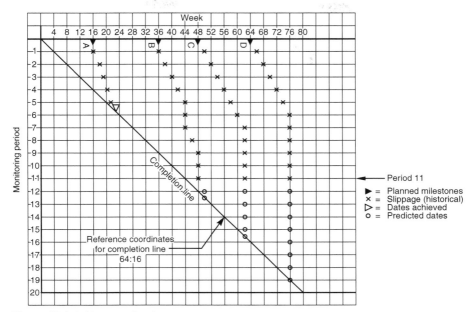

Figure 25.2 Milestone slip chart.

a move to a new premises. When the overall project consists of a number of identical or batch operations, each of which may be a subproject in its own right, it may be of advantage to use a technique called *line of balance* (LoB).

The quickest way to explain how this planning method works is to follow a simple example involving the construction of four identical, small, single-storey houses of the type shown in Chapter 47, Fig. 47.1. For the sake of clarity, only the first five activities will be considered, and it will be seen from Fig. 47.2 that the last of the five activities, E − 'floor joists', will be complete in week 9.

Assuming one has sufficient resources and space between the actual building plots, it is possible to start work on every house at the same time and therefore finish laying all the floor joists by week 9. However, in real life this is not possible, so the gang laying the foundations to house No. 1 will move to house No. 2 when foundation No. 1 is finished. When foundation No. 2 is finished, the gang will start No. 3, and so on. The same procedure will be carried out by all the following trades until all the houses are finished.

Another practical device is to allow a time buffer between the trades to give a measure of flexibility and introduce a margin of error. Frequently such a buffer will occur naturally for such reasons as hardening time of concrete, setting time of adhesive, or drying time of plaster or paint.

Table 47.1 can now be partially redrawn showing the buffer time, which was originally included in the activity duration. The new table is now shown as Table 25.1.

Fig. 25.3 shows the relationship between the trades involved. Each trade (or activity) is represented by two lines. The distance between these lines is the duration of the

Table 25.1 Activity summary.

Activity letter	Activity description	Adjusted duration (weeks)	Dependency	Total float (weeks)	Buffer (weeks)
A	Clear ground	2.0	Start	0	0.0
B	Lay foundations	2.8	A	0	0.2
C	Build dwarf walls	1.9	B	0	0.1
D	Oversite concrete	0.9	B	1	0.1
E	Floor joists	1.8	C and D	0	0.2

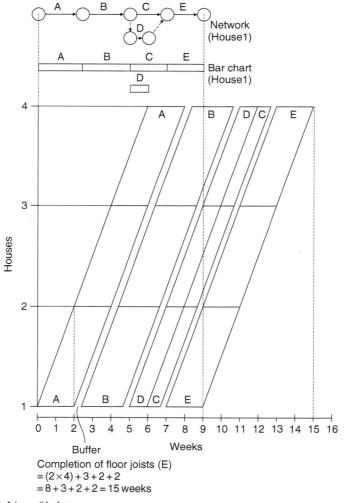

Completion of floor joists (E)
$= (2 \times 4) + 3 + 2 + 2$
$= 8 + 3 + 2 + 2 = 15$ weeks

Figure 25.3 Line of balance.

activity. The distance between the activities is the buffer period. As can be seen, all the work of activities A to E is carried out at the same rate, which means that for every house, enough resources are available for every trade to start as soon as its preceding trade is finished. This is shown to be the case in Fig. 25.3.

However, if only one gang is available on the site for each trade, for example if only one gang of concreters laying the foundations (activity B) is available, concreting on house 2 cannot start until ground clearance (activity A) has been completed. The chart would then be as shown in Fig. 25.4. If instead there were two gangs of concreters available on the site, the foundations for house 2 could be started as soon as the ground has been cleared.

Building the dwarf wall (activity C) requires only 1.9 weeks per house, which is a faster rate of work than laying foundations. To keep the bricklaying gang going smoothly from one house to the next, work can only start on house 1 in week 7.2, that is after the buffer of about 2.5 weeks following the completion of the foundations of house 1. In this way, by the time the dwarf walls are started on house 4, the foundations (activity B) of house 4 will just have been finished. (In practice of course there

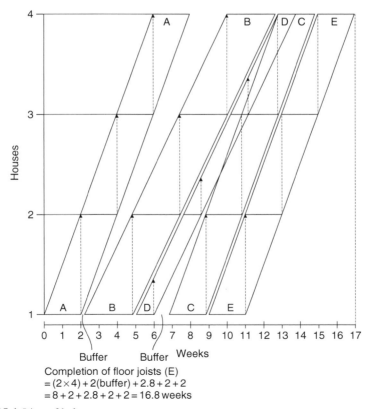

Figure 25.4 Line of balance.

would be a further buffer to allow the concrete to harden sufficiently for the brick-laying to start.)

As the oversite concreting (activity D) only takes 0.9 weeks, the one gang of labourers doing this work will have every oversite completed well before the next house is ready for them. Their start date could be delayed if necessary by as much as 3.5 weeks, since apart from the buffer, this activity (D) has also 1 week float.

It can be seen therefore from Fig. 25.4 that by plotting these operations with the time as the horizontal axis and the number of houses as the vertical axis, the following becomes apparent.

If the slope of an operation is less (i.e., flatter) than the slope of the preceding operation, the chosen buffer is shown at the *start* of the operation. If, on the other hand, the slope of the succeeding operation is *steeper*, the buffer must be inserted at the *end* of the previous operation, since otherwise there is a possibility of the trades clashing when they get to the last house.

What becomes very clear from these diagrams is the ability to delay the start of an operation (and use the resources somewhere else) and still meet the overall project programme.

When the work is carried out by trade gangs, the movement of the gangs can be shown on the LoB chart by vertical arrows as indicated in Fig. 25.4.

Simple examples

Chapter outline

To illustrate the principles set out in Chapter 20, let us now examine two simple examples.

Example 1

For the first example, let us consider the rather mundane operation of getting up in the morning, and let us look at the constituent activities between the alarm going off and boarding our train to the office.

The list of activities — not necessarily in their correct sequence — is roughly as follows:

		Time (minutes)
A	Switch off the alarm clock	0.05
B	Lie back and collect your thoughts	2.0
C	Get out of the bed	0.05
D	Go to the bathroom	0.10
E	Wash or shower	6.0
F	Brush teeth	3.0
G	Brush hair	3.0
H	Shave (if you are a man)	4.0
J	Boil water for tea	2.0
K	Pour tea	0.10
L	Make toast	3.0
M	Fry eggs	4.0

Continued

Project Management, Planning and Control. https://doi.org/10.1016/B978-0-12-824339-8.00026-2

		Time (minutes)
N	Serve breakfast	1.0
P	Eat breakfast	8.0
Q	Clean shoes	2.0
R	Kiss the wife goodbye	0.10
S	Don coat	0.05
T	Walk to the station	8.0
U	Queue and buy ticket	3.0
V	Board the train	1.0
	—	50.45

The operations listed earlier can be represented diagrammatically in a network. This would look something like that shown in Fig. 26.1.

It will be seen that the activities are all joined in one long string, starting with A (switch off the alarm) and ending with V (board the train). If we give each activity a time duration, we can easily calculate the total time taken to perform the complete operation by simply adding up the individual durations. In the example given, this total time — or project duration — is 50.45 minutes. In theory, therefore, if any operation takes a fraction of a minute longer, we will miss our train. Consequently, each activity becomes critical and the whole sequence can be seen to be on the critical path.

In practice, however, we will obviously try to make up the time lost on an activity by speeding up a subsequent one. Thus, if we burn the toast and have to make a new

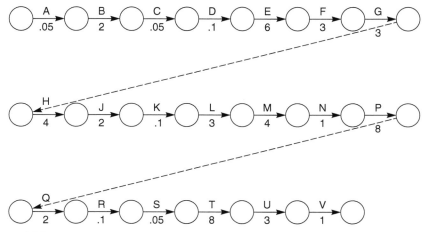

Figure 26.1

piece, we can make up the time by running to the station instead of walking. We know that we can do this because we have a built-in margin or float in the journey to the station. This float is, of course, the difference between the time taken to walk and run to the station. In other words, the path is not as critical as it might appear, that is we have not pared each activity down to its minimum duration in our original sequence or network. We had something up our sleeve.

However, let us suppose that we cannot run to the station because we have a bad knee; how can then we make up for the lost time? This is where network analysis comes in. Let us look at the activities succeeding the making of toast (L) and see how we can make up the lost time of, say, 2 minutes. The remaining activities are as follows:

		Times (minutes)
M	Fry eggs	4.0
N	Serve breakfast	1.0
P	Eat breakfast	8.0
Q	Clean shoes	2.0
R	Kiss the wife goodbye	0.10
S	Don coat	0.05
T	Walk to the station	8.0
U	Queue and buy ticket	3.0
V	Board the train	1.0
–	–	27.15

The total time taken to perform these activities is 27.15 minutes.

Therefore, the first question is, do we have any activity which is unnecessary? Yes. We need not kiss the wife goodbye. But this only saves us 0.1 minute, and the saving is of little benefit. Besides, it could have serious repercussions.

The second question must therefore be, are there any activities which we can perform simultaneously? Yes. We can clean our shoes while the eggs fry. The network shown in Fig. 26.2 can thus be redrawn as demonstrated in Fig. 26.3. The total now from M to V adds up to 25.15 minutes. We have, therefore, made up our lost 2 minutes without apparent extraeffort. All we have to do is to move the shoe-cleaning box to a position in the kitchen where we can keep a sharp eye on the eggs while they fry.

Figure 26.2

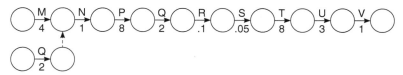

Figure 26.3

Encouraged by this success, let us now re-examine the whole operation to see how else we can save a few minutes, as a few moments extra in bed are well worth saving. Let us therefore see what other activities can be performed simultaneously:

1. We could brush our teeth under the shower.
2. We could put the kettle on before we shaved so that it boils while we shave.
3. We could make the toast while the kettle boils or while we fry the eggs.
4. We could forget about the ticket and pay the ticket collector at the other end.
5. We could clean our shoes while the eggs fry as previously discussed.

Having considered the earlier list, we eliminate (1) since it is not nice to spit into the bathtub, and (4) is not possible because we have an officious guard on our barrier. So we are left with (2), (3) and (5). Let us see what our network looks like now (Fig. 26.4). The total duration of the operation or programme is now 43.45 minutes, a saving of 7 minutes or over 13% for no additional effort. All we did was to resequence the activities. If we moved the washbasin near the shower and adopted the 'brush your teeth while you shower' routine, we could save another 3 minutes, and if we bought a season ticket we would cut another 3 minutes off our time. It can be seen, therefore, that by a little careful planning we could well spend an extra 13 minutes in bed − all at no extra cost or effort.

If a saving of over 25% can be made on such a simple operation as getting up, it is easy to see what tremendous savings can be made when planning complex manufacturing or construction operations.

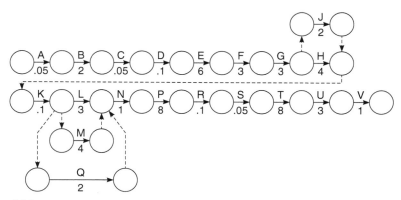

Figure 26.4

Let us now look at our latest network again. From A to G, the activities are in the same sequence as on our original network. H and J (shave and boil water) are in parallel. H takes 4 minutes and J takes 2. We therefore have 2 minutes float on activity J in relation to H. To get the total project duration, we must, therefore, use the 4 minutes of H in our adding-up process, that is the *longest* duration of the parallel activities.

Similarly, activities L, M and Q are being carried out in parallel and we must, therefore, use M (fry eggs) with its duration of 4 minutes in our calculation. Activity L will, therefore, have 1 minute float while activity Q has 2 minutes float. It can be seen, therefore, that activities H, L and Q could all be delayed by their respective floats without affecting the overall programme. In practice, such a float is absorbed by extending the duration to match the parallel critical duration or left as a contingency for disasters. In our example, it may well be prudent to increase the toast-making operation from 3 minutes to 4 by reducing the flame on the grill in order to minimize the risk of burning the bread.

Example 2

Let us now look at another example. Suppose we decide to build a new room into the loft space of our house. We decide to coordinate the work ourselves because the actual building work will be carried out by a small jobbing builder, who has little idea of planning, while the drawings will be prepared by a freelance architect who is not concerned with the meaning of time. If the start of the programme is brief to the architect and the end is the fitting of carpets, let us draw up a list of activities we wish to monitor to ensure a speedy completion of the project. The list would be as follows:

		Days
A	Brief architect	1
B	Architect produces plans for planning permission	7
C	Obtain the planning permission	60
D	Finalize the drawings	10
E	Obtain tenders	30
F	Adjudicate bids	2
G	Builder delivers materials	15
H	Strip roof	2
J	Construct dormer	2
K	Lay floor	2
L	Tile the dormer walls	3
M	Felt the dormer roof	1
N	Fit window	1

Continued

		Days
P	Move the CW tank	1
Q	Fit doors	1
R	Fit the shelves and cupboards	4
S	Fit the internal lining and insulation	4
T	Run electric cables	2
U	Cut a hole in existing ceiling	1
V	Fit stairs	2
W	Plaster walls	2
X	Paint	2
Y	Fit the carpets	1
	–	156

Rather than drawing out all these activities in a single long string, let us make a preliminary analysis on which activities can be carried out in parallel. The following immediately spring to mind:

1. Final drawings can be prepared while planning permission is obtained.
2. It may even be possible to obtain tenders during the planning permission period, which is often extended.
3. The floor can be laid while the dormer is being tiled.

The preliminary network would, therefore, be as shown in Fig. 26.5.

If all the activities were carried out in series, the project would take 156 days. As drawn in Fig. 26.5, the duration of the project is 114 days. This already shows a considerable saving by utilizing the planning permission period for finalizing drawings and obtaining tenders.

However, we wish to reduce the overall time even further, so we call the builder in before we start work and go through the job with him. The first question we ask is, how many men he will employ. He says between two and four. We then make the following suggestions:

1. Let the electrician run the cables while the joiners fit the stairs.
2. Let the plumber move the tank while the roof of the dormer is being constructed.
3. Let the glazier fit the windows while the joiner fits the shelves.
4. Let the roofer felt the dormer while the walls are being tiled.
5. Fit the doors while the cupboards are being built.

The builder may object that this requires too many men, but you can tell him that his overall time will be reduced and he will probably gain in the end. The revised network is shown in Fig. 26.6. The total project duration is now reduced to 108 days. The same network in precedence format [activity on node, (AoN)] is shown in Fig. 26.7.

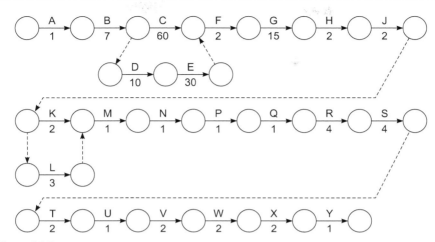

Figure 26.5

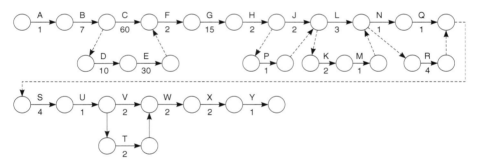

Figure 26.6

If we now wish to reduce the period even further, we may have to pay the builder a little extra. However, let us assume that time is of the essence, as our rich old uncle will be coming to stay and an uncomfortable night on the sofa in the sitting room might prejudice our chances in his will. It is financially viable, therefore, to ensure that the room will be complete.

Suppose we have to cut the whole job to take no longer than 96 days. Somehow, we have to save another 12 days. First, let us look at those activities that have float. N and Q together take 2 days while R takes 4. N and Q have, therefore, 2 days float. We can utilize this by splitting the operation S (fit internal lining) and doing 2 days' work while the shelves and cupboards are being built. The network of this section would, therefore, appear as in Fig. 26.8. We have saved 2 days provided that labour can be made available to start insulating the rafters.

If we adjudicate the bids (F) before waiting for planning permission, we can save another 2 days. This section of the network will, therefore, appear as in Fig. 26.9.

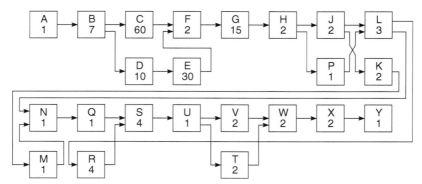

Figure 26.7 Precedence network.

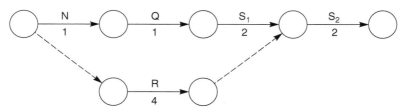

Figure 26.8

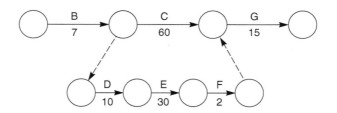

Figure 26.9

Total saving to this stage is $2 + 2 = 4$ days. We have to find another 8 days, so let us look at the activities that take the longest: C (obtaining planning permission) cannot be reduced since it is outside our control. It is very difficult to hurry a local authority. G (builder delivers materials) is difficult to reduce since the builder will require a reasonable mobilization period to buy materials and allocate resources. However, if we select the builder before planning permission has been received, and we do, after all, have 18 days float in loop D–E–F, we may be able to get him to place preliminary orders for the materials required first, and thus enable work to be started a little earlier. We may have to guarantee to pay the cost for this material if planning permission is not granted, but as time is of the essence we are prepared to take the risk. The saving could well be anything from 1 to 15 days.

Let us assume we can realistically save 5 days. We have now reduced the programme by $2 + 2 + 5 = 9$ days. The remaining days can now only be saved by reducing the actual durations of some of the activities. This means more resources and hence more money. However, the rich uncle cannot be put off, so we offer to increase the contract sum if the builder can manage to reduce V, T, W and X by 1 day each, thus saving 3 days altogether. It should be noted that we only save 3 days although we have reduced the time of four activities by 1 day each. This is, of course, because V and T are carried out in parallel, but our overall period — for very little extra cost — is now 96 days, a saving of 60 days or 38%.

Example 3

This example from the IT industry uses the AoN (precedence) method of network drafting. This is now the standard method for this industry, probably because of the influence of MS Project and because networks in IT are relatively small compared to the very large networks in construction, which can have between 200 and several thousand activities. The principles are of course identical.

A supermarket requires a new stock-control system linked to a new check-out facility. This involves removing the existing checkout, designing and manufacturing new hardware and writing new software for the existing computer, which will be retained.

The main activities and durations (all in days) for this project are as follows:

		Days
A	Obtain a brief from the client (the supermarket owner)	1
B	Discuss the brief	2
C	Conceptual design	7
D	Feasibility study	3
E	Evaluation	2
F	Authorization	1
G	System design	12
H	Software development	20
J	Hardware design	40
K	Hardware manufacture	90
L	Hardware delivery (transport)	2
M	Removal of existing checkout	7
N	Installation of new equipment	6
P	Testing on site	4
Q	Handover	1
R	Trial operation	7
S	Close out	1

The network for this project is shown in Fig. 26.10 from which it can be seen that there are virtually no parallel activities, so only two activities, M (removal of existing checkout) and H (software development), have any float. However, the float of M is only 1 day, so for all intents and purposes it is also critical. It may be possible, however, to start J (hardware design) earlier after G (system design) is 50% complete. This change is shown on the network in Fig. 26.11. As a result of this change, the overall project period has been reduced from 179 to 173 days. It could be argued that the existing checkout (M) could be removed earlier, but the client quite rightly wants to make sure that the new equipment is ready for dispatch before removing the old one. As the software developed under H is only required at the time of the installation (N), there is still plenty of float (106 days) even after the earlier start of hardware design (J) to make sure everything is ready for the installation of the new equipment (N).

In practice, this means that the start of software development (H) could be delayed if the resources allocated to H are more urgently required by another project.

Summary of operation

The three examples given are, of course, very small, simple programmes, but they do show the steps that have to be taken to get the best out of network analysis. These are as follows:

1. Draw up a list of activities and anticipated durations.
2. Make as many activities as possible run in parallel.
3. Examine new sequences after the initial network has been drawn.
4. Start a string of activities as early as possible and terminate as late as possible.
5. Split activities into two or more steps if necessary.
6. If time is vital, reduce durations by paying more for extra resources.
7. Always look for new techniques in the construction or operation being programmed.

It is really amazing what savings can be found after a few minutes' examination, especially after a good night's sleep.

Example 4 (using manual techniques)

An example of how the duration of a small project can be reduced quite significantly using manual techniques is shown by following the stages shown in Fig. 26.13.

The project involves the installation of a pump, a tank and the interconnecting piping, which has to be insulated. Fig. 26.12 shows the diagrammatic representation

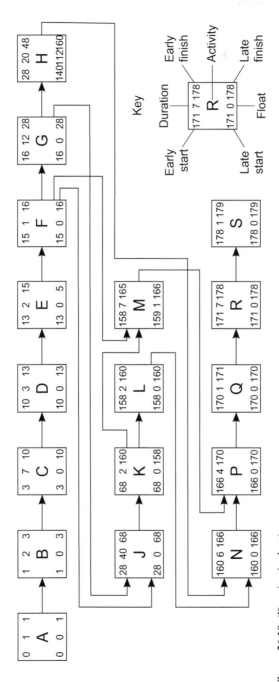

Figure 26.10 (Duration in days).

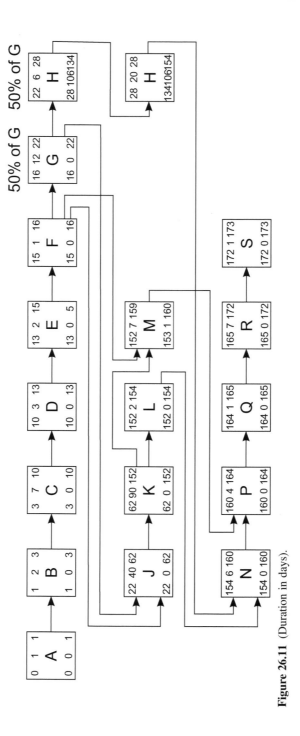

Figure 26.11 (Duration in days).

of the scheme, which does not include the erection of the pipe bridge over which the line has to run. All the networks in Fig. 26.13 are presented in activity on arrow, AoN and bar chart format, which clearly show the effect of overlapping activities. Fig. 26.13A illustrates all the five operations in sequence. This is quite a realistic procedure, but it takes the maximum amount of time — 16 days. By erecting the tank and pump at the same time (Fig. 26.13B), the overall duration has been reduced to 14 days. Fig. 26.13C shows a further saving of 3 days by erecting the pipe over the bridge while also erecting the pump and tank, giving an overall time of 11 days. When the pipe laying is divided into three sections (D_1, D_2 and D_3), it is possible to weld the last two sections at the same time, thus reducing the overall time to 10 days (Fig. 26.13D). Further investigation shows that while the last two sections of pipe are being welded it is possible to insulate the already completed section. This reduces the overall duration to 8 days (Fig. 26.13E).

It can be argued, of course, that an experienced planner can foresee all the possibilities right from the start, and produce the network and bar chart shown in Fig. 26.13E without going through all the previous stages. However, most mortals tend to find the optimum solution to a problem by stages, using the logical thought processes as outlined earlier. A sketch pad and pocket calculator are all that is required to run through these steps. A computer at this stage would certainly not be necessary.

It must be pointed out that although the example shown is only a very small project, such problems occur almost daily and valuable time can be saved by just running through a number of options before the work actually starts. In many cases, the five activities will be represented by only one activity, for example 'install lift pump system' on a larger construction network, and while this master network may be computerized, the small 'problem networks' are far more easily analysed manually.

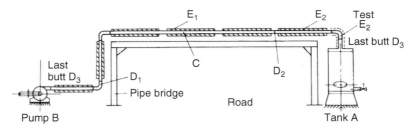

Figure 26.12 Pipe bridge.

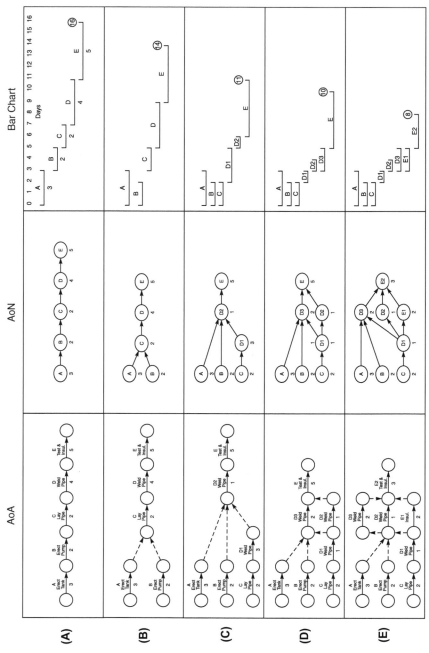

Figure 26.13 Small pipeline project.

Progress reporting

27

Chapter outline

Feedback   204

Having drawn the network programme, it is now necessary to develop a simple but effective system of recording and reporting progress. The conventional method of recording progress on a bar (Gantt) chart is to thicken up or hatch in the bars, which are purposely drawn 'hollow' to allow this to be done. When drafting the network, activities are normally represented by single solid lines (Fig. 27.1), but the principle of thickening up can still be applied. The simplest way is to thicken up the activity line and black in the actual node point (Fig. 27.2). If an activity is only partially complete (say 50%), it can be easily represented by only blacking in 50% of the activity (Fig. 27.2). It can be seen, therefore, that in the case of the string of activities shown in Fig. 27.2, the first activity is complete while the second one is half complete. By rights, therefore, the week number at that stage should be 4 + 50% of 6 = 7. However, this presupposes that the first activity has not been delayed and finished on week 4 as programmed.

How, then, can one represent the case of the first activity finishing, say, 2 weeks late (week 6)? The simple answer is to cross out the original week number (4) and write the revised week number next to it, as shown in Fig. 27.3. If the duration of the second activity cannot be reduced, that is, if it still requires 6 weeks as programmed, it will be necessary to amend all the subsequent week numbers as well (Fig. 27.4).

This operation will, of course, have to be carried out every time there is a slippage, and it is prudent, therefore, to leave sufficient space over the node point to enable this to be done. Alternatively, it may be more desirable to erase the previous week numbers and insert the new ones, provided, of course, the numbers are written in pencil and not ink. At first sight, the job of erasing some 200 node numbers on a network may appear to be a tedious and time-consuming exercise. However, in practice, such an updating function poses no problems. A reasonably experienced planner can update a complete network consisting of about 200 activities in less than 1 hour. When one remembers that in most situations only a small proportion of the activities on a network require updating, the speed of the operation can be appreciated.

Figure 27.1 Weeks.

Project Management, Planning and Control. https://doi.org/10.1016/B978-0-12-824339-8.00027-4

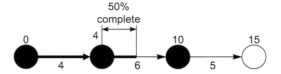

Figure 27.2

Figure 27.3

Figure 27.4

Naturally, only the earliest dates are calculated, as this answers the most important questions, that is,

1. When can a particular activity start?
2. When will the whole project be completed?

At this stage, there is no need to calculate floats since these can be ascertained rapidly as and when required, as explained in Chapter 21.

Precedence (activity on node) networks can be updated as shown in Chapter 19, Figs 19.2 and 19.3.

Feedback

Apart from reporting progress, it is also necessary to update the network to reflect logic changes and delays. This updating, which has to be on a regular basis, must reflect two main types of information:

1. What progress, if any, has been achieved since the last update or reporting stage?
2. What logic changes have to be incorporated to meet the technical or programme requirements?

To enable planners to incorporate this information on a revised or updated network, they must be supplied with data in an organized and regular manner. Many schemes — some very complex and some very simple — have been devised to enable this to happen. Naturally, the simpler the scheme, the better, and the less the paper used, the more the information on the paper will be used.

The ideal situation is, therefore, one where no additional forms whatsoever are used, and this ideal can indeed be reached by using the latest IT systems.

However, unless the operatives in the field have the facilities to electronically transmit the latest updated position direct to the planning engineer's computer, a paper copy of the updated network will still have to be produced on site and sent back to the planning engineer. Provided that:

1. The networks have been drawn on small sheets, that is, A3 or A4, or have been photographically reduced to these sizes.
2. If a photocopier is available, updating the network is merely a question of thickening the completed or partially completed activities, amending any durations where necessary and taking a photostat copy. This copy is then returned to the planner, preferably electronically. When a logic change is necessary, the amendment is made on a copy of the last network and this too is returned to the planner. If all the disciplines or departments do this and return their feedback regularly to the planner, a master network incorporating all these changes can be produced and the effects on other disciplines calculated and studied.

There may be instances where a department manager may want to change a sequence of activities or add new items to his or her particular part of the network. Such logic changes are most easily transmitted to the planner electronically, or, in the absence of such facilities, by placing an overlay over that portion of the network that has to be changed and sketching in the new logic freehand.

When logic changes have been proposed − for this is all a department can do in isolation at this stage − the effect on other departments only becomes apparent when a new draft network has been produced by the planner. Before accepting the situation, the planner must either inform the project manager or call a meeting of all the interested departments to discuss the implications of the proposed logic changes. In other words, the network becomes what it should always be − a focal point for discussion, a means by which the job can be seen graphically, so that it can be amended to suit any new restraints or requirements.

In many instances, it will be possible for the planner to visit the various departments and update the programme by asking a few pertinent questions. This reduces the amount of paper even more and has, of course, the advantage that logic changes can be discussed and provisionally agreed right away. On a site, where the contract has been divided into a number of operational areas, this method is particularly useful as area managers are notorious for shunning paperwork − especially reports. Even very large projects can be controlled in this manner, and the personal contact again helps to generate the close relationship and involvement so necessary for good morale.

Where an efficient cost-reporting system is in operation, and provided that this is geared to the network, the feedback for the programme can be combined with the weekly cost-report information issued in the field or shop.

A good example of this is given in Chapter 32, which describes the earned value analysis (EVA) cost-control system. In this system, the cost-control and cost-reporting procedures are based on the network so that the percentage complete of an operation can be taken from the site returns and entered straight onto the network.

The application of EVA is particularly interesting, as the network can be either electronically or manually analysed while the cost report is produced by a computer, both using the same database.

One of the greatest problems found by main contractors is the submission of updated programmes from sub-suppliers or subcontractors. Despite clauses in the purchase order or subcontract documents, requiring the vendor to return a programme within a certain number of weeks of order date and update it monthly, many suppliers just do not comply. Even if programmes are submitted as requested, they vary in size and format from a reduced computer printout to a crude bar chart, which shows activities possibly useful to the supplier but quite useless to the main contractor or client.

One reason for this production of unsatisfactory information is that the main contractor (or consultant) was not specific enough in the contract documents setting out exactly what information is required and when it is needed. To overcome this difficulty, the simplest way is to give the vendor a pre-printed bar chart form as part of the contract documents, together with a list of suggested activities which *must* appear on the programme.

A pre-printed table, as drawn in Fig. 27.5, shows by the letter X which activities are important for monitoring purposes, for typical items of equipment or materials. The list can be modified by the supplier if necessary, and obviously each main contractor can draw up his own requirements depending on the type of industry he is engaged in, but the basic requirements from setting out drawings to final test certificates are included. The dates by which some of the key documents are required should, of course, be given in the purchase order or contract document, as they may be linked to stage payments and/or penalties such as liquidated damages.

The advantages of the main contractor requesting the programme (in the form of a bar chart) to be produced to his or her own format, a copy of which is shown in Fig. 27.6, are as follows:

1. All the returned programmes are of the same size and type and can be more easily interpreted and filed by the main contractor's staff.
2. Where the supplier is unsophisticated, the main contractor's programme is of educational value to the supplier.
3. As the format is ready-made, the supplier's work is reduced and the programme will be returned by him earlier.
4. As all the programmes are on A4 size paper, they can be reproduced and distributed more easily and speedily.

To ensure that the supplier has understood the principles and uses the correct method for populating the completed bar chart, an instruction sheet, as shown in Fig. 27.7, should be attached to the blank bar chart.

Activity	Pumps	Heat exchanger	Air fins	Compress and turbines	Vessels towers	Valves	Struct. steel	Instr. panels	Large motors	Switchgear MCC invertors	Transformers	Fans	Pipe work
Drawings A – Setting plans	X	X	X	X	X	X	X	X	X	X	X	X	X
Drawings B – As specified	X	X	X	X	X	X	X	X	X	X	X	X	X
Drawings C – (Final)	X	X	X	X	X	X	X	X	X	X	X	X	X
Foster Wheeler Eng. cut-off	X	X	X	X	X	X	X	X	X	X	X		X
Place sub-orders	X	X		X	X	X	X					X	X
Receive forgings		X											X
Receive plate		X			X								
Receive seals	X			X		X	X		X				
Receive couplings	X		X	X								X	
Receive gauges/instrum.		X		X	X			X		X	X		
Receive tubes/fittings			X		X								
Receive bearings	X		X	X					X			X	
Receive motor/actuator	X		X						X			X	
Casting of casing	X					X			X			X	
Casting impeller	X			X		X						X	
Casting bedplate	X			X								X	
Machine casting	X			X								X	
Machine impeller	X		X	X								X	
Machine flanges	X	X	X	X	X	X			X			X	
Machine gears	X			X					X			X	
Machine shaft	X			X	X	X			X			X	
Assemble rotor	X		X	X					X			X	
Assemble equipment	X	X	X	X		X		X	X	X	X	X	X
Weld frame/supports		X	X	X	X		X	X	X	X	X	X	
Roll and weld shell		X	X	X	X	X	X						
Drill tube plate		X			X								
Form dished ends		X			X								
Weld/roll tubes		X	X										
Weld nozzles	X	X	X	X	X	X							
Fit internals			X		X						X		
Access platforms		X					X		X	X			
Light presswork/guards	X		X					X	X			X	
Heat treatment		X			X					X	X		X
Wiring									X		X		
Windings								X	X				
Lube-oil system	X												
Control system								X				X	
Galvanizing/plating	X	X	X	X	X	X		X	X	X	X	X	X
Painting/priming	X	X	X	X	X	X		X	X	X	X	X	X
Testing pressure/mech.	X	X	X	X	X	X		X	X	X	X	X	X
Testing witness/perform.	X	X	X	X	X	X		X		X	X	X	X
Prepare despatch	X	X	X	X	X	X	X	X	X	X	X	X	X
Data books/oper. instructions	X	X	X	X	X	X		X	X	X	X		X
Weld procedures		X			X								X
Spares schedules	X	X	X	X	X	X		X	X	X	X	X	X
Test certs	X	X	X	X	X	X		X	X	X	X	X	X

Figure 27.5 Suggested activities for a manufacturer's bar chart.

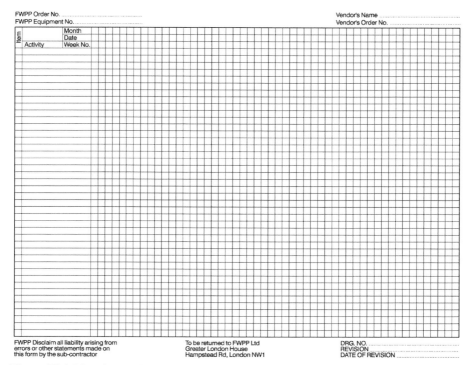

Figure 27.6 Manufacturer's bar chart.

Foster Wheeler Power Products Ltd

Instructions to vendors for completing
FWPP's standard programme format

1 Vendors are required to complete a Manufacturing Programme using the FWPP Standard Bar Chart form enclosed herewith.

2 The block on the top at the page given the FWPP Order Number, FWPP Equipment Number, Vendor's Name and Vendor's Order Number will be filled in by FWPP Purchasing Department at time of order issue.

3 Where a starting date is not known, Vendors must give the programme in week numbers with Week 1 as the date of the order. Subsequently, after order has been placed, the correct FWPP Week Number must be substituted together with the corresponding calendar date.

4 The left-hand column headed 'Activity' must be filled in by the Vendor showing the various stages of the manufacturing process. This should start with production of the necessary drawings requested in the Purchase Order document and continue through various stages of materials arriving at the Vendor's works, manufacuturing stages, assembly stages, testing stages and ending with actual delivery date.

5 For the benefit of vendors the attached Table shows some typical stages which FWPP Expeditors will be monitoring but it must be emphasized that these are for guidance only and must be amended or augmented by the Vendor to suit his method of production.

The Table consists of eleven (11) common items of equipment normally associated with Petrochemical Plants and where an item of equipment does not fall into one of these categories, vendors are required to build up their own detailed lists.

6 Activities with a duration of one (1) week or more should be represented by a thick line

    thus: ■■■■■■■■■■■■■■■■■■■

while shorter activities or specific events such as cut-off dates or despatch dates should be shown by a triangle

          despatch
    thus: ▽

7 This programme must be returned to FWPP within three (3) weeks of receiving the Purchase Order.

Figure 27.7

Project management and network planning

Chapter outline

Responsibilities of the project managers

It is not easy to define the responsibilities of a project manager, mainly because the scope covered by such a position varies not only from industry to industry but also from one company to another. However, three areas of responsibility are nearly always part of the project manager's brief:

1. To build the job to specification and to satisfy the operational (performance, quality and safety) requirements.
2. To complete the project on time.
3. To build the job within previously established budgetary constraints.

The last two are, of course, connected; generally, it can be stated that if the job is on schedule, the cost has either not exceeded the budget, or additional resources have been supplied by the contractor to rectify his own mistakes or good grounds exist for claiming any extra costs from the client. It is far more difficult to obtain extra cash if the programme has been exceeded and the client has also suffered loss due to the delay.

Time, therefore, is vitally important, and the control of time, whether at the design stage or the construction stage, should be a matter of top priority with the project

Project Management, Planning and Control. https://doi.org/10.1016/B978-0-12-824339-8.00028-6

manager. It is surprising, therefore, that so few project managers are fully conversant with the mechanics of network analysis and its advantages over other systems. Even if it had no other function but to act as a polarizing communication document, it would justify its use in preference to other methods.

Information from network

A correctly drawn and regularly updated network can be used to give vital information and has the following beneficial effects on the project.

1. It enables the interaction of the various activities to be shown graphically and clearly.
2. It enables spare time or float to be found where it exists so that advantage can be taken to reduce resources, if necessary.
3. It can pinpoint potential bottlenecks and trouble spots.
4. It enables conflicting priorities to be resolved in the most economical manner.
5. It gives an up-to-date picture of progress.
6. It acts as a communication document between all disciplines and stakeholders.
7. It shows all interested parties the intent of the method of construction.
8. It acts as a focus for discussion at project meetings.
9. It can be expanded into subnets showing greater detail, or contracted to show the chief over-all milestones.
10. If updated in coloured pencil, it can act as a spur between rival gangs of workers.
11. It is very rapid and cheap to operate and is a base for earned value analysis (EVA).
12. It is quickly modified if circumstances warrant it.
13. It can be used when formulating claims, as evidence of disruption due to late decisions or delayed drawings and equipment.
14. Networks of past jobs can be used to draft proposal networks for future jobs.
15. Networks stimulate discussion provided everyone concerned is familiar with them.
16. It can assist in formulating a cash-flow chart to minimize additional funding.

To get the maximum benefit from networks, a project manager should be able to read them as a musician reads music. He should feel the slow movements and the crescendos of activities and combine these into a harmonious flow until the grand finale is reached.

To facilitate the use of networks at discussions, the sheets should be reduced photo-graphically to A3 (approximately 42 × 30 cm). In this way, a network can be folded once and kept in a standard A4 file, which tends to increase its usage. Small networks can, of course, be drawn on A3 or A4 size sheets in the first place, thus saving the cost of subsequent reduction in size.

It is often stated that networks are not easily understood by the man in the field, the area manager or the site foreman. This argument is usually supported by statements that the field men were brought up on bar charts and can, therefore, understand them fully, or that they are confused by all the computer printouts, which take too long to digest. Both statements are true. A bar chart is easy to understand and can easily be updated by hatching or colouring in the bars. It is also true that computer output sheets can be overwhelming by their complexity. Even if the output is restricted to a discipline report, only applicable to the person in question, confusion is often caused by the mass of data. As is so often the case, network analysis and computerization are regarded as being synonymous, and the drawbacks of the latter are then invoked (often quite unwittingly) to discredit the former.

The author's experience, however, contradicts the argument that site people cannot or will not use networks. On the contrary, once the foreman or chargehand understands and appreciates what a network can do, he will prefer it to a bar chart. This is illustrated by the following example, which describes an actual situation on a contract.

Site-preparation contract

The job described was a civil-engineering contract comprising the construction of oversite base slabs, roads, footpaths and foul and stormwater sewers for a large municipal housing scheme consisting of approximately 250 units. The main contractor, who confined his site activities to the actual house building, was anxious to start work as soon as possible to get as much done before the winter months. It was necessary, therefore, to provide him with good roads and a fully drained site.

The contract award was in June and the main contractor was programmed to start building operations at the end of November the same year. To enable this quite short civil-engineering stage to be completed on time, it was decided to split the site into four main areas that could be started at about the same time. The size and location of these areas were dictated by such considerations as access points, site clearance (including a considerable area of woodland), natural drainage and house-building sequence.

Once this principle was established by management, the general site foreman was called in to assist in the preparation of the network, although it was known that he had never even heard of, let alone worked to, a critical path programme.

After explaining the basic principles of network techniques, the foreman was asked where he would start work, what machines he would use, which methods of excavation and construction he intended to follow, etc. As he explained his methods, the steps were recorded on the back of an old drawing print by the familiar method of lines and node points (arrow diagram). Gradually, a network was evolved which grew before his eyes and his previous fears and scepticism began to melt away.

When the network of one area was complete, the foreman was asked for the anticipated duration of each activity. Each answer was religiously entered on the network without query, but when the forward pass was made, the overall period exceeded the contract period by several weeks. The foreman looked worried, but he was now involved. He asked to be allowed to review some of his durations and reassess some of the construction methods. Without being pressurized, the man, who had never used network analysis before, began the process that makes network analysis so valuable, that is he reviewed and refined the plan until it complied with the contractual requirements. The exercise was repeated with the three other areas, and the following day the whole operation was explained to the four chargehands who were to be responsible for those areas.

Four separate networks were then drawn, together with four corresponding bar charts. These were pinned on the wall of the site hut with the instruction that one of the programmes, either network or bar chart, to be updated daily. Great stress was laid on the need to update regularly, since it is the monitoring of the programme that is so often neglected once the plan has been drawn. The decision on which of the programmes was used for recording progress was left to the foreman, and it is interesting to note that the network proved to be the format he preferred.

Since each chargehand could compare the progress in his area with that of the others, a competitive spirit developed quite spontaneously to the delight of management. The result was that the job was completed 4 weeks ahead of schedule without additional cost. These extra weeks in October were naturally extremely valuable to the main contractor, who could get more units weatherproof before the cold period of January to March. The network was also used to predict cash flow, which proved to be remarkably accurate. (The principles of this are explained in Chapter 31).

It can be seen, therefore, that in this instance a manual network enabled the project manager to control both the programme (time) and the cost of the job with minimum paperwork. This was primarily because the men who actually carried out the work in the field were involved and were convinced of the usefulness of the network programme.

Confidence in plan

It is vitally important that no one, but no one, associated with a project must lose faith in the programme or the overall plan. It is one of the prime duties of a project manager to ensure that this faith exists. When small cracks do appear in this vital bridge of understanding between the planning department and the operational departments, the project manager must do everything in his power to close them before they become chasms of suspicion and despondency. It may be necessary to re-examine the plan, or change the planner or hold a meeting explaining the situation to all parties, but a plan in which the participants have no faith is not worth the paper it is drawn on.

Having convinced all parties that the network is a useful control tool, the project manager must now ensure that it is kept up to date and the new information transmitted to all the interested parties as quickly as possible. This requires exerting a constant pressure on the planning department, or planning engineer, to keep to the 'issue deadlines', and equally learning on the operational departments to return the feedback documents regularly. To do this, the project manager must use a combination of education, indoctrination, charm and rank pulling, but the feedback *must* be returned as regularly as the issue of the company's pay cheque.

The returned document might only say 'no change', but if this vital link is neglected, the network ceases to be a live document. The problem of feedback for the network is automatically solved when using the EVA cost control system (explained in Chapter 32), since the man-hour returns are directly related to activities, thus giving a very accurate percentage completion of each activity.

It would be an interesting and revealing experience to carry out a survey amongst project managers of large projects to obtain their unbiased opinion on the effectiveness of networks. Most of the managers, with whom this problem was discussed, felt that there was some merit in network techniques, but, equally, most of them complained that too much paper was being generated by the planning department.

Network and method statements

More and more clients and consultants require contractors to produce method statements as part of their construction documentation. Indeed, a method statement for certain complex operations may be a requirement of ISO 9000 Part I. A method statement is basically an explanation of the sequence of operations augmented by a description of the resources (i.e. cranes and other tackle) required for the job. It must be immediately apparent that a network can be of great benefit, not only in explaining the sequence of operations to the client but also for concentrating the writer's mind when the sequence is committed to paper. In the same way, the designer produces a freehand sketch of his ideas, a construction engineer will also be able to draw a freehand network to crystallize his thoughts.

The degree of detail will vary with the complexity of the operation and the requirements of the client or consultant, but it will always be a clear graphical representation of the sequences, which can replace pages of narrative. Any number of activities can be 'extracted' from the network for further explanation or in-depth discussion in the accompanying written statement.

The network, which can be produced manually or by computer, will mainly follow conventional lines and can, of course, be in arrow diagram or precedence format. For certain operations, however, such as structural steelwork erection, it may be advantageous to draw the network in the form of a table, where the operations (erect column, erect beam, plumb, level, etc.) are in horizontal rows. In this way, a highly organized, easy-to-read network can be produced. Examples of such a procedure are shown in Figs 28.1 and 28.2. There are doubtless other situations where this system can be adopted, but the prime objective must always be clarity and ease of understanding. Complex networks only confuse clients and reflect a lack of appreciation of the advantages of method statements.

Integrated systems

The trend is to produce and operate integrated project management systems. By using the various regular inputs generated by the different operating departments, these systems can, on demand, give the project manager an up-to-date status report of the job in terms of time, cost and resources. This facility is particularly valuable once the project has reached the construction stage. The high cost of mainframe machines and the unreliability of regular feedback — even with the use of terminals — have held back

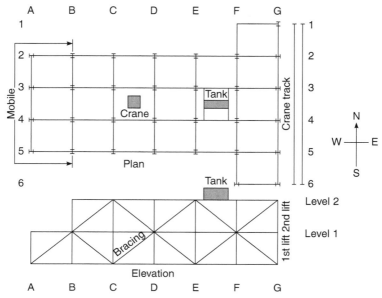

Figure 28.1 Structural framing plan.

the full utilization of computing facilities in the field, especially in remote sites. The PCs, with their low cost, mobility and ease of operation, have changed all this so that effective project-control information can be generated on the spot.

The following list shows the types of management functions that can be successfully carried out either in the office, in the workshop or on site by a single computer installation:

• Cost accounting
• Material control
• Plant movement
• Machine loading
• Man-hour and timesheet analysis
• Progress monitoring
• Network analysis and scheduling
• Risk analysis
• Technical design calculations, etc.

Additional equipment is available to provide presentation in a graphic form such as bar charts, histograms, S-curves and other plots. If required, these can be in a number of colours to aid in identification.

The basis of all these systems is, however, still a good planning method based on well-defined and realistic networks and budgets. If this base is deficient, all comparisons and controls will be fallacious. Therefore, the procedures described in Chapters 20 and 21 still apply. In fact, the more sophisticated the analysis and data processing,

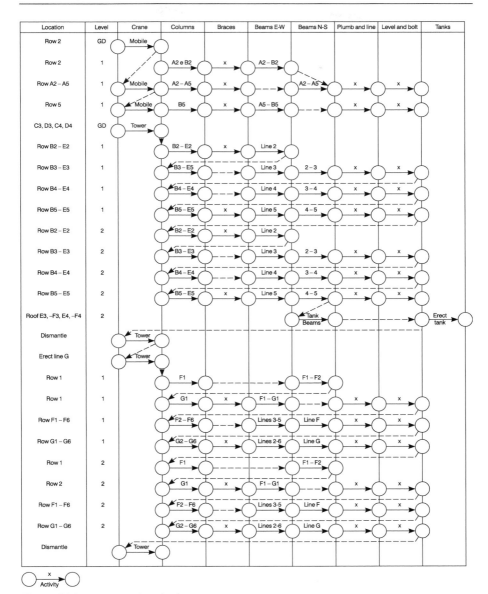

Figure 28.2 Network of method statement.

the more accurate and meaningful the base information has to be. This is because the errors tend to be multiplied by further manipulation, and the wider dissemination of the output will, if incorrect, give more people the wrong data on which to base management decisions.

Networks and claims

From the contractor's point of view, one of the most useful (and lucrative) applications of network presentation arises when it is necessary to formulate claims for extension of time, disruption to anticipated sequences or delays of equipment deliveries. There is no more convincing system than a network to show a professional consultant how his late supply of design information has adversely affected progress on site, or how a late delivery has disrupted the previously anticipated and clearly stated method of construction.

It is, of course, self-evident that to make the fullest use of the network for claim purposes, the method of construction *must* have been previously stated, preferably also in network form. A wise contractor will include a network showing the anticipated sequences with his tender, and indicate clearly the deadlines by which drawings, details and equipment are required.

In most cases, the network will be accepted as a fair representation of the construction programme, but it is possible that the client or consultant will try to indemnify themselves by such statements as that they (the consultant) do not necessarily accept the network as the only logical sequence of operations. Therefore, it is up to the contractor to use his skills and experience to construct the works in light of circumstances prevailing at the time.

Such vague attempts to forestall genuine claims for disruption carry little weight in a serious discussion amongst reasonable people, and count even less should the claim be taken to arbitration or adjudication. The contractor is entitled to receive his access, drawings and free issue equipment in accordance with his stated method of construction, as set out in his tender, and all the excuses or disclaimers by the client or consultant cannot alter this right. Those contractors who have appreciated this facility have undoubtedly profited handsomely by making full use of network techniques, but these must, of course, be prepared accurately.

To obtain the maximum benefit from the network, the contractor must show the following:

1. The programme was reasonable and technically feasible.
2. It represented the most economical construction method.
3. Any delays in the client's drawings or materials will either lengthen the overall programme or increase costs, or both.
4. Any acceleration carried out by him to reduce the delay caused by others resulted in increased costs.
5. Any absorption of float caused by the delay increased the risk of completion on time and had to be countered by acceleration in other areas or by additional costs.

The last point is an important one, since 'float' belongs to the contractor. It is the contractor who builds it into his programme. It is the contractor who assesses the risks and decides which activities require priority action. The mere fact that a delayed component only reduces the float of an activity, without affecting the overall programme, is not a reason for withholding compensation if the contractor can show increased costs were incurred.

Examples of claims for delays

The following examples show how a contractor could incur (and probably reclaim) costs by late delivery of drawings or materials by the employer.

Example 1

To excavate a foundation, the network in Fig. 28.3 was prepared by the contractor. The critical path obviously runs through the excavation, giving the path through the reinforcing steel supply and fabrication a float of 4 days. If the drawings are delayed by 4 days, both paths become critical and, in theory, no delays occur. However, in practice, the contractor may now find that the delay in the order for reinforcing steel has lost him his place in the queue of the steel supplier, since he had previously advised the supplier that information would be available by day 10. Now that the information was only given to the supplier on day 14, labour for the cages was diverted to another contract and, to meet the new delivery of day 29, overtime will have to be worked. These overtime costs are claimable.

In any case, the 4-day float, which the contractor built in as an insurance period, has now disappeared, so even if the steel had arrived by day 29 and the cage fabrication took longer than 3 days, a claim would have been justified.

Example 2

The network in Fig. 28.4 shows a sequence for erecting and connecting a set of pumps. The first pump was promised to be delivered by the client on a 'free issue' basis in week 0. The second pump was scheduled for delivery in week 4. In the event, both pumps were delivered together in week 4. The client argued that since there was a float of 4 days on pump 1, there was no delay to the programme since handover could still be effected by week 16.

What the programme does *not* show, and what it *need not show*, are the resource restraints imposed by the contractor to give him economical working. A network submitted as a contractual document need only show the logic from an *operational* point of view. Resource restraints are *not* logic restraints since they can be overcome by merely supplying additional resources.

The contractor rightly pointed out that he always intended to utilize the float on the first set of pumps to transfer the pipe fitters and electricians to the second pump as soon

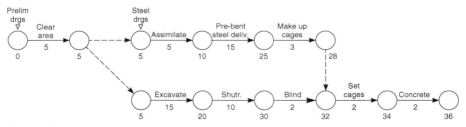

Figure 28.3

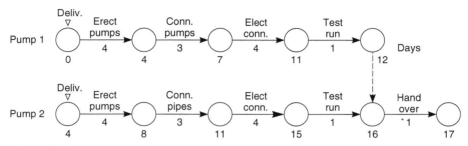

Figure 28.4

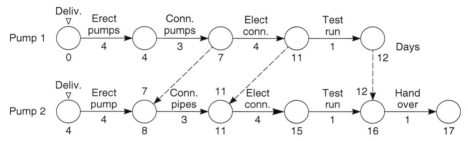

Figure 28.5

as the first pump was piped up and electrically connected. The *implied* network, utilizing the float economically, was therefore as shown in Fig. 28.5.

Now, to meet the programme, the contractor has to employ two teams of pipefitters and electricians, which may have to be obtained at additional cost from another site and certainly requiring additional supervision if the two pumps are geographically far apart. Needless to say, if the contractor shows the *resource* restraints in his contract network, his case for a reimbursement of costs will be that much easier to prove.

Force majeure claims

The causes giving rise to force majeure claims are usually specified in the contract, and there is generally no difficulty in claiming an extension of time for the period of a strike or (where permitted) the duration of extraordinary bad weather. What is more difficult to prove is the loss of time caused by the *effect* of a force majeure situation. It is here where a network can help the contractor to state his case.

Example 3

A boiler manufacturer has received two orders from different clients and has programmed the two contracts through his shops in such a way that as one boiler leaves the assembly area, the parts of the second boiler can be placed into position ready for assembly. The simplified network is shown in Fig. 28.6. Because the factory

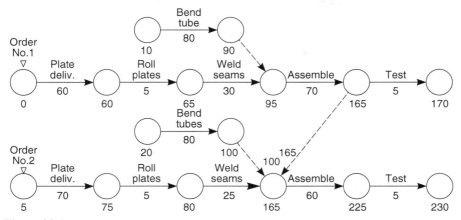

Figure 28.6

had only one assembly bay, boiler no. 2 assembly had to await completion of boiler no. 1, and the delivery promises of boiler no. 2 reflected this.

Unfortunately, the plate for the drum of boiler no. 1, which was ordered from abroad, was delayed by a national dock strike that lasted 15 days. The result was that both boilers were delayed by this period, although the plate for boiler no. 2 arrived as programmed.

The client of boiler no. 2 could not understand why his boiler should be delayed because of the late delivery of a plate for another boiler, but when shown the network, he appreciated the position and granted an extension. Had the assembly of boiler no. 2 started first, boiler no. 1 would have been delayed 70 days instead of only 15, while boiler no. 2 would have incurred storage costs for 60 days. Clearly, such a situation was seen to be unrealistic by all parties. The revised network is shown in Fig. 28.7.

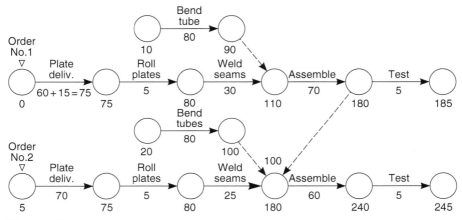

Figure 28.7

Example 4

The contract for large storage tanks covered supply, erection and painting. Bad weather was a permissible force majeure claim. During the erection stage, high winds slowed down the work because the cranes would not handle the large plates safely. The winds delayed the erection by four weeks, but by the time the painting stage started, the November mists set in and the inspector could not allow painting to start on the damp plate. The contractor submitted a network with the contract to show that the painting would be finished *before* November. Because of the high winds, the final coat of paint was, in fact, delayed until March, when the weather permitted painting to proceed.

Fig. 28.8 illustrates the network submitted which, fortunately, clearly showed the non-painting month, so that the client was aware of the position before contract award. The same point could obviously have been made on a bar chart, but the network showed that no acceleration was possible after the winds delayed the erection of the side plates. To assist in relating the week numbers to actual dates, a week number/calendar date table should be provided on the network.

The above examples may appear to be rather negative, that is it looks as if network analysis is advocated purely as a device with which the contractor can extract the maximum compensation from the client or his advisers. No doubt, in a dispute both sides will attempt to field whatever weapons are at their disposal, but a more positive interpretation is that network techniques surely put *all* parties on their mettle. Everyone can see graphically the effects of delays on other members of the construction team and the cost or time implications that can develop. The result is, therefore, that all parties will make sure that they will not be responsible for the delay, so that in the end everyone − client, consultant and contractor − will benefit: the client, because he gets his job on time; the consultant, because his reputation is enhanced; and the contractor, because he can make a fair profit.

Fortunately, the trend is for claims to be reduced due to the introduction of partnering. In these types of contracts, which are usually a mixture of firm price and reimbursable costs, an open book policy by the contractor allows the employer to see how and where his money is being expended, so that there are no hidden surprises at the end of the contract ending up as a claim.

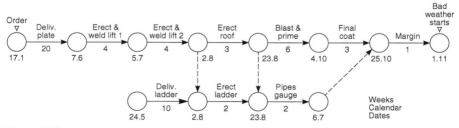

Figure 28.8

Frequently, any cost savings are shared by a predetermined ratio so that all parties are encouraged to minimize delays and disruptions as much as possible. In such types of contracts, network analysis can play an important part, in that, provided the network is kept up to date and reflects the true and latest position of the contract, all parties can jointly see graphically where the problem lies and can together hammer out the most economical solution.

Network applications outside the construction industry

Chapter outline

Most of the examples of network analysis in this book are taken from the construction industry, mainly because network techniques are particularly suitable for planning and progressing the type of operations found in either the design office or on a site. However, many operations outside the construction industry that comprise a series of sequential and/or parallel activities can benefit from network analysis − indeed, the Polaris project is an example of such an application.

The following examples are included, therefore, to show how other industries can make use of network analysis, but as can be seen from Chapter 26, even the humble task of getting up in the morning can be networked. When network analysis first came into existence, a men's magazine even published a critical path network of seduction!

Bringing a new product onto the market

The operations involved in launching a new product require careful planning and coordination. The example shows how network techniques were used to plan the development, manufacture and marketing of a new type of water metre for use in countries where these are installed on all premises.

The list of operations is first grouped into five main functions:

A	Management
B	Design and development
C	Production
D	Purchasing and supply
E	Sales and marketing

Project Management, Planning and Control. https://doi.org/10.1016/B978-0-12-824339-8.00029-8

Each main function is then divided into activities that have to be carried out in defined sequences and by specific times. The management function would therefore include the following of product:

A—1	Definition of product	Size, range, finish, production rate, etc.
2	Costing	Selling price, manufacturing costs
3	Approvals for expenditure	Plant materials, tools and jigs, storage advertising, training, etc.
4	Periodic reviews	—
5	Instruction to proceed with stages	—

The design and development function would consist of the following:

B—1	Product design brief
2	Specification and parts list
3	Prototype drawings
4	Prototype manufacture
5	Testing and reports
6	Preliminary costing

Once the decision has been made to proceed with the water metre, the production department will carry out the following activities:

C—1	Production planning
2	Jig tool manufacture
3	Plant and machinery requisition
4	Production schedules
5	Materials requisitions
6	Assembly line installation
7	Automatic testing
8	Packing bay
9	Inspection procedures
10	Labour recruitment and training
11	Spares schedules

The purchasing and supply function involves the procurement of all the necessary raw materials and bought-out items, and includes the following activities:

D—1	Material enquiries
2	Bought-out items enquiries
3	Tender documents
4	Evaluation of bids
5	Long-delivery orders
6	Short-delivery orders
7	Carton and packaging
8	Instruction leaflets, etc.
9	Outside inspection

The sales and marketing function will obviously interlink with the management function and consists of the following activities:

E—1	Sales advice and feedback	
2	Sales literature	Photographs, copying, printing, films, displays, packaging
3	Recruitment of sales staff	
4	Sales campaign and public relations	
5	Technical literature	Scope and production
6	Market research	

Obviously, the above breakdowns are only indicative and the network shown in Fig. 29.1 gives only the main items to be programmed. The actual programme for such a product would be far more detailed and would probably contain about 120 activities.

The final presentation for those who prefer it could then be in a bar chart form covering a time span of approximately 18 months from conception to main production run.

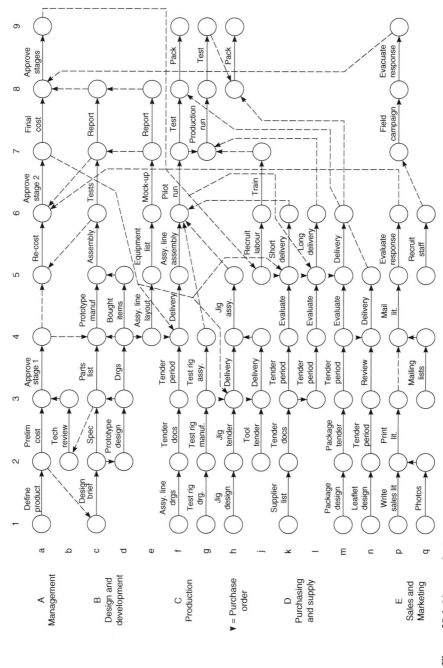

Figure 29.1 New product.

Moving a factory

One of the main considerations in moving the equipment and machinery of a manufacturing unit from one site to another is to carry out the operation with the minimum loss of production. Obviously, at some stage manufacturing must be halted unless certain key equipment is duplicated, but if the final move is carried out during the annual works' holiday period the loss of output can be minimized.

Consideration must therefore be given to the following points:

1. Identifying equipment or machines which can be temporarily dispensed with;
2. Identifying essential equipment and machines;
3. Dismantling problems of each machine;
4. Re-erection on the new site;
5. Service connections;
6. Transport problems − weight, size, fragility, route restrictions;
7. Orders in pipeline;
8. Movement of stocks;
9. Holiday periods;
10. Readiness of new premises;
11. Manpower availability;
12. Overall cost;
13. Announcement of move to customers and suppliers;
14. Communication equipment (telephone, e-mail, fax);
15. Staff accommodation during and after the move;
16. Trial runs; and
17. Recruitment and staff training.

By collecting these activities into main functions, a network can be produced that will facilitate the organization and integration of the main requirements. The main functions would therefore be the following:

A	Existing premises and transport
B	New premises − commissioning
C	Services and communications
D	Production and sales
E	Manpower, staffing

The network for the complete operation is shown in Fig. 29.2. It will be noticed that, as with the previous example, horizontal banding (as described in Chapter 19) is of considerable help in keeping the network disciplined.

By transferring the network onto a bar chart, it will be possible to arrange for certain activities to be carried out at weekends or holidays. This may require a rearrangement

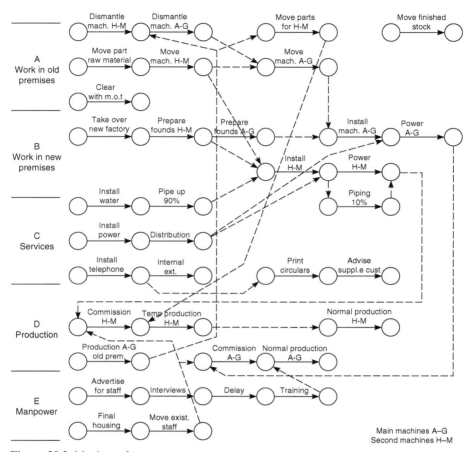

Figure 29.2 Moving a factory.

of the logic, which, though not giving the most economical answer in a physical sense, is still the best overall financial solution when production and marketing considerations are taken into account.

Centrifugal pump manufacture

The following network shows the stages required for manufacturing centrifugal pumps for the process industry. The company providing these pumps has no foundry, so the un-machined castings have to be bought in.

Assuming that the drawings for the pump are complete and the assembly line set-up, a large order for a certain range of pumps requires the following main operations:

1. Order castings – bodies, impellers, etc.;
2. Order raw materials for shafts, seal plates, etc.;
3. Order seals, bearings, keys, bolts;
4. Machine castings, impellers, shafts;
5. Assemble;
6. Test;
7. Paint and stamp;
8. Crate and dispatch; and
9. Issue installation and maintenance instructions and spares list.

Fig. 29.3 shows the network of the various operations complete with coordinate node numbers, durations and earliest start times. The critical path is shown by a thickened line and total float can be seen by inspection. For example, the float of all the activities on line C is 120−48 = 72 days. Similarly, the float of all activities on line D is 120−48 = 72 days.

Fig. 29.4 is the network redrawn in bar chart form, on which the floats have been indicated by dotted lines. It is apparent that the preparation of documents such as maintenance manuals, spares lists and quotes can be delayed without an ill effect for a considerable time, thus releasing these technical resources for more urgent work such as tendering for new enquiries.

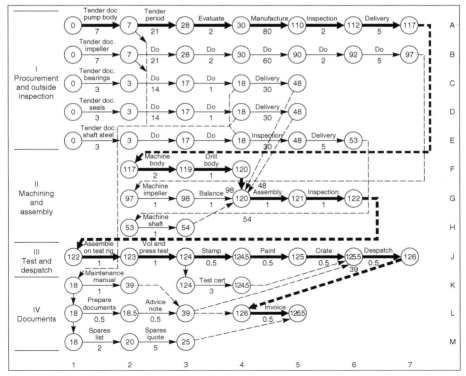

Figure 29.3 Pump manufacture (duration in days).

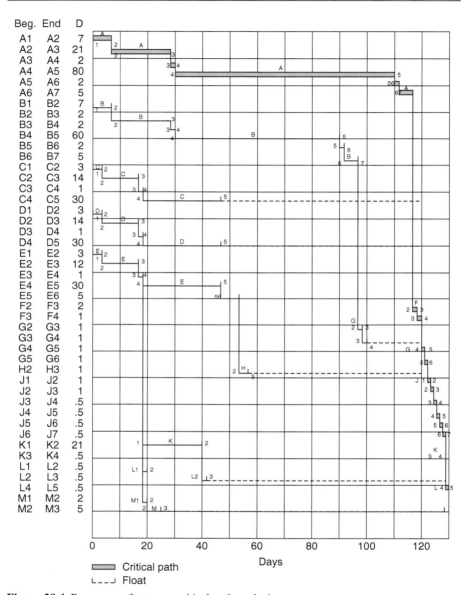

Figure 29.4 Pump manufacture − critical path analysis.

Planning a mail order campaign

When a mail order house decides to promote a specific product, a properly coordinated sequence of steps has to be followed to ensure that the campaign will have the maximum impact and success. The following example shows the activities required for promoting a new set of record albums and involves both the test campaign and the main sales drive.

The two stages are shown separately on the network (Fig. 29.5) as they obviously occur at different times, but in practice intermediate results could affect the management decisions on packaging and text on the advertising leaflet. At the end of the test, shot management will have to decide on the percentage of records to be ordered to meet the initial demand.

In practice, the test shot will consist of three or more types of advertising leaflets and record packaging, and the result of each type will have to be assessed before the final main campaign leaflets are printed.

Depending on the rate of return of orders, two or more record-ordering and dispatch stages will have to be allowed for. These are shown on the network as b1 and b2.

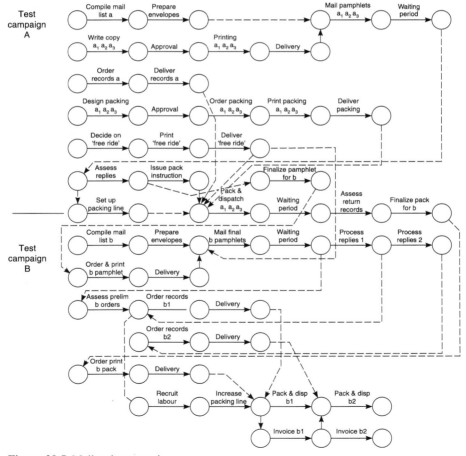

Figure 29.5 Mail order campaign.

Manufacture of a package boiler

The programme in this example covers the fabrication and assembly of a large package boiler of about 75,000 kg of superheated steam per hour at 30 bar g and a temperature of 300°C. The separate economizer is not included.

The drum shells, drum ends, tubes, headers, doors and nozzles are bought out, leaving the following manufacturing operations:

1. Weld drums (longitudinal and circumferential seams);
2. Weld on drum ends;
3. Weld on nozzles and internal supports;
4. Drill drum for tubes;
5. Stress relieve top and bottom drums;
6. Bend convection bank tubes;
7. Fit and expand tubes in drums – set up erection frame;
8. Weld fins to furnace tubes, pressure test;
9. Produce water wall panels;
10. Gang bend panels;
11. Erect wall panels;
12. Weld and drill headers; stress relieve;
13. Weld panels to headers;
14. Weld on casing plates;
15. Attach peepholes, access doors, etc.;
16. Pressure test;
17. Seal-weld furnace walls;
18. Fit burners and seals;
19. Air test – inspection;
20. Insulate;
21. Prepare for transport; and
22. Dispatch.

There are four main bands in the manufacturing programme:

A	Drum manufacture
B	Panel and tube manufacture
C	Assembly
D	Insulation and preparation for dispatch

The programme assumes that all materials have been ordered and will be available at the right time. Furthermore, in practice, sub-programmes would be necessary for panel fabrication, which includes blast cleaning the tubes and fin bar, automatic welding, interstage inspection, radiography and stress relieving. Fig. 29.6 shows the main production stages covering a period of approximately 7 months.

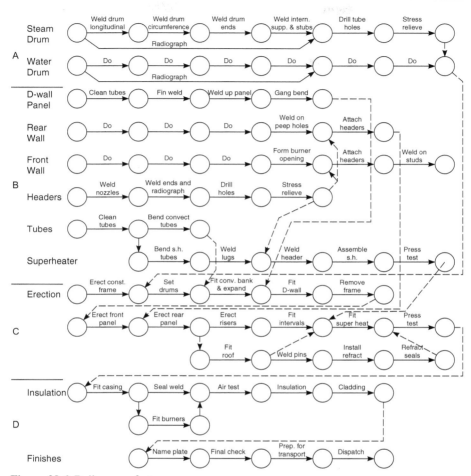

Figure 29.6 Boiler manufacture.

Manufacture of a cast machined part

The casting, machining and finishing of a steel product can be represented in network form as shown in Fig. 29.7. It can be seen that the total duration of the originally planned operation is 38 hours, but the aim was to reduce this manufacturing period to make this product more competitive. By incorporating the principle, that efficiency can be increased if some of the operations on a component can be performed while it is on the move between workstations, it is obviously possible to reduce the overall manufacturing time. The obvious activities that can be carried out while the component is actually being transported (usually on a conveyor system) is cooling off, painting and paint drying. As can be seen from Fig. 29.8 such a change in the manufacturing procedure saves 3 hours.

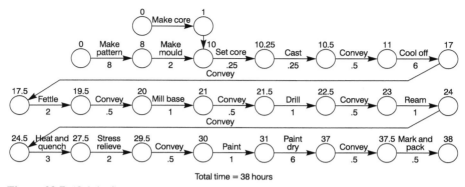

Figure 29.7 (Original).

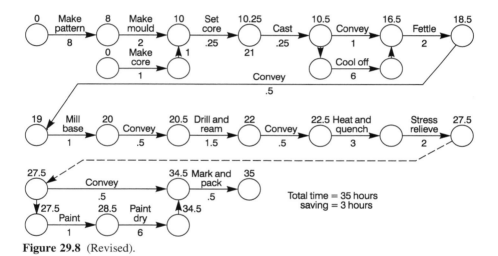

Figure 29.8 (Revised).

Any further time savings now require a reduction in the duration of some of the individual activities. The first choice must obviously be those with the longest durations, that is

1. Make pattern (8 hours);
2. Cool off (6 hours); and
3. Dry paint (8 hours).

These operations require new engineering solutions. For example, in activity (1), the pattern may have to be split, with each component being made by a separate pattern maker. It may also be possible to subcontract the pattern to a firm with more resources. Activity (2) can be reduced in time by using forced-draught air to cool the casting before fettling. Care must, of course, be taken not to cool it at such a rate that it causes cracking or other metallurgical changes. Conversely to activity (2), the paint drying in

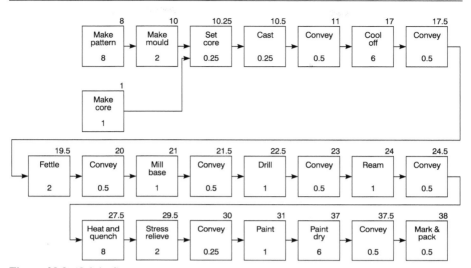

Figure 29.9 (Original).

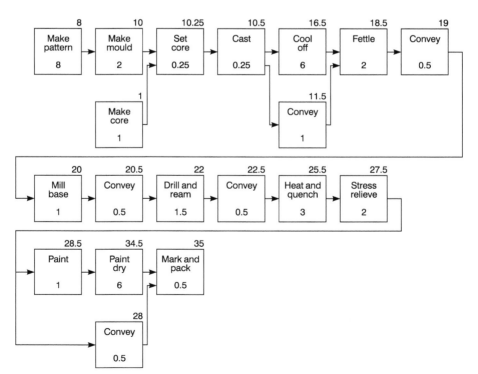

Figure 29.10 (Revised).

activity (3) can be speeded up by blowing warm air over the finished component. If the geographical layout permits, it may be possible to take the heated air from the cooling process, pass it through a filter and use it to dry the paint.

Further time reductions are possible by reducing the machining time of the milling and drilling operations. This may mean investing in cutters or drills that can withstand higher cutting speeds. It may also be possible to increase the speed of the different conveyors, which, even on the revised network, make up 1 hour of the cycle time.

For those planners who are familiar with manufacturing flow charts, it may be an advantage to draw the network in precedence (activity on node) format (see Chapter 19). Such a representation of the initial and revised networks is shown in Figs. 29.9 and 29.10, respectively.

It is important to remember that the network itself does not reduce the overall durations. Its first function is to show in a graphic way the logical interrelationship of the production processes and the conveying requirement between the manufacturing stages. It is then up to the production engineer or controller to examine the network to see where savings can be made. This is, in fact, the second function of the network — to act as a catalyst for the thought processes of the user to give him the inspiration to test a whole series of alternatives, until the most economical or fastest production sequence has been achieved.

The use of a PC at this stage will, of course, enable the various trial runs to be carried out quite rapidly, but, as can be seen, even a manual series of tests takes no longer than a few minutes. As explained in Chapter 21, the first operation is to calculate the shortest forward pass — a relatively simple operation — leaving the more complex calculations of float to the computer when the final selection has been made.

Resource loading

Chapter outline

Most modern computer programs incorporate facilities for resource loading, resource allocation and resource smoothing. Indeed, the Primavera P6 program shown in Chapter 51 features such a capability.

In principle, the computer aggregates a particular resource in any time period and compares this with a previously entered availability level for that resource. If the availability is less than the required level, the program will either:

1. Show the excess requirement in tabular form, often in a different colour to highlight the problem; or
2. Increase the duration of the activity requiring the resource to spread the available resources over a longer period, thus eliminating the unattainable peak loading.

The more preferable action by the computer is option (1), that is the simple report showing the overrun of resources. It is then up to management to make the necessary adjustments by either extending the time period — if the contractual commitments permit — or mobilizing additional resources. In practice, of course, the problem is complicated by such issues as available access or working space as well as financial, contractual or even political restraints. Often it may be possible to make technical changes that alter the resource mix. For example, a shortage of carpenters used for formwork erection may make it necessary to increase the use of precast concrete components with a possible increase in cost but a decrease in time. Project management is more than just writing and monitoring programs. The so-called project-management systems marketed by software companies are really only there to present to the project manager, on a regular basis, the position of the project to date and the possible consequences unless some form of remedial action is taken. The type of action and the timing of it rest fairly and squarely on the shoulders of management.

The options by management are usually quite wide, provided sufficient time is taken to think them out. It is in such situations that the 'what if' scenarios are a useful facility on a computer. However, the real implication can only be seen by 'plugging' the various alternatives into the network on paper and examining the downstream effects in company with the various specialists, who, after all, have to do the actual work. There is no effective substitute for good teamwork.

Project Management, Planning and Control. https://doi.org/10.1016/B978-0-12-824339-8.00030-4

The alternative approach

Resource smoothing can, of course, be done very effectively without a computer — especially if the program is not very large. Once a network has been prepared, it is very easy to convert it into a bar chart, as all the 'thinking' has already been completed. Using the earliest starting and finishing times, the bars can be added to the gridded paper in minutes. Indeed, the longest operation in drawing a bar chart (once a network has been completed) is writing down the activity descriptions on the left-hand side of the paper. By leaving sufficient vertical space between the bars and dividing the grid into week (or day) columns, the resource levels for each activity can be added. Generally, there is no need to examine more than two types of resources per chart, as only the potentially restrictive or quantitatively limited ones are of concern. When all the activity bars have been marked with the resource value, each time period is added up vertically and the total entered in the appropriate space. The next step is to draw a histogram to show the graphical distribution of the resources. This will immediately highlight the peaks and troughs, and trigger off the next step — resource smoothing.

Manual resource smoothing is probably the most practical method, as unprogrammable factors such as access, working space, hard-standing for cranes, and personality traits of foremen, can only be considered by a human when the smoothing is carried out. Nevertheless, the smoothing operation must still follow the logical pattern given below:

1. Advantage should be taken of float. In theory, activities with free float should be the first to be extended, so that a limited resource can be spread over a longer time period. In practice, however, such opportunities are comparatively rare, and for all normal operations, all activities with total float can be used for the purpose of smoothing. The floats can be indicated on the bars by dotted line extensions, again read straight off the network by subtracting the earliest from the latest times of the beginning node of the activity.
2. When the floats have been absorbed and the resources are distributed over the longer activity durations, another vertical addition is carried out from which a new histogram can be drawn. A typical network, bar chart and histogram are shown in Fig. 30.1.
3. If the peaks still exceed the available resources for any time period, logic changes will be required. These changes are usually carried out on the network, but it may be possible to make some of them by 'sliding' the bars on the bar chart. For example, a common problem when commissioning a process or steam-raising plant is a shortage of suitably qualified commissioning engineers. If the bars of the bar chart are cut out and pasted onto cardboard, with the resources written against each time period on the activity bar, the various operations can be moved on the time-scaled bar chart until an acceptable resource level is obtained. The reason it is not always necessary to use the network is that in a commissioning operation there is often considerable flexibility as to which machine is commissioned first. Whether pump A is commissioned before or after compressor B is often a matter of personal choice rather than logical necessity. When an acceptable solution has been found, the strips of bar can be held on to the backing sheet with an adhesive putty (Blu-Tack) and (provided the format is of the necessary size) photocopied for distribution to interested parties.
4. If the weekly (or daily) aggregates are totalled cumulatively it is sometimes desirable to draw the cumulative curve (usually known as the S-curve, because it frequently takes the shape of

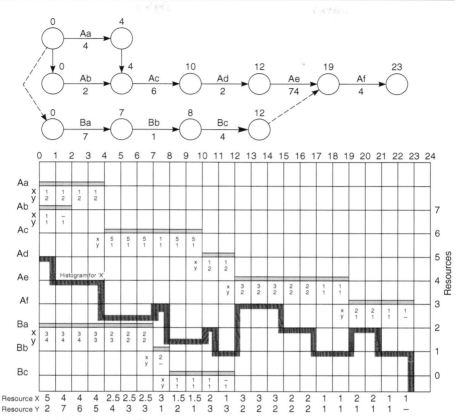

Figure 30.1 Bar chart and histogram.

an elongated letter S), which gives a picture of the build-up (and run-down) of the resources over the period of the project. This curve is also useful for showing the cumulative cash flow, which, after all, is only another resource. An example of such a cash-flow curve is given in Chapter 45.

The following example shows the above steps in relation to a small construction project where there is a resource limitation. Fig. 30.2 shows the activity on arrow configuration, and Fig. 30.3 shows the same network in activity on node configuration. Fig. 30.4 shows their translation into a bar (or Gantt) chart where the bars are in fact a string of resource numbers. For simplicity, all the resources shown are of the same type (e.g. welders). By adding up the resources of each week, a totals table can be drawn, from which it can be seen that in week 9 the resource requirement is 14. This amount exceeds the availability, which is only 11 welders, and an adjustment is therefore necessary. Closer examination of the bar chart reveals a low resource requirement of only 6 in week 12. A check on the network (Fig. 30.2) shows that there is

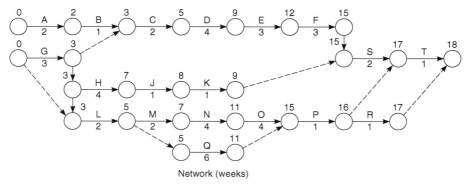

Network (weeks)

Figure 30.2

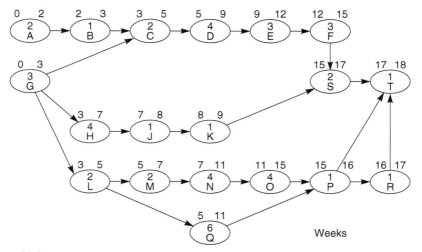

Figure 30.3

$15 - 9 = 6$ weeks' float on activity K. This activity can therefore be used to smooth the resources. By delaying activity K by 3 weeks, the resource requirement is now:

Week 9, −10
Week 12, −10

A histogram and a cumulative resource curve are drawn from the revised totals table. The latter can also be used as a planned performance curve, as the resources (if men) are directly proportional to man-hours. It is interesting to note that any 'dip' or 'peak' in the cumulative resource curve indicates a change of resource requirement, which should be investigated. A well-planned project should have a smooth resource curve following approximately the shape of letter S.

The method described may appear to be lengthy and time consuming, but the example given by Fig. 30.2 or 30.3 and Fig. 30.4, including the resource smoothing

Time taken to
draw table and curve:

Start 11.38 ⎱ 4 min
Finish 11.42 ⎰

R = Resource per week   MAX R = 11   Time 19 acts = 4 min
X = Float   Resource table (early start)   Time 200 acts = 45 min

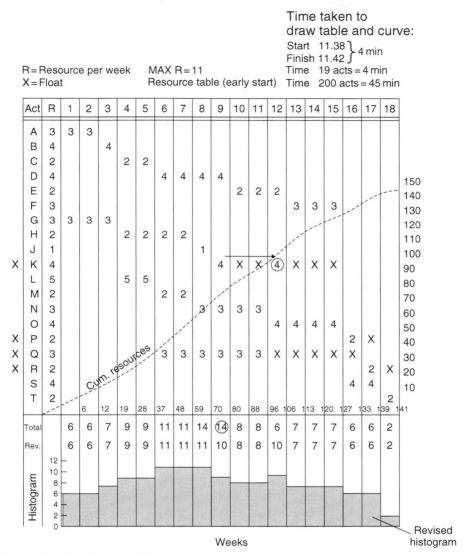

X	Act	R	1	2	3	4	5	6	7	8	9	10	11	12	13	14	15	16	17	18	
	A	3	3	3																	
	B	4				4															
	C	2					2	2													
	D	4						4	4	4	4										150
	E	2										2	2	2							140
	F	3													3	3	3				130
	G	3	3	3	3																120
	H	2						2	2	2	2										110
	J	1									1										100
X	K	4									4	X	X	(4)	X	X	X				90
	L	5						5	5												80
	M	2						2	2												70
	N	3									3	3	3	3							60
	O	4												4	4	4	4				50
X	P	2																2	X		40
X	Q	3						3	3	3	3	3	3	X	X	X	X	X			30
X	R	2																	2	X	20
	S	4															4	4			10
	T	2																		2	
			6	12	19	28	37	48	59	70	80	88	96	106	113	120	127	133	139	141	
	Total		6	6	7	9	9	11	11	14	(14)	8	8	6	7	7	7	6	6	2	
	Rev.		6	6	7	9	9	11	11	11	10	8	8	10	7	7	7	6	6	2	

Cum. resources

Histogram (12, 10, 8, 6, 4, 2, 0)

Weeks

Revised histogram

Figure 30.4 Histogram and 'S' curve.

and curve plotting, took exactly 6 minutes. Once the activities and resources have been listed on graph paper, the bar chart drafting and resource smoothing of a practical network of approximately 200 activities can usually be carried out in about 1 hour.

Most of the modern computers' project-management programs have resource smoothing facilities, which enable the base to be re-positioned on the screen to give the required resource total for any time period.

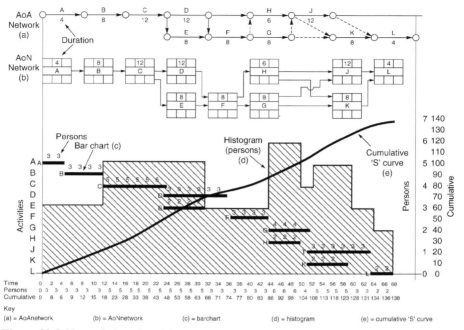

Figure 30.5 Network, bar chart, histogram and 'S' curve.

However, it is advisable not to do this automatically as the machine cannot make allowances for congestion of work area, special skills of operators, clients' preferences and other factors only apparent to the people on the job.

Fig. 30.5 shows the relationship between the networks, bar chart, histogram and cumulative S-curve.

It should be noted that the term used for redistributing the resources was 'resource smoothing'. Some authorities also use the term 'resource levelling', by which they mean flattening out the histogram to keep the resource usage within the resource availability for a particular time period. However, to do this without moving the position of some of the activities is just about impossible. Whether the resources are 'levelled' to reduce the unacceptable peaks due to resource restraints or 'smoothed' to produce a more even resource usage pattern, activities have to be readjusted along the time scale by utilizing the available float. To ascribe different meanings to the terms smoothing and levelling is therefore somewhat hair splitting, as in both cases the operations to be carried out are identical. If the resource levels are so restricted that even the critical activities have to be extended, the project completion will inevitably be delayed.

Further reading

Schwindt, C. (2005). *Resource allocation in project management*. Springer.

Cash flow forecasting

31

Chapter outline

It has been stated in Chapter 30 that it is very easy to convert a network into a bar chart, especially if the durations and week (or day) numbers have been inserted. Indeed, the graphical method of analysis actually generates the bar chart as it is developed.

If we now divide this bar chart into a number of time periods (say, weeks or months) it is possible to see, by adding up vertically, what work has to be carried out in any time period. For example, if the time period is in months, then in any particular month we can see that one section is being excavated, other is being concreted and another is being scaffolded and shuttered, etc.

From the description, we can identify the work and can then find the appropriate rate (or total cost) from the bills of quantities. If the total period of that work takes 6 weeks and we have used up 4 weeks in the time period under consideration, then approximately two-thirds of the value of that operation has been performed and could be certificated.

By this process, it is possible to build up a fairly accurate picture of anticipated expenditure at the beginning of the job, which in itself might well affect the whole tendering policy. Provided the job is on programme, the cash flow can be calculated, but, naturally, due allowance must be made for the different methods and periods of retentions, billing and reimbursement. The cost of the operation must therefore be broken down into six main constituents:

- Labour
- Plant
- Materials and equipment
- Subcontracts
- Site establishment
- Overheads and profit

By drawing up a table of the main operations as shown in the network, and splitting up the cost of these operations (or activities) into the six constituents, it is possible to calculate the average percentage what each constituent contains in relation to the value. It is very important, however, to deduct the values of the subcontracts from any operation and treat these subcontracts separately. The reason for this is, of course, that a subcontract is self-contained and is often of a specialized nature. To break up a

Project Management, Planning and Control. https://doi.org/10.1016/B978-0-12-824339-8.00031-6

subcontract into labour, plant, materials, etc., would not only be very difficult (as this is the prerogative of the subcontractor) but would also seriously distort the true distribution of the remainder of the project.

Example of cash flow forecasting

The simplest way to explain the method is to work through the example described in Figs 31.1−31.6. This is a hypothetical construction project of three identical simple unheated warehouses with a steel framework on independent foundation blocks,

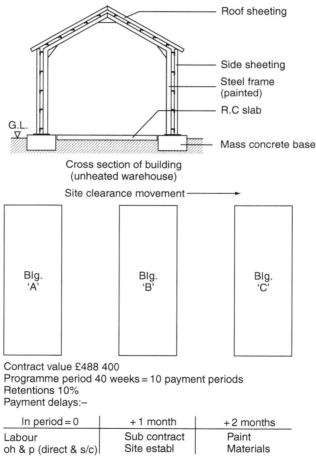

Figure 31.1 Warehouse building.

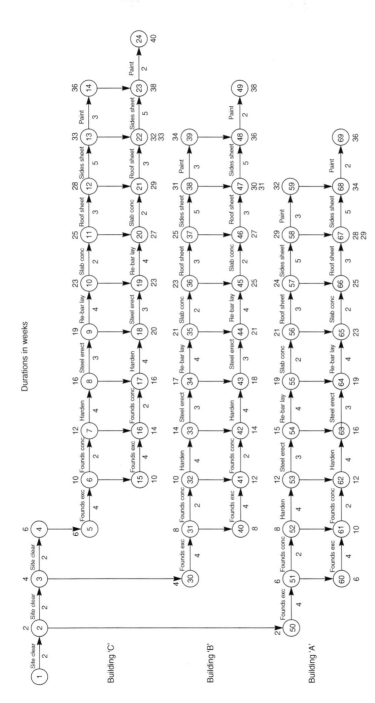

Durations in weeks

Figure 31.2 Construction network.

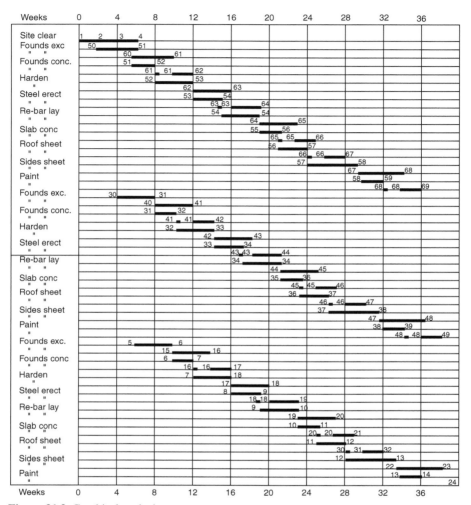

Figure 31.3 Graphical analysis.

profiled steel roof and side cladding and a reinforced-concrete ground slab. It has been assumed that as once an area of site has been cleared excavation work can start, and the sequences of each warehouse are identical. The layout is shown in Fig. 31.1 and the network for the three warehouses is shown in Fig. 31.2.

Fig. 31.3 shows the graphical analysis of the network separated for each building. The floats can be easily seen by inspection, for example there is a 2-week float in the first paint activity (58–59) because there is a gap between the following dummy 59–68 and activity 68–69. The speed and ease of this method soon become apparent after a little practice.

One building Units in $ × 100

Activity	Duration weeks	Total value	Labour		Plant		Materials		Sub Contr		Site Estb.		OH & P	
			Value	%	Value	%	Value	%	Value	%	Value	%	Value	%
Clear site	2	62	30	48	20	32	–		3	5	4	7	5	8
Founds exc.	4	94	40	43	40	43	–		–		6	6	8	8
" "	4	94	40	43	40	43	–		–		6	6	8	8
Founds conc.	2	71	20	28	10	14	30	42	–		5	8	6	8
" "	2	71	20	28	10	14	30	42	–		5	8	6	8
Steel erect	3	220	–		–		–		200	91	–		20	9
" "	3	220	–		–		–		200	91	–		20	9
Re-bar lay	4	106	30	28	–		60	56	–		7	7	9	9
" "	4	106	30	28	–		60	56	–		7	7	9	9
Slab conc.	2	71	20	28	10	14	30	42	–		5	8	6	8
" "	2	71	20	28	10	14	30	42	–		5	8	6	8
Roof sheet	3	66	–		–		–		60	91	–		6	9
" "	3	66	–		–		–		60	91	–		6	9
Sides sheet	5	100	–		–		–		90	90	–		10	10
" "	5	100	–		–		–		90	90	–		10	10
Paint	3	66	–		–		–		60	91	–		6	9
"	2	44	–		–		–		40	91	–		4	9
Total direct		743	250	34	140	19	240	32			50	7	63	8
Total sub-contr.		885							803	91			82	9
Grand total		1628												
For 3 blgs		4884												

Figure 31.4 Earned value table.

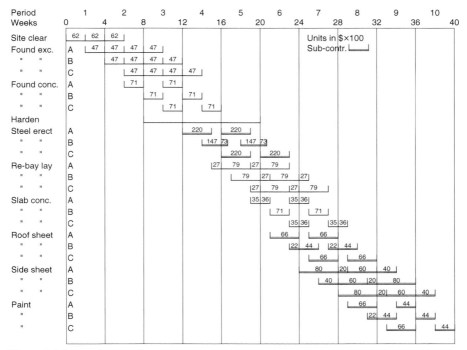

Figure 31.5 Bar charts and costs.

Group	Item	% / Delay	1	2	3	4	5	6	7	8	9	10	11
	Period		1	2	3	4	5	6	7	8	9	10	11
	Week	0	4	8	12	16	20	24	28	32	36	40	44
	Total S/C		—	—	—	367	660	381	318	438	354	128	
	S/C	91				334	600	347	289	399	322	116	
	OH & P	9				33	60	34	29	39	32	12	
$×100	**Direct**	%	171	368	448	216	247	368	284	36			
	Labour	34	58	125	153	74	84	159	97	12			
	Plant	19	33	70	85	41	47	89	54	7			
	Material	32	55	118	143	69	79	150	91	11			
	Site est.	7	12	26	31	15	17	33	20	3			
	OH & P	8	13	29	36	17	20	37	22	3			
	Total value		171	368	448	583	907	849	602	474	354	128	
	Outflow	Delay											
	Labour	0	58	125	153	74	84	159	97	12			
	Plant	2			33	70	85	41	47	89	54	7	
$×100	Material	2			55	118	143	69	79	150	91	11	
	S/C	1					334	600	347	289	399	322	116
	Site est.	1		12	26	31	15	17	33	20	3		
	OH & P	0	13	29	36	17	20	37	22	3			
	S/C OH&P	0				33	60	34	29	39	32	12	
	Out		71	166	303	343	741	957	654	602	579	352	116
$×100	In 90%	1		154	331	403	525	816	764	542	427	319	115
	Net flow		(71)	(12)	28	60	(216)	(141)	110	(60)	(152)	(33)	(1)
–	Cumul. out		71	237	540	883	1624	2581	3235	3837	4416	4768	4884
+	Cumul. in			154	485	888	1413	2229	2993	3535	3962	4281	4396
	Cumul. net		−71	−83	−55	+5	−211	−352	−242	−302	−454	−487	−488

Figure 31.6 Cash flow chart.

The bar chart in Fig. 31.5 has been drawn straight from the network (Fig. 31.2) and the costs in $100 units added from Fig. 31.4. For example, in Fig. 31.4 the value of foundation excavation for any one building is $9400 per 4-week activity. As there are two 4-week activities, the total is $18,800. To enable the activity to be costed in the corresponding measurement period, it is convenient to split this up into two-weekly periods of $4700. Hence, in Fig. 31.5, foundation excavation for building A is shown as follows:

47 in period 1
47 + 47 = 94 in period 2
47 in period 3

The summation of all the costs in any period is shown in Fig. 31.6.

Fig. 31.6 clearly shows the effect of the anticipated delays in payment of certificates and settlement of contractor's accounts. For example, material valued at 118 in period 2 is paid to the contractor after 1 month in period 3 (part of the 331, which is 90% of 368, the total value of period 2), and is paid to the supplier by the contractor in period 4 after the 2-month delay period.

From Fig. 31.6, it can be seen that it has been decided to extract overhead and profit monthly as the job proceeds, but this is a policy that is not followed by every company. Similarly, the payment delays may differ in practice, but the principle would be the same.

Fig. 31.6 shows the total outflow and inflow for each time period, and the net differences between the two. When these values are plotted on graphs as in Figs 31.7 and 31.8, it can be seen that there are only three periods of positive cash flow, that is periods 3, 4 and 7. However, while this shows the actual periods when additional moneys have to be made available to fund the project, it does not show, because the gap between the outflow and inflow is so large for most of the time, that for all intents and purposes the project has a negative cash flow throughout its life.

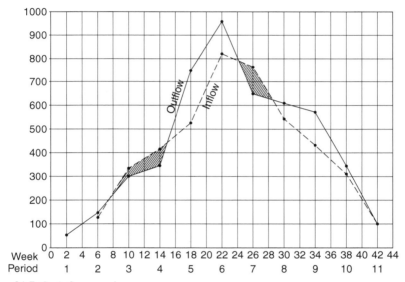

Figure 31.7 Cash flow graph.

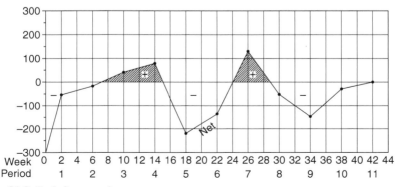

Figure 31.8 Cash flow graph.

This becomes apparent when the cumulative outflows and inflows, tabulated in the last three lines of Fig. 31.6, are plotted on a graph as in Figs 31.9 and 31.10. From these, it can be seen that cumulatively, a positive cash flow (a mere $500) is in period 4.

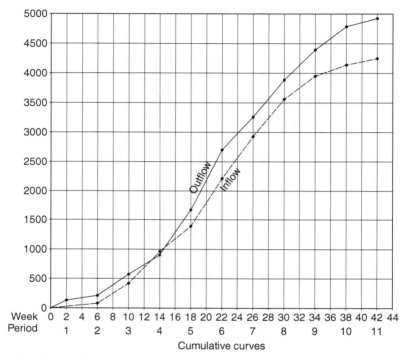

Figure 31.9 Cumulative curves.

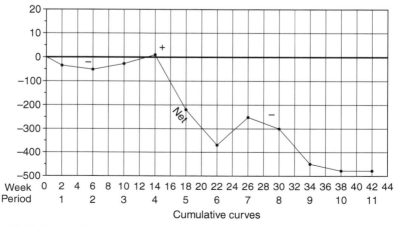

Figure 31.10 Cumulative net flow.

This example shows that the project is not self-financing and will possibly only show a profit when the 10% retention moneys have been released. To restore the project to a positive cash flow, it would be necessary to negotiate a sufficiently large mobilization fee at the start of the project to ensure that the contract is self-financing.

Cost control and EVA

Chapter outline

Apart from ensuring that their project is completed on time, all managers, whether in the office, workshop, factory or on-site, are concerned with the cost. There is little consolation in finishing on time, when, from a cost point of view, one wished the job had never started.

Cost control has been a vital function of management since the days of the pyramids, but only too frequently is the term confused with mere cost reporting. The cost report is usually part of every manager's monthly report to his superiors, but an account of the past month's expenditure is only stating historical facts. What the manager needs is a regular and up-to-date monitoring system that enables him to identify the expenditure with specific operations or stages, determine whether the expenditure was cost-effective, plot or calculate the trend and then take immediate action if the trend is unacceptable.

Network analysis forms an excellent base for any cost-control system, as the activities can each be identified and costed so that the percent completion of an activity can also give the proportion of expenditure, if that expenditure is time related. Therefore, the system is ideal for construction sites, drawing offices or factories where the basic unit of control is the man-hour.

SMAC – man-hour control

Site man-hours and cost (SMAC)[1] is a cost-control system developed in 1978 specifically on a critical path network base for either manual or computerized cost and progress monitoring, which enables performance to be measured and trends to be

---

[1] SMAC is the proprietary name given to the cost-control program developed by Foster Wheeler.

Project Management, Planning and Control. https://doi.org/10.1016/B978-0-12-824339-8.00032-8

evaluated, thus providing the project manager with an effective instrument for further action. The system, which is now known as earned value analysis (EVA), can be used for all operations where man-hours or costs have to be controlled, and as most functions in an industrial (and now more and more commercial) environment are based on man-hours and can be planned with critical path networks, the utilization of the system is almost limitless.

The following operations or activities could benefit from the system:

1. Construction sites
2. Fabrication shops
3. Manufacturing (batch production)
4. Drawing offices
5. Removal services
6. Machinery commissioning
7. Repetitive clerical functions
8. Road maintenance

The criteria laid down when the system was first mooted were:

1. *Minimum site (or workshop) input.* Site staff should spend their time managing the contract and not filling in unnecessary forms.
2. *Speed.* The returns should be monitored and analysed quickly so that the action can be taken.
3. *Accuracy.* The man-hour expenditure must be identifiable with specific activities that are naturally logged on time sheets.
4. *Value for money.* The useful man-hours on an activity must be comparable with the actual hours expended.
5. *Economy.* The system must be inexpensive to operate.
6. *Forward looking.* Trends must be seen quickly so that remedial action can be taken when necessary.

The final system satisfied all these criteria with the additional advantage that the percent complete returns become a simple but effective feedback for updating the network programme.

One of the most significant differences between EVA and the conventional progress-reporting systems is the substitution of 'weightings' given to individual activities, by the concept of 'value hours'. If each activity is monitored against its budget hours (or the hours allocated to that activity at the beginning of the contract) then the 'value hour' is simply the percent complete of that activity multiplied by its budget hours. In other words, it is the useful hours against the actual hours recorded on the time sheets.

If all the value hours of a project are added up and the total divided by the total budget hours, the overall percent complete of the project is immediately seen.

The advantage of this system over the weighting system is that activities can be added or eliminated without having to 're-weight' all the other activities. Furthermore, the value hours are a tangible parameter, which, if plotted on a graph against actual hours, budget hours and predicted final hours, give the manager a 'feel' of the progress of the job that is second to none. The examples in Tables 32.1 and 32.2 show the difference between the two systems.

Table 32.1 Weighting system.

1	2	3	4	5	6	7
Activity No.	**Activity**	**Budget × 100**	**Weighting**	**% Complete**	**% Weighted**	**Actual hours × 100**
1	A	1000	0.232	100	23.2	1400
2	B	800	0.186	50	9.3	600
3	C	600	0.140	60	8.4	300
4	D	1200	0.279	40	11.2	850
5	E	300	0.070	70	4.9	250
6	F	400	0.093	80	7.4	600
Total		4300	1.000		64.4	4000

Overall % complete = 64.4%

Predicted final hours = $\dfrac{4000}{0.644}$ = 6211 × 100 hours

Efficiency = $\dfrac{4300 \times 0.644}{4000}$ = 69.25

Table 32.2 Value hours (earned value) system.

1	2	3	4	5	6
Activity No.	Activity	Budget × 100	% Complete	Value hours × 100	Actual hours × 100
1	A	1000	100	1000	1400
2	B	800	50	400	600
3	C	600	60	360	300
4	D	1200	40	480	850
5	E	300	70	210	250
6	F	400	80	320	600
Total		4300		2770	4000

Overall % complete $= \dfrac{2770}{4300} = 64.4$

Predicted final hours $= \dfrac{4000}{0.644} = 6211 \times 100$ hours

Efficiency $= \dfrac{2770}{4000} = 69.25$

Summary of advantages

Comparing the weighting and value-hours systems, the following advantages of the latter are immediately apparent:

1. The basic value-hours system requires only six columns against the weighting system's seven.
2. There is no need to carry out a preliminary time-consuming 'weighting' at the beginning of the job.
3. Activities can be added, removed or have the durations changed without the need to recalculate the weightings of each activity. This saves hundreds of man-hours on a large project.
4. The value hours are easily calculated and can even, in many cases, be assessed by inspection.
5. Errors are easily seen, as the value can never be more than the budget.
6. Budget hours, actual hours, value hours and forecast hours can all be plotted on one graph to show the trend.
7. The method is ideal for assessing the value of work actually completed for progress payments to main and subcontractors. Since it is based on man-hours, it truly represents construction progress independent of material or plant costs, which so often distort the assessment.

The efficiency (output/input) for each activity is obtained by dividing the value hours by the actual hours. This is also known as the cost performance index (CPI).

The analysis can be considerably enhanced by calculating the efficiency and forecast final hours for each activity and adding these to the table.

The forecast final hours are obtained by either:

1. Dividing the budget hours by the efficiency; or
2. Dividing the actual hours by the % complete.

Both these methods give the same answer as the following proof (using the same abbreviations) shows:

1. $\text{Final hours} = \dfrac{\text{budget}}{\text{efficiency}} = \dfrac{B}{G}$

   $\text{Efficiency (CPI)} = \dfrac{\text{value}}{\text{actual}} = \dfrac{\text{earned value}}{\text{actual}} = \dfrac{E}{C}\,(\text{value is always the numerator})$

   $\text{Hence, final hours} = \dfrac{\text{budget}}{\text{value/actual}} = \dfrac{\text{budget}}{\text{value}} \times \text{actual} = \dfrac{B \times C}{E}$

   But $\text{value} = \text{budget} \times \text{complete} = B \times D$

2. $\text{Hence, final hours} = \dfrac{\text{budget} \times \text{actual}}{\text{budget} \times \text{complete}} = \dfrac{\text{actual}}{\text{complete}} = \dfrac{B \times C}{B \times D} = \dfrac{C}{D}$

Example 1 shows the earned value table for a small project consisting of three activities where there was reasonable progress.

The overall percentage complete of the work can be obtained by adding all the value hours in column E and dividing them by the total budget hours in column B, that is, E/B.

$$\text{Thus, overall percentage complete} = \frac{\text{total value}}{\text{total budget}} = \frac{E}{B} = \frac{540}{1800}$$

$$= 0.3 \text{ or } 30\%$$

$$\text{Thefore cast final hours } F = \frac{\text{total actual}}{\text{overall\%}} = \frac{600}{0.3} = 2000\frac{C}{D}$$

$$\text{As total efficiency of the project (CPI)} = \frac{\text{value}}{\text{actual}} = \frac{540}{600} = 0.9 \text{ or } 90\frac{E}{C}$$

$$\text{Alternatively, the forecast final hours } F = \frac{\text{budget}}{\text{efficiency}} = \frac{1800}{0.9} = 2000\frac{B}{G}$$

It can be seen that the difference between the calculated final hours of 2000, and the sum of the values of column F of 1950, is only 50 hours or 2.5%, and this tends to be the variation on projects with a large number of activities.

When an analysis is carried out after a period of poor progress as shown in the table of Example 2, the increase in the forecast final hours and the decrease in the efficiency become immediately apparent. An examination of the table shows that this is due to the abysmal efficiencies (column G) of activities 1 and 2.

In this example, the overall % complete is as follows:

$E/B = 310/1800 = 0.17222$ or 17.222

The efficiency (CPI) is $E/C = 310/600 = .5167$ or 52 approx

The forecast final hours are $C/D = 600/0.17222 = 3484$

or $B/G = 1800/0.5167 = 3484$

This is still a large overrun, but it is considerably less than the massive 8750 hours produced by adding up the individual forecast final hours in column F.

Clearly, such a discrepancy of 5266 hours in Example 2 calls for an examination. The answer lies in the offending activities 1 and 2, which need to be restated so that the actual hours reflect the actual situation on the job. For example, if it is found that activities 1 and 2 required rework to such an extent that the original work was completely wasted and the job had to be started again, it is sensible to restate the actual hours of these activities to reflect this, that is, all the abortive work is 'written off' and a new assessment of 0% complete is made from the starting point of the rework. There is little virtue in handicapping the final forecast with the gross inefficiency caused by unforeseen rework problems. Such a restatement is shown in Example 2a.

Comparing Examples 2 and 2a, it will be noted that:

1. The total budget hours are the same, that is, 1800.
2. The total actual hours are now only 350 in Example 2a because 180 hours have been written off for activity 1A and 70 hours have been written off for activity 2A.
3. The value hours are the same, that is, 310.
4. The overall % complete is the same.
5. The forecast final hours are now only 1700 because although the 250 aborted hours had to be included, the efficiency of the revised activities 1 and 2B has improved;
6. The overall efficiency is 310/350 = 0.885 or 88.5%.

The forecast final hours calculated by dividing the budget hours by the efficiency comes to $1800/0.885 = 2033$ hours. This is more than the 1700 hours obtained by adding all the values in column F, but the difference is only because the percent complete assessment of activities is so diverse.

In practice, such a difference is both common and acceptable because:

1. On medium or large projects, wide variations of % complete assessments tend to follow the law of 'swings and roundabouts', and cancel each other out.
2. In most cases, therefore, the sensible method of forecasting the final hours is to either:
   a. Divide the budget hours by the efficiency, that is, B/G or
   b. Divide the actual hours by the % complete, that is, C/D. Both, of course, give the same answer.
3. The column F (forecast final hours) is in most cases not required, but should it be necessary to find the forecast final hours of a specific activity, this can be done at any stage by simply dividing the actual hours of that activity by its percent complete.

4. It must be remembered that comparing the forecast final hours with the original budget hours is only a reporting function and its use should not be given too much emphasis. A much more important comparison is that between the actual hours and the value hours as this is a powerful and essential control function.

As stated earlier, two of the criteria of the system were the absolute minimum amount of form-filling for reporting progress and the accurate assessment of percent complete of specific activities. The first requirement is met by cutting down the reporting items to three essentials:

1. The activity numbers of the activities worked on in the reporting period (usually 1 week).
2. The *actual* hours spent on each of these activities, taken from the time cards.
3. The assessment of the percent complete of each reported activity. This is made by the 'man on the spot'.

The third item is the most likely one to be inaccurate, as any estimate is a mixture of fact and opinion. To reduce this risk (and thus comply with the second criterion, i.e., accuracy), the activities on the network have to be chosen and 'sized' to enable them to be estimated, measured or assessed in the field, shop or office by the foreman or supervisor in charge. This is an absolute prerequisite of success, and its importance cannot be over-emphasized.

Individual activities must not be so complex or long (in time) that further breakdown is necessary in the field, nor should they be so small as to cause unnecessary paperwork. For example, the erection of a length of ducting and supports (Fig. 32.1) could be split into the activities shown in Figs. 32.2 and 32.3.

Any competent supervisor can see that if the two columns of frame 1 (Activity 1) have been erected and stayed, the activity is about 50% complete. He may be conservative and report 40% or optimistic and report 60%, but this ±20% difference is not important in the light of the total project. When all these individual estimates are

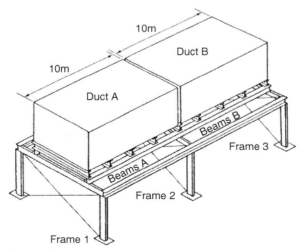

Figure 32.1 Duct support.

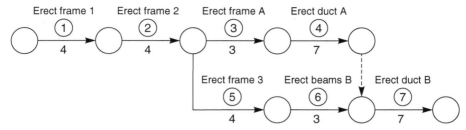

Figure 32.2 Duct and support network.

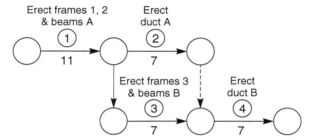

Figure 32.3 Duct and support network (condensed).

summated, the discrepancies tend to cancel out. What is important is that the assessment is realistic and checkable. Similarly, if 3 m of the duct between frames 1 and 2 have been erected, it is *about* 30% complete. Again, a margin on each side of this estimate is permissible.

However, if the network were prepared as shown in Fig. 32.3 the supervisor may have some difficulty in assessing the percent complete of activity 1 when he had erected and stayed the columns of frame 1. He now has to mentally compute the man-hours to erect and stay two columns in relation to four columns and four beams. The percent complete could be between 10% and 30%, with an average of 20%. The ± percentage difference is now 50%, which is more than double the difference in the first network. It can be seen therefore that the possibility of error and the amount of effort to make an assessment or both is greater.

Had the size of each activity been *reduced* to each column, beam or brace, the clerical effort would have been increased and the whole exercise would have been less viable. It is important therefore to consult the men in the field or on the shop floor before drafting the network and fixing the sequence and duration of each activity.

EVA for civil engineering projects

Most civil engineering contracts have an in-built earned value system, because the monthly re-measure is in fact a valuation of the work done. By using the composite rates in the bill of quantities or the schedule of rates, the monetary value of the work done to date can be easily established. This can then be translated into a curve and the value at any time period can then be divided by the corresponding value of the cash flow curve (which is in effect the planned work) to give the approximate % complete [schedule performance index (SPI)]. However, these values do not give a true picture of the work actually done, as the rates in the bills include overheads and profit as well as contingency allowance.

In order to compare actual costs with earned value, it is necessary to take the contingency, profit and overhead portion out of the unit rates so that only true labour, material and plant costs remain. This reduction could be between 5% and 10%. Re-measuring of the completed work can then take place as normal in the conventional physical units of m, m^2, m^3, tonne, etc. The measured quantities are then multiplied by the new (reduced) rates to give the useful work done in monetary terms and become in effect the earned value.

Planned costs

These can be taken from the S-curve of the histogram or cash flow curve. These will include labour, materials and plant costs, but they must again be at the reduced rate, that is, without overheads and profit.

Budget costs	
Labour	Total assessed time for activity × average labour rate
Materials	Estimated purchase price of materials required for this activity
Plant	Where plant is exclusive to the activity, for example, scaffolding
	• plant hire cost (external or internal) for the planned period of the activity • for example, £/day × anticipated number of days the plant is required
	Where plant is shared with other activities, for example, dumper truck
	• an approximate % assessment of usage must be made for each activity
Actual costs	
Labour	Time sheet hours × average labour rate
Materials	Invoice cost of material quantities used to date, or unit rate of material × quantity used to date
Plant	Where plant is exclusive to activity
	• invoiced plant hire rate, for example, £/day × number of days worked to date
	When plant is shared with other activities:
	• assessed % of plant hire rate, for example, £/day × number of days worked to date

Site overheads

Establishment and indirect labour costs cannot be a part of the EVA system. However, they must be recorded separately. The indirect labour costs must be plotted on the bottom of the EVA set of curves to ensure that they have not been inflated to off-set overruns of direct labour costs.

As with the normal EVA system, at any particular point in time.

Earned value (EV)	= Measured work in £, $, euros, etc.
Overall % complete	= EV/total original budget
Efficiency (CPI)	= EV/actual costs
SPI	= EV/planned costs

Where there are no bills of quantities, labour costs can still be used for monitoring progress using the EVA system. Material and plant costs can never be used for monitoring progress.

To operate such an EVA system, the budget costs, planned costs and actual costs (and the calculated earned value) must be broken down into labour, materials and plant for each activity, as each is measured differently. For any particular point in time, it is quite possible to have installed the planned materials and expended the associated plant costs and yet have an overrun or underrun of labour costs, depending on the effectiveness of supervision, method of working, climatic conditions or a myriad of other factors.

It can be seen, therefore, that this breakdown of each activity into three cost items can be very time consuming and, for this reason, the conventional monthly measurement of completed work based on the rates in the bills of quantities is the preferred method used by most companies in the civil engineering and building industries.

Example

A large storage tank consists of 400 steel plates, which, at $150 each, gives a material cost of $60,000 (for simplicity, other material costs have been ignored).

The duration for erecting and testing is 6 weeks.

The labour (mostly welding) man-hours are 1200, which at $20/hour, gives a labour cost of $24,000.

Again, for simplicity, plant costs (cranes and welding sets) are regarded as site overheads and can be ignored.

All the plates arrive on site on day 1 and have to be paid for on day 28.

Work starts as soon as the plates have been unloaded (by others).

By day 7 (1 week later), 60 plates have been erected and welded.

Therefore, at day 7, the percent complete is $60/400 = 15\%$.

The EV calculations are carried out weekly, and the time sheets show that after the first week, the men have booked a total of 200 man-hours.

Therefore, the earned value is 15% of 1200 = 180 so that the efficiency (CPI) = 180/200 = 90%.

If the men are paid production bonuses, they would not get a bonus for this week as the productivity bonus (as agreed with the unions) only starts at 97% efficiency.

The costs incurred to date are therefore only labour costs and are 200 × $20 = $4000.

At day 28 (after 4 weeks), 300 plates have been erected.

The % complete is now 300/400 = 75%.

The total man-hours booked to date are now 850.

The earned value can be seen to be 75% of 1200 = 900.

As the efficiency (CPI) is now 900/850 = 105%, the men get their bonuses.

The cost to date is now 850 × $20 = $17,000 plus the *total* material cost of $60,000 = $77,000.

Note that *all* the material has to be paid for − not just the material erected.

Alternative payment schedule

Supposing the terms of the contract were that the tank contractor had to be paid *weekly* for material and labour. He would therefore be paid as follows:

At the end of week 1:

Labour: 180 man-hours = $3600 (earned value, not actual cost).

Material: 60 plates at $150/plate = 150 × 60 = $9000 (plates erected).

The total payment is therefore $3600 + 9000 = $12,600 (ignoring retentions).

It can be seen therefore that the materials can be a useful aid in assessing percent complete, and although in order to obtain the total costs, they must be added to the labour costs, they *cannot* be part of the EVA. In other words, with the exception of the individual percentage complete assessment, the earned value, CPI, SPI, anticipated final cost and anticipated final completion time can only be calculated from the labour data.

The types of work that lend themselves to a similar treatment as the storage tank are the following:

- Pipework measured in metres
- Cable runs measured in metres
- Insulation measured in metres or square metres
- Steelwork measured in tonnes
- Refractory work measured in square metres
- Equipment measured in number of pieces, etc.

All labour must of course be measured in man-hours or money units.

If plant costs have to be booked to the work package, they can be treated in a similar way to equipment, except that payments (when the plant is hired) are usually made monthly.

Example 1. Reasonable progress.

A	B	C	D	E	F	G
Activity	Budget hours	Actual hours	% Complete	Value hours $B \times D$	Forecast final hours C/D	Efficiency (CPI) E/C
1	1000	200	20	200	1000	1.00
2	200	100	50	100	200	1.00
3	600	300	40	240	750	0.80
Total	1800	600		540	1950	

Example 2. Very poor progress due to rework.

A	B	C	D	E	F	G
Activity	Budget hours	Actual hours	% Complete	Value hours $B \times D$	Forecast final hours C/D	Efficiency (CPI) E/C
1	1000	200	5	50	4000	0.25
2	200	100	10	20	4000	0.20
3	600	300	40	240	750	0.80
Total	1800	600		310	8750	

Example 2a. Very poor progress due to rework.

A	B	C	D	E	F	G
Activity	Budget hours	Actual hours	% Complete	Value hours $B \times D$	Forecast final hours C/D	Efficiency (CPI) E/C
1A	0	0	100	0	180	0
1B	1000	20	5	50	400	2.5
2A	0	0	100	0	70	0
2B	200	30	10	20	300	0.67
3	600	300	40	240	750	0.80
Total	1800	350		310	1700	
1A or 2A are the works that have been written off.						

Further reading

APM. (2002). *Earned value management: APM guidelines.* APM.
Lewis, J. P. (2005). *Project planning scheduling and control.* McGraw-Hill.

Control graphs and reports

Chapter outline

In addition to the numerical report shown in Fig. 33.5, two very useful management control graphs can be produced:

1. Showing budget hours, actual hours, value hours and predicted final hours, all against a common time base; and
2. Showing percent planned, percent complete and efficiency, against a similar time base.

The actual shape of the curves on these graphs gives the project manager an insight into the running of the job, enabling the appropriate action to be taken.

Fig. 33.1 shows the site returns of man-hours of a small project over a 9-month period, and, for convenience, the table of percent complete and actual and value hours is drawn on the same page as the resulting curves. In practice, a greater number of activities would not make such a compressed presentation possible.

A number of interesting points are ascertainable from the curves:

1. There was obviously a large increase in site labour between the fifth and sixth months, as shown by the steep rise of the actual hours curve.
2. This has resulted in increased efficiency.
3. The learning curve given by the estimated final hours has flattened in month 6 making the prediction both consistent and realistic.
4. Month 7 showed a divergence of actual and value hours (indicated also by a loss of efficiency), which was corrected (probably by management action) by month 8.
5. It is possible to predict the month of actual completion by projecting all the curves forward. The month of completion is then given:
   a. When the value hours curve intersects the budget line; and
   b. When the actual hours curve intersects the estimated final hours curve.

In this example, one could safely predict the completion of the project in month 10.

It will be appreciated that this system lends itself ideally to computerization, giving the project manager the maximum information with very minimum site input. The sensitivity of the system is shown by the immediate change in efficiency when the value hours diverged from the actual hours in month 7. This alerts management to investigate and apply corrections.

Project Management, Planning and Control. https://doi.org/10.1016/B978-0-12-824339-8.00033-X

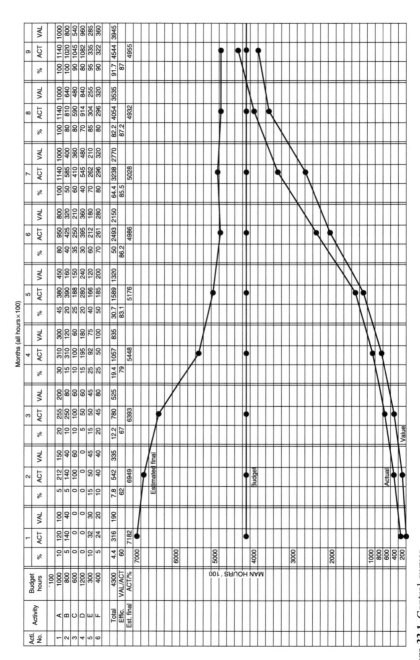

Figure 33.1 Control curves.

For maximum benefit, the returns and calculations should be carried out weekly. By using the normal weekly time cards very little additional site effort is required to complete the returns, and with the aid of a good computer program the results should be available 24 hours after the returns are received.

An example of the application of a manual earned value analysis (EVA) is shown in Figs. 33.2−33.9. The site-construction network of a package boiler installation is given in Fig. 33.2. Although the project consisted of three boilers, only one network, that of boiler no. 1 is shown. In this way, it was possible to control each boiler construction separately and compare performances. The numbers above the activity description are the activity numbers, while those below are the durations. The reason for using activity numbers for identifying each activity, instead of more conventional beginning and end event numbers, is that the identifier must always be uniquely associated with the activity description.

If the event numbers (in this case the coordinates of the grid) were used, the identifier could change if the logic was amended or other activities were inserted. In a sense, the activity number is akin to the node number of a precedence diagram, which is always associated with its activity. The use of precedence diagrams and computerized EVA is therefore a natural marriage, and to illustrate this point, a precedence diagram is shown in Fig. 33.3.

Once the network has been drawn, the man-hours allocated to each activity can be represented graphically on a bar chart. This is shown in Fig. 33.4. By adding up the man-hours for each week, the totals, cumulative totals and each week's percentage of the total man-hours can be calculated. If these percentages are then plotted as a graph, the planned percent-complete curve can be drawn. This is shown in Fig. 33.7.

All the work described up to this stage can be carried out before work starts on site. The only other operation necessary before the construction stage is to complete the left-hand side of the site-returns analysis sheet. This is shown in Fig. 33.5, which covers only periods 4−9 of the project. The columns to be completed at this stage are as follows:

1. The activity number
2. The activity title
3. The budget hours

Once work has started on site, the construction manager reports weekly on the progress of each activity worked on during that week. All he has to state is the following:

1. The activity number
2. The actual hours expended *in that week*
3. The percent complete of that activity to date

If the computation is carried out manually, the figures are entered on the sheet (Fig. 33.5) and the following values are calculated weekly:

1. Total man-hours expended this week (W column)
2. Total man-hours to date (A column)
3. Percent complete of project (% column)
4. Total value hours to date (V column)
5. Efficiency
6. Estimated final hours

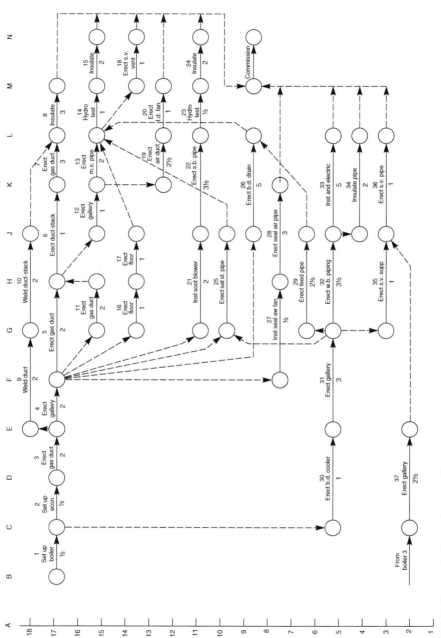

Figure 33.2 Boiler no. 1. Network arrow diagram.

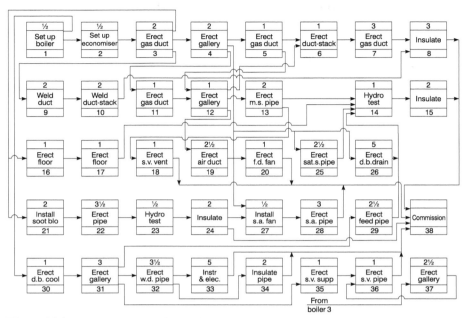

Figure 33.3 Boiler no. 1. Precedence diagram.

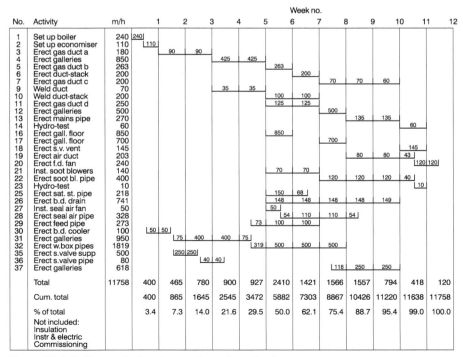

No.	Activity	m/h	1	2	3	4	5	6	7	8	9	10	11	12
1	Set up boiler	240	240											
2	Set up economiser	110	110											
3	Erect gas duct a	180		90	90									
4	Erect galleries	850				425	425							
5	Erect gas duct b	263						263						
6	Erect duct-stack	200							200					
7	Erect gas duct c	200								70	70	60		
9	Weld duct	70				35	35							
10	Weld duct-stack	200						100	100					
11	Erect gas duct d	250						125	125					
12	Erect galleries	500								500				
13	Erect mains pipe	270									135	135		
14	Hydro-test	60											60	
16	Erect gall. floor	850						850						
17	Erect gall. floor	700								700				
18	Erect s.v. vent	145											145	
19	Erect air duct	203								80	80	43		
20	Erect f.d. fan	240											120	120
21	Inst. soot blowers	140						70	70					
22	Erect soot bl. pipe	400								120	120	120	40	
23	Hydro-test	10											10	
25	Erect sat. st. pipe	218						150	68					
26	Erect b.d. drain	741						148	148	148	148	149		
27	Inst. seal air fan	50						50						
28	Erect seal air pipe	328						54	110	110	54			
29	Erect feed pipe	273						73	100	100				
30	Erect b.d. cooler	100		50	50									
31	Erect galleries	950			75	400	400	75						
32	Erect w.box pipes	1819						319	500	500	500			
35	Erect s.valve supp	500		250	250									
36	Erect s.valve pipe	80				40	40							
37	Erect galleries	618								118	250	250		
	Total	11758	400	465	780	900	927	2410	1421	1566	1557	794	418	120
	Cum. total		400	865	1645	2545	3472	5882	7303	8867	10426	11220	11638	11758
	% of total		3.4	7.3	14.0	21.6	29.5	50.0	62.1	75.4	88.7	95.4	99.0	100.0

Not included:
Insulation
Instr & electric
Commissioning

Figure 33.4 Boiler no. 1. Bar chart and man-hour loadings.

No.	Title	Orig	Rev	Period-4				Period-5				Period-6				Period-7				Period-8				Period-9			
				W	A	%	V	W	A	%	V	W	A	%	V	W	A	%	V	W	A	%	V	W	A	%	V
1	Set up boiler	240			233	100	240		230	100	240		230	100	240		230	100	240		230	100	240		230	100	240
2	Set up economiser	110			90	100	110		90	100	110		90	100	110		90	100	110		90	100	110		90	100	110
3	Erect gas duct a	180			155	100	180		155	100	180		155	100	180		155	100	180		155	100	180		155	100	180
4	Erect galleries	850		352	352	60	510	389	741	80	450	69	810	100	850		810	100	850		810	100	850		810	100	850
5	Erect gas duct b	263										200	200	60	158	55	255	100	263		255	100	263		255	100	263
6	Erect duct-stack	200																		180	180	80	160	5	185	100	200
7	Erect gas duct c	200																		62	62	30	60	88	150	75	150
9	Weld duct	70		32	32	50	35	33	65	100	70		65	100	70		65	100	70		65	100	70		65	100	70
10	Weld duct-stack	200																		110	110	60	120	70	180	100	200
11	Erect gas duct d	250										42	92	60	150	18	70	80	200	5	175	100	250	5	175	100	250
12	Erect galleries	500																		405	405	95	475	5	410	100	500
13	Erect mains pipe	270																						105	105	40	108
14	Hydro-test	60																									–
16	Erect gall. floor	850	–									420	420	60	510	360	780	90	765	5	785	100	850		785	100	850
17	Erect gall. floor	700																		340	340	45	315	310	650	100	700
18	Erect s.v. vent	145	–																								–
19	Erect air duct	203																						75	75	30	81
20	Erect f.d. fan	240	–																								–
21	Inst. soot blowers	140										65	65	55	77	65	130	95	133	5	135	100	140		135	100	140
22	Erect soot bl. pipe	400																		100	100	20	80	110	210	40	160
23	Hydro-test	10	–																								
25	Erect sat. st. pipe	218										125	125	60	131	85	210	100	218		210	100	218		210	100	218
26	Erect b.d. drain	741										130	130	20	148	132	212	45	333	136	398	60	445	122	520	80	598
27	Inst. seal air fan	50											40	100	50		40	100	50		40	100	50		40	100	50
28	Erect seal air pipe	328										45	45	20	66	37	82	40	131	128	210	85	279	10	220	90	293
29	Erect feed pipe	273										93	145	60	164	100	245	100	273		245	100	273		245	100	273
30	Erect b.d. cooler	100			105	100	100	52	105	100	100		105	100	100		105	100	100		105	100	100		105	100	100
31	Erect galleries	950		390	810	75	713	30	840	80	760	25	865	100	950		865	100	950		865	100	950		865	100	950
32	Erect w.box pipes	1819						300	300	20	364	460	760	30	546	445	1205	65	1182	405	1610	80	1455	340	1950	100	1819
35	Erect s.valve supp	500			440	100	500		460	100	500		460	100	500		460	100	500		460	100	500		460	100	500
36	Erect s.valve pipe	80		40	80	100	80		80	100	80		80	100	80		80	100	80		80	100	80		80	100	80
37	Erect galleries	618																						120	120	30	185
							106%				101%				104%				106%				105%				106%
	Totals:-	11758					11019				11548				11260				11141				11277				11095
				814	2314	21	2468	804	3118		3152	1724	4842	43	5030	1397	6239	56	6628	1881	8120	72	8513	1360	9480	86	10095

Legend

A – Actual accumulated manhours used to end of period
% – Estimate of percentage completion of activity
V – Value hours = budget x percentage completion
W – Hours worked this period

Figure 33.5 Earned value analysis sheet.

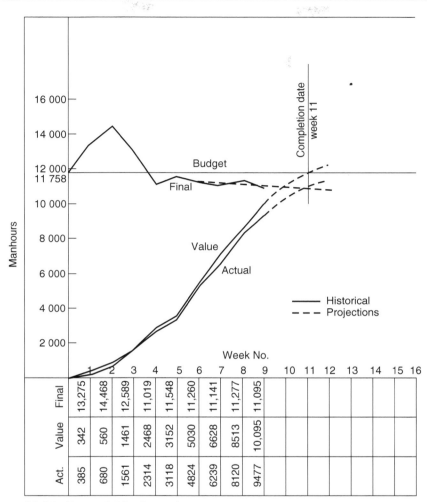

The table within the figure:

		2	3	4	5	6	7	8	9	10	11	12	13	14	15	16
Final		13,275	14,468	12,589	11,019	11,548	11,260	11,141	11,277	11,095						
Value		342	560	1461	2468	3152	5030	6628	8513	10,095						
Act.		385	680	1561	2314	3118	4824	6239	8120	9477						

Figure 33.6 Boiler no. 1. Man hour–time curves.

Alternatively, the site returns can be processed by computer and the resulting printout of part of a project is shown in Fig. 33.8. Whether the information is collected manually or electronically, the return can be made on a standard time sheet with the only addition being a % complete column. In other words, no additional forms are required to collect information for EVA. There are in fact only three items of data to be returned to give sufficient information:

1. The activity number of the activity actually being worked on in that time period;
2. the actual hours being expended on each activity worked on in that time period; and
3. The cumulative % complete of each of these activities.

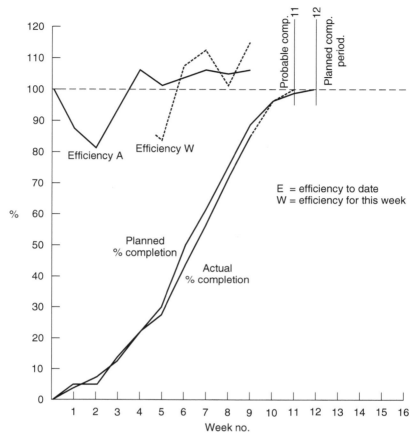

Figure 33.7 Boiler no. 1. Percentage—time curves.

All the other information required for computation and reporting (such as activity titles and activity man-hour budgets) will already have been inputted and stored in the computer. A typical modified time sheet is shown in Fig. 33.9.

A complete set of printouts produced by a modern project-management system are shown in Figs. 33.10—33.14. It will be noted that the network in precedence format has been produced by the computer, as have the bar chart and curves. In this program, the numerical EVA has been combined with the normal critical path analysis from one database, so that both outputs can be printed and updated at the same time on one sheet of paper. The reason the totals of the forecast hours are different from the manual analysis is that the computer calculates the forecast hours for each activity and then adds them up, while in the manual system the total forecast hours are obtained by simply dividing the actual hours by the percent complete rounded off to the nearest 1%.

Foster Wheeler Power Products Ltd.

Contract No. 2-322-04298
Construction at Suamprogetti

Site manhours and costing system (EVA)
Standard report
Manhours report

Events prec succ	Description	No off unit 0-rate	Hrs/ unit C-rate	Budgets original/ current	Period this accum.	% com	Cimp. value	Est. to compl.	Forecast last rep total	Var. from last rep total	Extra	Remarks
	Setup boiler											
0001-0001-01	Setup boiler	BLR 1	240.00	240.00 240.00	0.00 55.00	100	240	0	55 55	0 185		
	Setup econ											
0001-0002-01	Setup economizer	ECON 1	110.00	110.00 110.00	0.00 52.00	100	110	0	52 52	0 58		
	Erect ducts											
0001-0003-01	Erect ducts blr/econ	DUCT 1	180.00	180.00 180.00	0.00 257.00	100	180	0	257 257	0 −77		
	Erect b/d cooler											
0001-0004-01	Erect b/d cooler	VESSL 1	100.00	100.00 100.00	0.00 128.00	100	100	0	128 128	0 −28		
	Erect galleries											
0001-0005-01	Erect galls for blr	GALLS 1	850.00	850.00 850.00	0.00 651.00	850	850	0	651 651	0 199		
	Erect duct											
0001-006-01	Erect duct chimney dampers	DUCT 1	250.00	250.00 250.00	0.00 169.00	98	245	3	172 172	0 78		
	Erect galleries											

Figure 33.8 Standard E.V. Report printout.

WEEKLY TIMESHEET

| Name | | | Staff no. | | | | Week ending | | |

Project no.	Activity/ document no.	Mon.	Tues.	Wed.	Thu.	Fri.	Sat.	Sun.	Total	% complete	Remarks

| Signed | | Date | | Approved | | | Date | |

Figure 33.9 Weekly time sheet.

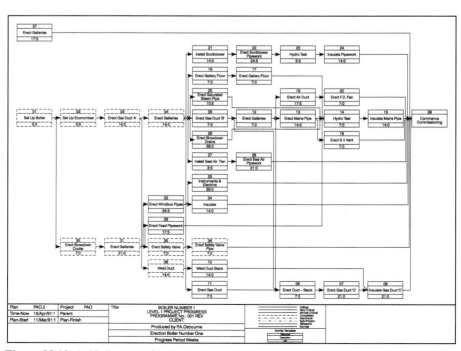

Figure 33.10 AoN diagram of boiler.

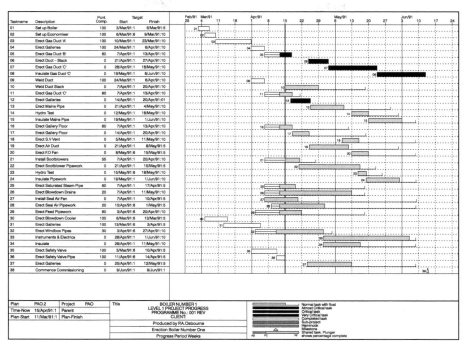

Figure 33.11 Bar chart.

As mentioned earlier, if the budget hours, actual hours, value hours and estimated final hours are plotted as curves on the same graph, their shape and relative positions can be extremely revealing in terms of profitability and progress. For example, it can be seen from Fig. 33.6 that the contract was potentially running at a loss during the first 3 weeks, as the value hours were less than the actual hours. Once the two curves crossed, profitability returned and in fact increased, as indicated by the diverging nature of the value- and actual-hour curves. This trend is also reflected by the final-hours curve dipping below the budget-hour line.

The percentage—time curves in Fig. 33.7 enable the project manager to compare the actual percent complete with the planned. This is a better measure of performance than comparing actual hours expended with planned hours expended. There is no virtue in spending the man-hours in accordance with a planned rate. What is important is the percent complete in relation to the plan and whether the hours spent were *useful* hours. Indeed, there should be every incentive to spend *less* hours than planned, provided that the *value* hours are equal or greater than the actual and the percent complete is equal or greater than the planned.

The efficiency curve in Fig. 33.7 is useful, as any drop is a signal for management action. Curve 'A' is based on the efficiency calculated by dividing the cumulative value hours by the cumulative actual hours for every week. Curve 'W' is the efficiency

Activity Number	Description	Planned to date	Budget hours	Actual hours	Complete %	Estimate to comp	Forecast hours	Variance -/+	EFF 1	Orig durn	Rem durn	Early start	Early Finish	Late start	Late finish	Target start	Target finish	Actual start	Actual finish	Float remain	Programme status against target	Action required
01	Set up boiler	240	240	230	100%	0	230	10	104	3.5	0.0	3MAR91	6MAR91	3MAR91	6MAR91	3MAR91	6MAR91	3MAR91	6MAR91	0.0	1 day (s) slippage	Complete
02	Set up economiser	110	110	90	100%	0	90	20	122	3.5	0.0	6MAR91	9MAR91	6MAR91	9MAR91	6MAR91	9MAR91	6MAR91	9MAR91	0.0	On target	Complete
03	Erect gas duct 'A'	180	180	155	100%	0	155	25	116	14.0	0.0	10MAR91	23MAR91	10MAR91	23MAR91	10MAR91	23MAR91	10MAR91	23MAR91	0.0	On target	Complete
04	Erect galleries	850	850	810	100%	0	810	40	105	14.0	0.0	24MAR91	6APR91	24MAR91	6APR91	24MAR91	6APR91	24MAR91	6APR91	0.0	On target	Complete
05	Erect gas duct 'B'	263	263	200	60%	133	133	-70	79	7.0	2.8	7/APR91	17/APR91	7/APR91	13/APR91	7/APR91	13/APR91	7/APR91	/	-3.8	4 day (s) slippage	Yes
06	Erect duct – stack	200	200	0	0%	200	200	0	0	7.0	7.0	24/APR91	1/MAY91	21/APR91	27/APR91	21/APR91	27/APR91	/	/	3.8	4 day (s) slippage	Yes
07	Erect gas duct 'C'	200	200	0	0%	200	200	0	0	21.0	21.0	1/MAY91	22/MAY91	28/APR91	18/MAY91	28/APR91	18/MAY91	/	/	3.8	4 day (s) slippage	Yes
08	Insulate gas duct 'C'	0	0	0	0%	0	0	0	0	21.0	21.0	22/MAY91	12/JUN91	19/MAY91	8/JUN91	19MAY91	8/JUN91	/	/	-3.8	4 day (s) slippage	Yes
09	Weld duct	70	70	65	100%	0	65	5	108	14.0	0.0	6/APR91	8/APR91	5/MAY91	18/MAY91	7/APR91	20/APR91	24/MAR91	6/APR91	0.0	On target	Complete
10	Weld duct stack	108	200	0	0%	200	200	0	0	14.0	14.0	15/APR91	28/APR91	5/APR91	18/MAY91	7/APR91	20/APR91	/	/	20.0	8 day (s) slippage	
11	Erect gas duct 'C'	125	250	92	60%	61	153	97	163	2.8	2.8	7/APR91	17/APR91	7/APR91	7/APR91	7/APR91	13/APR91	7/APR91	/	3.2	4 day (s) slippage	Yes
12	Erect galleries	0	500	0	0%	500	500	0	0	7.0	7.0	17/APR91	24/APR91	14/APR91	20/APR91	14/APR91	20/APR91	/	/	-3.8	4 day (s) slippage	
13	Erect mains pipe	0	270	0	0%	270	270	0	0	14.0	14.0	24/APR91	8/MAY91	21/APR91	18/MAY91	21/APR91	4/MAY91	/	/	10.2	4 day (s) slippage	
14	Hydro test	0	60	0	0%	60	60	0	0	7.0	7.0	8/MAY91	19/MAY91	2/MAY91	25/MAY91	12/MAY91	18/MAY91	/	/	6.0	1 day (s) slippage	
15	Insulate mains pipe	0	0	0	0%	0	0	0	0	14.0	14.0	19/MAY91	2/JUN91	15/MAY91	8/JUN91	19/MAY91	1/JUN91	/	/	6.0	1 day (s) slippage	Complete
16	Erect galley floor	850	850	420	60%	280	700	150	121	7.0	2.8	7/APR91	17/APR91	7/APR91	11/MAY91	7/APR91	13/APR91	7/APR91	/	24.2	4 day (s) slippage	Complete
17	Erect galley floor	0	700	0	0%	700	700	0	0	7.0	7.0	14/APR91	24/APR91	14/APR91	8/JUN91	14/APR91	20/APR91	/	/	24.2	4 day (s) slippage	
18	Erect S.V. vent	0	145	0	0%	145	145	0	0	7.0	7.0	5/MAY91	15/MAY91	5/MAY91	8/JUN91	5/MAY91	11/MAY91	/	/	24.2	4 day (s) slippage	
19	Erect air duct	0	203	0	0%	203	203	0	0	17.5	17.5	21/APR91	12/MAY91	1/JUN91	1/JUL91	21/APR91	8/MAY91	/	/	20.7	4 day (s) slippage	
20	Erect F.D. fan	0	240	0	0%	240	240	0	0	7.0	2.8	8/MAY91	19/MAY91	8/MAY91	8/JUL91	8/MAY91	15/MAY91	/	/	20.7	4 day (s) slippage	
21	Install sootblowers	70	140	65	55%	53	118	22	118	14.0	6.3	7/APR91	21/APR91	7/APR91	27/APR91	7/APR91	20/APR91	7/APR91	/	6.7	0 day (s) slippage	
22	Erect sootblower pipework	100	400	0	0%	400	400	0	0	24.5	24.5	21/APR91	15/MAY91	28/APR91	22/MAY91	21/APR91	15/MAY91	/	/	6.7	0 day (s) slippage	
23	Hydro test	0	10	0	0%	10	10	0	0	3.5	3.5	15/MAY91	19/MAY91	22/MAY91	25/MAY91	15/MAY91	18/MAY91	/	/	6.7	0 day (s) slippage	
24	Insulate pipework	0	0	0	0%	0	0	0	0	14.0	14.0	26/MAY91	2/JUN91	26/MAY91	8/JUN91	26/MAY91	1/JUN91	/	/	6.7	0 day (s) slippage	
25	Erect saturated steam pipe	150	218	125	60%	83	208	10	105	10.5	4.2	7/APR91	19/APR91	7/APR91	18/MAY91	7/APR91	17/MAY91	7/APR91	/	29.8	2 day (s) slippage	
26	Erect blowdown drains	148	741	130	20%	520	650	91	114	35.0	28.0	7/APR91	12/MAY91	7/APR91	18/MAY91	7/APR91	11/MAY91	7/APR91	/	6.0	1 day (s) slippage	
27	Install seal air fan	50	50	0	0%	50	50	0	0	3.5	3.5	7/APR91	18/APR91	7/APR91	8/JUN91	7/APR91	10/APR91	7/APR91	/	51.5	8 day (s) slippage	
28	Erect seal air pipework	54	328	45	20%	180	225	103	146	21.0	16.8	10/APR91	1/MAY91	10/APR91	8/JUN91	10/APR91	1/MAY91	10/APR91	/	38.2	0 day (s) slippage	
29	Erect feed pipework	173	273	145	60%	97	242	31	113	17.5	7.0	3/APR91	21/APR91	3/APR91	18/MAY91	3/APR91	20/APR91	3/APR91	/	27.0	1 day (s) slippage	Complete
30	Erect blowdown cooler	100	100	105	100%	0	105	-5	95	7.0	7.0	6/APR91	13/MAY91	6/APR91	13/MAY91	6/APR91	13/MAR91	6/MAR91	13MAR91	0.0	1 day (s) slippage	Complete
31	Erect galleries	950	950	865	100%	0	865	85	110	21.0	0.0	13/MAR91	3/APR91	13/MAR91	3/APR91	13/MAR91	3/APR91	13/MAR91	3/APR91	0.0	1 day (s) slippage	
32	Erect windbox pipes	819	1819	760	30%	1773	2533	-714	72	24.5	17.2	3/APR91	2/MAY91	3/APR91	4/MAY91	3/APR91	27/APR91	3/APR91	/	2.8	1 day (s) slippage	
33	Instruments & electrics	0	0	0	0%	0	0	0	0	35.0	35.0	28/APR91	6/JUN91	28/APR91	8/JUN91	28/APR91	1/JUN91	/	/	2.8	4 day (s) slippage	
34	Insulate	0	0	0	0%	0	0	0	0	14.0	14.0	28/APR91	16/MAY91	28/APR91	8/JUN91	28/APR91	11/MAY91	/	/	23.8	4 day (s) slippage	
35	Erect safety valve	500	500	460	100%	0	460	40	109	7.0	7.0	3/APR91	10/APR91	3/APR91	10/APR91	3/APR91	10/APR91	3/APR91	10/APR91	0.0	1 day (s) slippage	Complete
36	Erect safety valve pipe	80	80	80	100%	0	80	0	100	7.0	0.0	11/APR91	14/APR91	11/APR91	14/APR91	11/APR91	14/APR91	11/APR91	14/APR91	0.0	1 day (s) slippage	Complete
37	Erect galleries	0	618	0	0%	618	618	0	0	17.5	17.5	25/APR91	12/MAY91	22/APR91	8/JUN91	25/APR91	12/MAY91	/	/	27.5	On target	
38	Commence commissioning	0	0	0	0%	0	0	0	0	17.5	0.0	9/JUN91	12/JUN91	9/JUN91	9/JUN91	9/JUN91	9/JUN91	/	/	-3.8	4 day (s) slippage	Yes
Total		5882	11758	4842	43%	6977	11819	-61	104		5029											

Figure 33.12 Combined CPA and EVA print out.

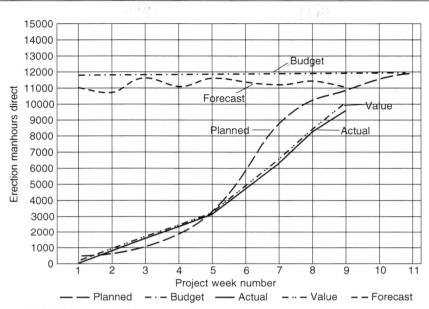

Figure 33.13 Boiler no. 1. Erection man-hours.

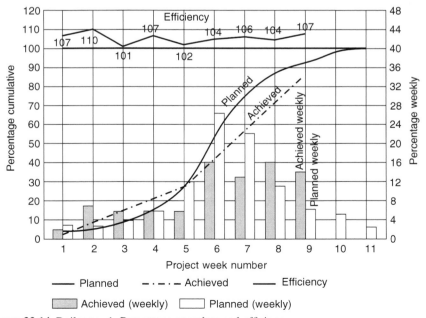

Figure 33.14 Boiler no. 1. Percentage complete and efficiency.

by dividing the value hours generated in a particular week by the actual hours expended in that week. It can be seen that Curve 'W' (shown only for periods 5—9) is more sensitive to change and is therefore a more dramatic warning device to management.

Finally, by comparing the curves in Figs. 33.6 and 33.7, the following conclusions can be drawn:

1. *Value hours exceed actual hours* (Fig. 33.6). This indicates that the site is efficiently run.
2. *Final hours are less than budget hours* (Fig. 33.6). This implies that the contract will make a profit.
3. *The efficiency is over 100% and rising* (Fig. 33.7). This bears out conclusion 1.
4. *The actual percent complete curve* (Fig. 33.7), although less than the planned, has for the last four periods been increasing at a greater rate than the planned (i.e. the line is at a steeper angle). Hence, the job may well finish *earlier* than planned (probably in week 11).
5. By projecting the value-hour curve forward to meet the budget-hour line, it crosses in week 11 (Fig. 33.6).
6. By projecting the actual-hour curve to meet the projection of the final-hour curve, it intersects in week 11 (Fig. 33.6). Hence, week 11 is the probable completion date.

The computer printout shown in Fig. 33.8 is updated weekly by adding the man-hours logged against individual activities. However, it is possible to show the *cost* of both the historical and current man-hours in the same report. This is achieved by feeding the average man-hour rate for the contract into the machine at the beginning of the job and updating it when the rate changes. Hence, the new hours will be multiplied by the current rates. A separate report can also be issued to cover the indirect hours such as supervision, inspection, inclement weather and general services.

As the value-hour concept is so important in assessing the labour content of a site or works operation, the following summary showing the computation in nonnumerical terms can be of help:

If

B	$=$	Budget hours (total)
C	$=$	Actual hours (total)
D	$=$	Percent complete
E	$=$	Value hours (earned value) (total)
F	$=$	Forecast final hours
G	$=$	Percent efficiency (CPI)

Then:

$$E = B \times D, \ D = E/B \times 100, \ G = E/C \times 100, F = C/D \text{ or } B/G$$

Overall project completion

Once the man-hours have been 'costed', they can be added to other cost reports of plant, equipment, materials, subcontracts, etc. so that an overall percent completion of a project can be calculated for valuation purposes on the only true common denominator of a project — money.

The total *value* to date divided by the revised budget × 100 is the percent complete of a job. The value-hour concept is entirely compatible with the conventional valuation of costing such as value of concrete poured, value of goods installed, cost of plant utilized — activities, which can, by themselves, be represented on networks at the planning stage.

Table 33.1 shows how the two main streams of operations, that is, those categories measured by cost and those measured by man-hours can be combined to give an overall picture of the percent complete in terms of cost and overall cost of a project. While the operations shown relate to a construction project, a similar table can be drawn for a manufacturing process, covering operations such as design, tooling, raw material purchase, machinery, assembly, testing, and packing.

Cost of overheads, plant amortization, licences, etc. can, of course, be added like any other commodity. An example giving quantities and cost values of a small job involving all the categories shown in Table 33.1 is presented in Table 33.2–33.4. It can be seen that in order to enable an overall percent complete to be calculated, all the quantities of the estimate (Table 33.2) have been multiplied by their respective rates — as in fact would be done as part of any budget — to give the estimated costs.

Table 33.3 shows the progress after a 16-week period, but in order to obtain the value hours (and hence the cost value) of category D it was necessary to break down the man-hours into work packages that could be assessed for percent completion. Thus, in Table 33.4, the pipelines A and B were assessed as 35% and 45% complete, and the pump and tank connections were found to be 15% and 20% complete, respectively. Once the value hours (3180) were found, they could be multiplied by the average cost per man-hour to give a cost value of $14628.

Table 33.5 shows the summary of the four categories. An adjustment should therefore also be made to the value of plant utilization category C as the two are closely related. The adjusted value total would therefore be as shown in column V.

With an expenditure to date true value of $104,048, the percent completion in terms of cost of the whole site is therefore as follows:

$$\frac{104,048}{202,000} \times 100 = 51.5$$

It must be stressed that the cost completed is not the same as the completion of construction work. It is only a valuation method when the material and equipment are valued (and paid for) in their month of arrival or installation.

Table 33.1

Basic method of measurement	Cost (money)			Man-hours
Method of measurement category	Bills of quantities A	Lump sum B	Rates C	Rate/hour D
Type of activity	Earth moving Civil work Painting Insulation Piping supply	Tanks Equipment (compressors, pumps, towers, etc.)	Mechanical plant Cranes Scaffolding Transport	Erection of: piping, electrical work, instrumentation, machinery, steelwork; testing, commissioning
Base for comparison of progress	Total of bills of quantities	Total of equipment items	Plant estimate	Man-hour budget
Periodic valuation	Measured quantities	Cost of items delivered	Cost of plant on-site	Value hours = % complete × budget
Method of assessment	Field measurement	Equipment count	Plant count	Physical % complete
Percent complete for reporting	Measured quantities × rates	Delivered cost	Cost of plant on-site	Value hours
Methods of measurement	Total in bill of quantities	Total equipment cost	Plant estimate	Man-hour budget

Total cost = measured quantity × rates + cost of items delivered + cost of plant on-site + actual hours × rate

$$\text{Total site percentage complete} = \frac{100 \,(\text{cost of A} + \text{cost of B} + \text{adjusted cost of C} + \text{value hours of D} \times \text{average rate})}{\text{Total budget}}$$

Table 33.2 Example showing effect of percent complete of different categories.

Estimate category	Item	Unit	Quantity	Rate ($/hour)	Cost $
A	Concrete	M^3	1000	25	25,000
	Pipe 6-inch	M	2000	3	6000
	Painting	M^2	2500	10	25,000
	–	–	–	–	56,000
B	Tanks	No	3	20,000	60,000
	Pumps	No	1	8000	8000
	Pumps	No	1	14,000	14,000
	–	–	–	–	82,000
C	Cranes	Hours	200	6015	12,600
	(hire)	Hours	400	–	6000
	Welding	–	–	–	18,000
D	Pipe fitters	Hours	4000	4} Av.	16,000
	Welders	Hours	6000	5} 4.6	30,000
	–	–	10,000	–	46,000

Table 33.3 Progress after 16 weeks.

Category	Item	Unit	Quantity	Rate ($/hour)	Cost $
A	Concrete poured	M^3	900	25	22,500
	Pipe 6-inch supplied	M	1000	3	3000
	Painting	M^2	500	10	5000
	Complete: $\dfrac{30,500}{56,000} \times 100 = 54.46$	–	–	30,500
B	Tanks delivered	No	2	20,000	40,000
	Pumps A	No	1	8000	8000
	Pumps B	No	1	–	–
	Complete: $\dfrac{48,000}{82,000} \times 100 = 58.53$	–	48,000

When the materials or equipment are paid for as they arrive on site (possibly a month before they are actually erected), or when they are supplied 'free issue' by the employer, they must not be part of the value or complete calculation.

It is clearly unrealistic to include materials and equipment in the complete and efficiency calculation as the cost of equipment is not proportional to the cost of installation.

Table 33.4

Category	Item	Unit	Quantity	Rate ($/hour)	Cost $
C	Cranes on-site	Hours	150	60	9000
	Welding plant	Hours	200	15	3000
	Complete: $\dfrac{12,000}{18,000} \times 100 = 66.66$		–	–	12,000
D	Pipe fitters	Hours	1800	4	7200
	Welders	Hours	2700	5	13,500
	–	–	–	–	20,700
Erection work		Budget M/H	Percent complete	Value hours	Actual hours
Pipeline A		3800	35	1330	1550
Pipeline B		2800	45	1260	1420
Pump connection		1800	15	270	220
Tank connection		1600	20	320	310
	$\left| \begin{array}{l} 10,000 \end{array} \right.$		3180	3500	

Complete: $\dfrac{3180}{10,000} \times 100 = 31.80$

Cost value (Av.) $= 3180 \times 4.6 = \$14,628$

Table 33.5 Total cost to date.

I	II	III	IV	V
Category	Budget	Cost	Value	Adjusted value
A	56,000	30,500	30,500	30,500
B	82,000	48,500	48,000	48,000
C	18,000	12,000	12,000	10,920
D	46,000	20,700	14,628	14,628
Total	$202,000	$111,200	$105,128	$104,048

For example, a carbon steel tank takes the same time to lift onto its foundations as a stainless steel tank, yet the cost is very different. Indeed, in some instances, an expensive item of equipment may be quicker and cheaper to install than an equivalent cheaper item, simply because the expensive item may be more 'complete' when it arrives on site.

All the items in the calculations can be stored, updated and processed by computer, so there is no reason why an accurate, up-to-date and regular progress report cannot be produced on a weekly basis, where the action takes place — on the site or in the workshop.

Clearly, with such information at one's fingertips, costs can truly be controlled — not merely reported!

It can be seen that the value hours for erection work are only 3180 against an actual man-hours usage of 3500. This represents an efficiency of only

$$\frac{3180}{3500} \times 100 = 91 \text{ approx.}$$

An adjustment should therefore also be made to the value of plant utilization, that is, $12,000 \times 91\% = 10,920$. The adjusted value total would therefore be as shown in column V.

The Site manhours and control (SMAC) system described on the previous pages was developed in 1978 by Foster Wheeler Power Products, primarily to find a quicker and more accurate method for assessing the % complete of multi-discipline, multi-contractor construction projects.

However, about 10 years earlier, the Department of Defense in the United States developed an almost identical system called cost, schedule, control system (CSCS), which was generally referred to as EVA. This was mainly geared to the cost control of defence projects within the United States, and apart from UK subcontractors to the American defence contractors, was not disseminated widely in the United Kingdom.

While the principles of SMAC and EVA are identical, inevitably a difference in terminology is developed, which has caused considerable confusion to students and practitioners. Fig. 33.16 lists these abbreviations and their meaning, and Fig. 33.17 shows the comparison between the now accepted EVA 'English' terms (shown in **bold**) and the CSCS jargon (shown in *italics*).

The CSCS also introduced four new parameters for cost efficiency and, for want of a better word, time efficiency:

1. The cost variance: this is the arithmetical difference at any point between the earned value and the actual cost.
2. The schedule variance: this is the arithmetical difference between the earned value and the scheduled (or planned) cost. However, comparing progress in time by subtracting the planned cost from the earned value is somewhat illogical as both are measured in monetary terms. It would make more sense to use parameters measured in time to calculate the time variance. This can be achieved by subtracting the actual duration [Actual time expended (ATE)] for a particular earned value from the originally planned duration (OD) for that earned value. This is shown clearly in Fig. 33.15. It can be seen that if the project is late, the result will be negative. There are therefore two schedule variances (SV):
   a. SV (cost), which is measured on the cost scale of the graph; and
   b. SV (time), which is measured on the time scale.

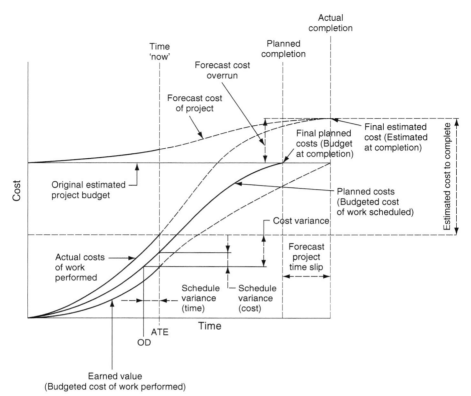

Figure 33.15 Earned value chart reproduced from BS 6079 'Guide to Project Management' by permission of British Standards Institution.

ECTC	is Estimated Cost To Complete
BAC	is Budget At Competition (current) (budget)
BCWS	is Budgeted Cost of Work Scheduled (current) (planned)
BCWP	is Budgeted Cost of Work Performed (earned value)
ACWP	is Actual Cost of Work Performed (actual)
OD	is Original Duration planned for the work to date
ATE	is the Actual Time Expended for the work to date
PTPT	is the planned Total Project Time
EAC	is Estimated Cost at Completion
ETPT	is Estimated Project Time
CPI	is Cost Performance Index = BCWP/ACWP = Efficiency
SPI	is Schedule Performance = BCWP/BCWS (cost based) = OD/ATE = % complete

Figure 33.16 Abbreviations used in EVA.

3. The cost efficiency is called the cost performance index (CPI) and is earned value/actual cost.
4. The 'time efficiency' is called the schedule performance index (SPI) and is earned value/scheduled (or planned) cost.

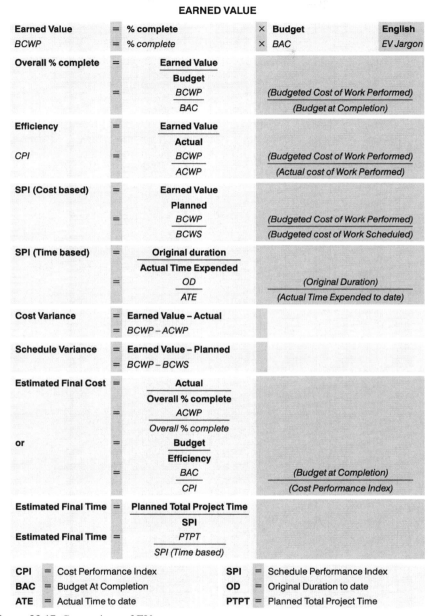

EARNED VALUE

Earned Value	=	% complete	×	Budget	**English**
BCWP	=	*% complete*	×	*BAC*	*EV Jargon*

Overall % complete	=	Earned Value		
		Budget		
	=	*BCWP*	*(Budgeted Cost of Work Performed)*	
		BAC	*(Budget at Completion)*	

Efficiency	=	Earned Value	
		Actual	
CPI	=	*BCWP*	*(Budgeted Cost of Work Performed)*
		ACWP	*(Actual cost of Work Performed)*

SPI (Cost based)	=	Earned Value	
		Planned	
	=	*BCWP*	*(Budgeted Cost of Work Performed)*
		BCWS	*(Budgeted cost of Work Scheduled)*

SPI (Time based)	=	Original duration	
		Actual Time Expended	
	=	*OD*	*(Original Duration)*
		ATE	*(Actual Time Expended to date)*

Cost Variance	=	Earned Value – Actual
	=	*BCWP – ACWP*

Schedule Variance	=	Earned Value – Planned
	=	*BCWP – BCWS*

Estimated Final Cost	=	Actual	
		Overall % complete	
	=	*ACWP*	
		Overall % complete	
or	=	Budget	
		Efficiency	
	=	*BAC*	*(Budget at Completion)*
		CPI	*(Cost Performance Index)*

Estimated Final Time	=	Planned Total Project Time
		SPI
Estimated Final Time	=	*PTPT*
		SPI (Time based)

CPI	=	Cost Performance Index	**SPI**	=	Schedule Performance Index
BAC	=	Budget At Completion	**OD**	=	Original Duration to date
ATE	=	Actual Time to date	**PTPT**	=	Planned Total Project Time

Figure 33.17 Comparison of EV terms.

Again, as with the schedule variance, measuring efficiency in time by dividing the earned value by the planned cost, which is measured in monetary terms or man-hours, is equally illogical and again it would be more sensible to use parameters measured in time.

Therefore, by dividing the planned duration for a particular earned value by the actual duration for that earned value, a more realistic index can be calculated. All time measurements must be in terms of hours, day numbers, week numbers, etc. − not calendar dates.

There are therefore now two SPIs:

a. SPI (cost) measured on the cost scale, that is, earned value/scheduled cost; and
b. SPI (time) measured on the time scale, that is, planned duration/actual duration.

In practice, the numerical difference between these two quotients is small, so that SPI (cost), which is easier to calculate, is sufficient for most purposes, bearing in mind that the result is still only a prediction based on historical data.

In 1996, the National Security Industrial Association of America published their own earned value management system, which dropped the terms such as actual cost of work performed, budgeted cost of work performed and budgeted cost of work scheduled (BCWS) used in CSCS, and adopted the simpler terms of earned value, actual and schedule instead.

Since then, the American Project Management Institute, the British Association for Project Management and the British Standards Institution have all discarded the CSCS abbreviations and also adopted the full English terms. In all probability, this will be the future universal terminology.

Fig. 33.17 clearly shows the earned value terms in both English (in **bold**) and EV jargon (in *italics*).

Earned Schedule

It has long been appreciated that schedule performance index (cost) (SPI_{cost}) based on the cost differences of the earned value and planned curves is somewhat illogical. An index reflecting schedule changes should be based on the time differences of a project. For this reason, schedule performance index (time) (SPI_{time}) is a more realistic approach and gives more accurate results, although in practice the numerical differences between SPI_{cost} and SPI_{time} are not very great. SPI_{time} for any point in time, or the current time, can be obtained by dropping a vertical line from the planned curve (BCWS) to the time baseline. This is time now (ATE). Next, dropping a vertical line from the point on the planned curve, where the planned value is equal to the earned value to the time baseline, gives the OD.

This duration from the start date to (OD) is referred to as 'Earned Schedule'

SPI_{time} is therefore OD/ATE, that is, the time efficiency. See Fig. 33.18.

Similar to budget cost/CPI = the final predicted cost, estimated duration/SPI_{time} = final completion time. It is important to remember that all units of durations on the time scale must be in day, week or month numbers, not in calendar dates.

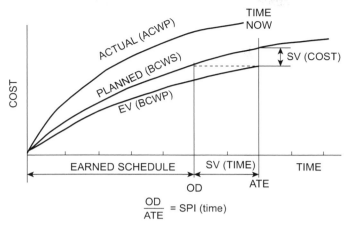

Figure 33.18 Control curves.

Integrated computer systems

Until 1992, the EVA system was run as a separate computer program in parallel with a conventional CPM system. Now, however, a number of software companies have produced project-management programs that fully integrate critical path analysis with EVA. One of the best programs of this type, Primavera P6, is fully described in Chapter 51.

The system can, of course, be used for controlling individual work packages, whether carried out by direct labour or by subcontractors, and by multiplying the total *actual* man-hours by the average labour rate, the cost to date is immediately available. The final results should be carefully analysed and can form an excellent base for future estimates.

As previously stated, apart from printing the EVA information and conventional CPM data, the program also produces a computer-drawn network. This is drawn in precedence format.

The information shown on the various reports include the following:

1. The man-hours spent on any activity or group of activities
2. The percent complete of any activity
3. The overall percent complete of the total project
4. The overall man-hours expended
5. The value (useful) hours expended
6. The efficiency of each activity
7. The overall efficiency
8. The estimated final hours for completion
9. The approximate completion date
10. The man-hours spent on extra work
11. The relationship between programme and progress
12. The relative performance of subcontractors or internal subareas of work

EVA % complete assessment

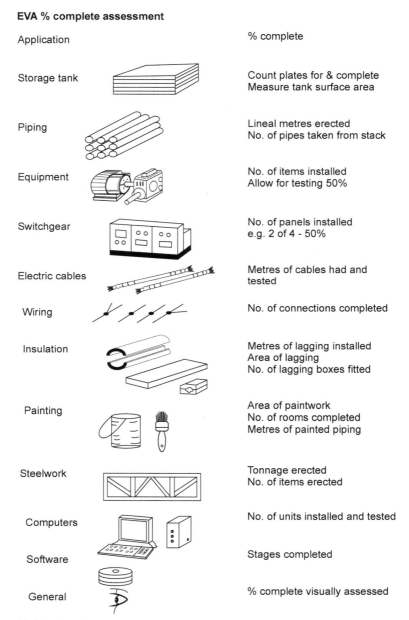

Application	% complete
Storage tank	Count plates for & complete Measure tank surface area
Piping	Lineal metres erected No. of pipes taken from stack
Equipment	No. of items installed Allow for testing 50%
Switchgear	No. of panels installed e.g. 2 of 4 - 50%
Electric cables	Metres of cables had and tested
Wiring	No. of connections completed
Insulation	Metres of lagging installed Area of lagging No. of lagging boxes fitted
Painting	Area of paintwork No. of rooms completed Metres of painted piping
Steelwork	Tonnage erected No. of items erected
Computers	No. of units installed and tested
Software	Stages completed
General	% complete visually assessed

Figure 33.19 EVA % complete assessment.

EVA % complete assessment

For updating purposes, it is necessary to assess the percent complete of each EVA-related activity on the schedule. This is in fact a measure of the work done to date and involves different units of measurement, depending on the type of work involved. For example, the unit for cables laid is the metres actually installed, while for paint-work, it is the area in m^2 of paint applied on date.

Fig. 33.19 gives an overview of some of the units used on a construction contract. This is only for assessment of percent complete. The EVA calculations must still be in man-hours or monetary units.

Procurement

34

Chapter outline

Project Management, Planning and Control. https://doi.org/10.1016/B978-0-12-824339-8.00034-1

Procurement is the term given to the process of acquiring goods or services.

The importance of procurement in a project can be appreciated by inspecting the pie chart (Fig. 34.1) from which it can be seen that for a typical capital project, procurement represents over 80% of the contract value.

The main functions involved in the procurement process are as follows:

1. Procurement strategy
2. Approved tender list
3. Pre-tender survey
4. Bidder selection
5. Request for quotation (RfQ)
6. Tender evaluation
7. Purchase order
8. Expediting, monitoring and inspection
9. Shipping and storage
10. Erection and installation
11. Commissioning and handover

These main functions contain a number of operations designed to ensure that the desired goods are correctly described and ordered, delivered when and where required and conform to the specified quality and performance criteria.

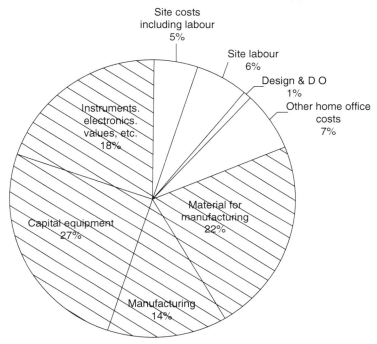

Figure 34.1 Total project investment breakdown. Procurement value $= 18 + 27 + 14 + 22 =$ 81%.

The functions are described further followed by a more detailed discussion of the types of contracts and supporting requirements.

Procurement strategy

Before any major purchasing operation is considered, a purchasing strategy must be drawn up which sets out the criteria to be followed. These include:

- The need and purpose of the proposed purchase.
- Should the items be bought or leased?
- Should the items be made in-house?
- Will there be a construction element involved? If yes, it will be necessary to draw up sub-contract documentation. If no construction or assembly element, a straight purchase order will suffice.
- How many companies will be invited to tender?
- Will there be open or selective tendering? This will often depend on the value and strategic or security restrictions. European Union regulations require contracts (subject to certain conditions) over a certain value to be opened up for bidding to competent suppliers in all member states.
- Are there prohibited areas for purchase or restrictive shipping conditions?

- What are the major risks associated with the purchase at every stage?
- What is the country of jurisdiction for settling disputes and what disputes procedures will be incorporated?
- What contract law is applicable, what language will be used, and in what currency will the goods be paid for?
- How will bids be opened and assessed and by whom? With public authorities tenders are usually opened by a committee.

Supply chain

Every purchaser must bear in mind that most items of equipment contain components that are outsourced by the manufacturer or subcontractor to specialist companies. These companies may themselves subcontract subcomponents to other specialists so that the original purchaser is utterly dependent on each subcontractor being technically and commercially rigorous and astute to ensure that the basic criteria of reliability, repeatability and sustainability are maintained through the whole supply chain.

The technical failure of any component, or the non-compliance with the relevant environmental and health and safety legislation, can have serious repercussions on the reputation and profitability of the purchaser. The problem has been exacerbated by the globalization of many manufacturing organizations where the inspection of the actual component or manufacturer's premises may be difficult and costly. A monitoring and reporting mechanism must therefore be set up right through the supply chain to ensure that technical and environmental standards are maintained.

This issue has to be confronted by every manufacturing industry whether it be construction, engineering, electronics, pharmaceuticals, textiles, toys, etc.

Approved tender list

When the procurement strategy has been agreed, a list of approved (or client-nominated) tenderers can be drawn up.

For many large infrastructure or power plant projects, it is beneficial to obtain the advice of specialist suppliers or contractors during the design stage. This will then inevitably lead to the possibility of reducing the number of competitive bids or even contravening (in Europe) EU procurement laws. In practice, many such projects will be constructed by consortia comprising civil, mechanical and electrical contractors and manufacturers, so that this risk has been transferred from the client/operator to the consortium.

Most major companies operate a register of the vendors who have carried out work on earlier contracts and have reached the required level of performance. These are generally referred to as 'approved vendor lists' and should only contain the names of those vendors who have been surveyed and whose capacities have been clearly established.

The use of computers enables such lists to be very sophisticated and almost certainly to contain information on company capacity and performance together with the details of all the work the contractor has carried out and the level of his performance in each case.

Lists are normally prepared on a commodity basis, but they are only of real use if they are constantly updated.

A typical entry against a vendor on such a list includes the following:

- Company name and address
- Telephone, telex and telefax no.
- Company annual turnover
- List of main products
- Value of last order placed
- Performance rating on:
  - ○ Adherence to price
  - ○ Adherence to delivery
  - ○ Adherence to quality requirements
- Ability to provide documents on time
- Cooperation during design stage
- Responses to emergency situations

Many organizations also keep a second list, comprising companies that have expressed their wish to be considered for certain areas of work, but are not on the approved vendors' list. When the opportunity arises for such companies to be considered, they should be sent the pre-qualification questionnaires shown in Fig. 34.2 and, if necessary, be visited by an inspector.

In this way, the list will be a dynamic document onto which new companies are added and from which unsatisfactory companies are removed.

Pre-tender survey

The approved vendor list is a useful tool, but it cannot be so comprehensive as to cover every commodity and service in every country of the world. For new commodities or new markets, therefore, a pre-tender survey is necessary. This is particularly important when a contractor is about to undertake work in a new country with whose laws and business practices he is not familiar.

The summary should be as comprehensive as possible and the buyer should draw information from every available source. The subject is discussed in more detail later in this chapter under 'Overseas bidder selection', but the principles are equally applicable to surveys carried out in the United Kingdom or United States.

Bidder selection

Before an enquiry can be issued, a list of bidders has to be compiled. Although the names of qualified companies will almost certainly be taken from the purchaser's approved vendor list, the actual selection of the prospective bidder for a particular enquiry requires careful consideration.

FOSTER WHEELER POWER PRODUCTS LTD
P.O. Box 160, Greater London House, Hampstead Road, London NW1 7QN

PRE-QUALIFICATION QUESTIONNAIRE

This questionnaire shall be completed by Vendors of direct materials.

1.0 General

1.1 Full legal Company name ....

1.2 Legal status of Vendor - whether a Private or Public Limited Liability Company, Partnership, Consortium, etc.
....

1.3 Registered office and/or other legal address ....
....
....
Telephone No: ....
Telex No: .... Fax No: ....
Telegraphic Address ....
Contact (English speaking) ....
Position Held by Contact ....
Vendor Type Manufacture/Distributor/Stockist/Agent

1.4 Full address of all Branch Offices ....
....
....
....

1.5 Full name & address of all Subsidiary & Associate Companies
....
....
....

1.6 Name & address of Parent or Holding Company ....
....
Telephone No:.................... Telex ....
Contact .... Position ....
Annual Turnover ....

Where there is insufficient space for a full reply to a question Please attach additional sheets.

PRE-QUALIFICATION QUESTIONNAIRE

a) Levels of turnover of Company for the last four financial years:-

198_
198_
198_
198_

3.2 Total value of orders waiting manufacture or in progress, in equivalent pounds sterling:

3.3 Companies expecting to supply goods totalling more than £25,000 sterling shall provide evidence of their financial standing. This shall include a statement of the situation for the current year to date (unaudited) and of the two previous years.

Please provide the following information:

a) Are shares quoted on Public Stock Exchange .... If so at which Exchange ....

b) Copy of Accounts of Company and of Group.

c) Liquidity Ratio i.e. Trade amounts to available cash.

d) Average Working Capital of Company and of Group.

3.4 Bank References

4.0 Work History

Client	Goods Supplied	Value	Date of Delivery

5.0 Number of Employees - Company Only

a) Administrative Staff -

b) Graduate Engineers -

c) Draughting Personnel -

d) Quality Assurance Personnel -

e) Inspection Personnel -

f) Procurement Personnel -

g) Qualified Welders -

h) Machine Shop Personnel -

i) General Works Staff -

6.0 Comments

Date Completed .... By ....

Signed ....

PRE-QUALIFICATION QUESTIONNAIRE

2.0 Description of Goods Supplied

2.1 Type of Goods & Ranges ....
....
....
....
....
....
....
....
....

2.2 Location & capacity of manufacturing & servicing facilities
....
....
....
....
....

2.3 List of design codes to which previous work has been produced
....
....

2.4 Licenses & monograms held ....
....

2.5 Test Facilities ....
....
....

2.6 Details of components normally sub-contracted to others
....
....
....

2.7 Quality Assurance Details in accordance with attached Quality Assurance Approval letter (2 sheets) and form A.

3.0 Financial

3.1 Levels of turnover of Vendor in equivalent pounds sterling:

PRE-QUALIFICATION QUESTIONNAIRE (REVIEW)   FWPP Ref:
(NOT TO BE ISSUED TO VENDOR)

Vendor Name ....
Location ....
Date Questionnaire sent to Vendor ....
              returned   ....

Recommendation of Review:

Additional Contacts with Vendor:

Name 1) .... Position 1) ....
Name 2) .... Position 2) ....

Reviewed by Project Procurement Manager & Manager of Quality Assurance or his nominee

Recommendation

1) Enter Vendor on FWPP list   YES/NO
2) Reject   YES/NO

Project Procurement Manager   Signature ....
                                  Date   ....

Manager of Quality Assurance or nominee

                                  Signature ....
                                  Date   ....

Recommendation Agreed by Procurement Manager   YES/NO

                                  Signature ....
                                  Date   ....

Figure 34.2

During the preparation stage of the equipment requisition, a number of suitable companies may well have been suggested to the buyer by the project manager, the engineering department, the construction department and, of course, the client. While all this, often unsolicited, advice must be given serious consideration, the final choice is the responsibility of the procurement manager. In most cases, however, the client's and project manager's suggestions will be included.

The number of companies invited to bid depends on the product, the market conditions and the 'imposed' names by the client. In no case, however, should the number of invited bidders be less than 3, or more than 10. In practice, a bid list of six companies gives a reasonable spread of prices, but if too many of the invited companies refuse to bid, further names can be added to ensure a return of at least three valid tenders.

To ensure that the minimum number of good bids is obtained, some purchasers tend to favour a long bidder's list. This practice is costly and time consuming, especially if the support documentation, such as specifications and drawings, is voluminous. A far better method is to telephone prospective bidders before the enquiry is sent out and, after describing the bid package in broad terms, obtain the assurance from the companies in question that they will indeed submit a bid when the documents are received. In this way, the list can be kept to a reasonable size. If a vendor subsequently refuses to bid, he runs the risk of being crossed off the next bid list.

While the approved vendor list is an excellent starting point for bidder selection, further research is necessary to ensure that the selected companies are able to meet the quality and programme requirements. The prospective vendor must be experienced in manufacturing the item in question, have the capacity to produce the quantity in time, be financially stable and meet all the necessary technical and contractual requirements.

The buyer who keeps in touch with the market conditions can often obtain useful information from colleagues in other companies, from technical journals and even the daily press on the workload or financial status of a particular supplier. An announcement in the financial columns that a company has obtained an order for 10,000 pumps to be delivered over 2 years should trigger off an investigation into the company's total capacity, before it is asked to bid for more pumps. Loading up a good supplier until he is overstretched and becomes a bad supplier is not in the best interest of either party.

When a purchaser lets it be known that a certain enquiry will be issued, the buyer will undoubtedly be contacted by many companies eager to be invited to bid. The information given by the salesman of one company must be compared with the, often contradictory, information from the opposition before a realistic decision can be taken. Only by being familiar with the product and the market forces at the time of the enquiry, can the buyer make objective judgements. On no account should vendors be included on a bid list to keep salesmen at bay or just to make up numbers.

For items not usually ordered by a purchaser, a different approach is necessary. In all probability, no approved vendors list exists for such a product and unless someone in the procurement department, or in one of the other technical departments, knows of suitable suppliers, a certain amount of research has to be carried out by the buyer.

The most obvious sources of suitable vendors are the various technical trade directories and technical indexes available on microfilm or microfiche. In addition, a telephone call to the relevant trade association will often yield a good crop of prospective suppliers. Whatever the source, and whoever is chosen for the preliminary list, it will be prudent to send a questionnaire to all the selected companies, requesting the following details:

- Full company name and address (postal and website)
- Telephone, fax no. and e-mail address
- Company annual turnover
- Name of bank
- Names of three references
- Confirmation that the specified product can be supplied
- What proportion will be subcontracted
- Name of parent company — if any

It is clear from the above that these questions must be asked well in advance of the enquiry date. It is necessary, therefore, for the buyer to scan the requisition schedule as soon as it is issued to extract such materials and items for which no vendors list has been compiled. In other words, the buyer will have to perform one of the most important functions of project management — forward planning.

Request for quotation

The procurement manager will first of all:

- Produce a final bid list of competent tenderers
- Decide on the minimum and maximum number of bids to be invited
- Ensure that the specification and technical description of goods or services to be purchased are complete
- Decide on the type of delivery and insurances required (ex-works, FAS, FOB, etc.)
- Agree to the programme with delivery dates for information and goods with the project manager
- Decide on the most appropriate general conditions of purchase or contract
- Draft the special conditions of contract (if required)
- Decide the type and number of document requirements (manuals)
- Agree with project manager on the amount of liquidated damages required against late delivery
- Draft and issue the RfQ letter
- Enclose with the RfQ the list of drawings and other data required for submitting bids

It is important to include any special requirements over and above the usual conditions of the contract with the enquiry documents. Failure to do so might generate claims for extras once the contract is underway. Such additional requirements may include special delivery needs and restrictions, access restrictions, monitoring procedures such as EVA, inspection facilities and currency restrictions, etc. Discounts and bonus payments are best discussed at the interviews with the preferred supplier/ contractor after the tenders have been opened.

The following information must be requested *from* tenderers in the RfQ:

- Price and delivery as specified
- Discounts and terms of payment required
- Date of issue of advance date technical data, layout drawings, setting plans, etc.
- Production and delivery programme
- Expediting schedule for suborders
- Spares lists and quotation for spares
- Guarantees and warranties offered
- Alternative proposal for possible consideration

Tender evaluation

The procurement manager must next:

- Decide on bid opening procedure and open bids
- Receive and log the bids as they arrive
- Assess previous experience and check reference contracts or sites
- Check financial stability of bidders
- Check list of past clients
- Set up bidders meetings
- Interview site or installation manager or foreman (if applicable)
- Discuss quantity discounts
- Negotiate early or other payment discounts
- Obtain parent company guarantees when the contract is with a subsidiary company
- Agree the maintenance (guarantee) period
- Discuss and agree bid, performance, maintenance and advance payment bonds
- Produce bid summary (bid tabulation) (Fig. 34.3)
- Carry out technical evaluation
- Carry out commercial evaluation

The following are the main items that have to be compared when assessing competing bids:

- Basic cost
- Extras
- Delivery and shipping cost
- Insurance
- Cost of testing and inspection
- Cost of documentation
- Cost of recommended spares
- Discounts
- Delivery period
- Terms of payment
- Retentions guarantees
- Compliance with purchase conditions

Figure 34.3 Bid summary.

All the vendors' prices must, of course, be compared with the estimator's budgets, which will also appear on the sheet.

If the value of the bids (or at least one bid) is within the budget, the bid evaluations can proceed as described later. If, on the other hand, all the bids are higher than the budget, a meeting has to be convened with the project manager at which one of the four decisions has to be taken, depending largely on the overall project programme:

1. If there is time, the enquiry can be reissued to a new list of vendors, spreading the area to include overseas suppliers, provided foreign suppliers are not excluded by the terms of the contract.
2. Review the original design or design standards to see whether savings can be made. While such savings may be difficult to make with equipment items, which have to meet specified technical requirements, it may be possible to effect cost savings in the type of finish, or the materials of construction. For example, a tank that was originally specified as galvanized may be acceptable with a good paint finish.
3. If there is no time to reissue the inquiry, it may be necessary to call in the two lowest bidders and either negotiate their prices down to the budget level or ask them to rebid within a few days on the basis of a few quickly ascertainable design changes.
4. If it is discovered that the original budget estimate was too low, the lowest bid (if technically and commercially acceptable) will have to be submitted for approval by the project manager.

These commercial comparisons must be carried out for every enquiry. However, an additional technical assessment must be produced when the material and equipment is other than a general commodity item, which only has to comply with a material specification.

Although the purpose of this evaluation process is to find the cheapest bidder who complies with the specification and contractual requirements, the technical capability of the bidder as well as their track record of timely completion and past relationships (good or bad) with other stakeholders should also be taken into consideration. This may of course be a problem in itself with government contracts, where there can always be accusations of favouritism or even nepotism if the least expensive bidder has not been selected.

Purchase order

When the selected bid has been agreed, the purchase order or contract document must be issued with all the same attachments, which were part of the RfQ. The only additional documents are those containing the terms and conditions agreed at the bidder's meeting or other written agreements made during the bid evaluation process. These will include the following:

- The procedures for payments as set out in contract
- The stages for issuing interim and final acceptance certificates

Any changes to the specification or drawings, etc. after the date of the purchase order will be issued as a variation order and controlled using established configuration management procedures.

As the contract proceeds, the procurement manager must:

- Set priorities from programme and revisions to programme, if necessary
- Set out bonded areas and marking when advance payments are considered
- Arrange to carry out regular expediting and check expediting reports
- Carry out stage inspection and specified tests, and check inspection and test reports
- Carry out route survey (if required)
- Issue final delivery instruction for documents and goods
- Issue packing instructions
- Issue shipping information, sailing times
- Advise on shipping restrictions (conference lines requirements, etc.)
- Advise on current site conditions for storage
- Gree and finalize close-out procedures
- Obtain details of after sales service

Expediting, monitoring and inspection

Between the time of issuing the purchase order or contract documents and the actual delivery of the goods or services, the progress of the production (manufacture) of the goods must be monitored if quality and delivery dates are to be maintained.

The contract documentation will have (or should have) included the appropriate portion of the overall project programme. From this, the contractor or supplier will be required to produce his own construction or manufacturing programme. This should be issued to the main contractor or client within two or three weeks of receipt of order and can be used to monitor progress.

It is beneficial if an expediter or inspector visits the works or offices of the supplier within 2 weeks of contract award to ensure that:

1. The contract documents have been received and understood
2. The contractor's or supplier's own programme has been started or is ready to be issued
3. The supplier or contractor has all the information he or she needs

At regular intervals an expediter will then visit the supplier and check that the supplier's own programme is being met and, in particular, that sub-orders for materials and components have been placed, bearing in mind the lead times for these items. Any slippage to either the purchaser's or supplier's programme must be reported immediately to the project manager so that appropriate action can be taken. Where time is of the essence, a number of options are available to bring the delivery of the goods back on schedule. These are as follows:

1. The purchaser's programme may have sufficient float (permissible delay) in the delivery string, which may make further action unnecessary, but pressure to deliver to the revised date must now be applied.
2. If there is no float available, the supplier must be urged to work overtime and/or weekends to speed up manufacture. It may be worthwhile for the purchaser to pay for these premium

man-hours in full or in part, as even if liquidated damages are imposed and obtained, the financial cost of delay, in addition to loss of prestige or reputation, is often very much greater than the value of any liquidated damages received.

3. If there is no liquidated damages clause in the purchase agreement or contract but delivery by a stated date is a fundamental requirement, the purchaser may threaten the supplier with legal proceedings, breach of contract or any other device, but while this may punish the supplier, it will not deliver the goods. For this reason, early warning of slippage is essential.

Because of the danger from the imposition of damages at large, due to possible delayed delivery, prudent suppliers, and especially contractors, should request that a liquidated damages clause be inserted into the contract, even if this has not been originally provided.

Similarly, an inspector will visit the supplier early on to ensure that the necessary materials and certificates of conformity are either on order or available for inspection.

This expediting and inspection process should continue until all the items are delivered or the specified services are commenced.

Shipping and storage

To ensure that the materials or equipment are delivered on time and at the correct location, checks should be carried out to ensure that, in the case of overseas procurement, there are no shipping, landing or customs problems. Overland deliveries of large components may require a route survey to ensure that bridges are high or strong enough, roads wide enough and the necessary local authority permits and police escorts are in place.

Special attention must be given to contracts when dealing with the construction industry. Here, the timing of deliveries has to be carefully calculated to ensure that there is sufficient unloading and storage space, adequate cranage is available, there is no interference with other site users and that adequate temporary protective materials are ready.

In the case of urban construction, interference with traffic or the general public must be minimized.

Erection and installation

Once on site, fabrication, erection and installation work should be regularly monitored using earned value techniques, which are related to the construction programme. Only in this way can the efficiency of a subcontractor be assessed and predictions as to final cost and completion can be made.

As the cost of delays to completion can be many times greater than the losses to production or usage of the facility being constructed, the imposition of liquidated damages (even if possible) will not recover the losses and are at best a deterrent. A far better safeguard against late completion is the insistence that realistic and updated construction programmes are produced, regularly (preferably weekly) monitored and, with the aid of network analysis and EVA techniques, kept on target. Although this may require additional resources and incentives, it may well be the best option.

As most construction contracts will have been placed on the basis of an agreed set of conditions of contracts (whether general or special), it will be necessary to check that the conditions are fully met, in particular those relating to quality, stage completion and health and safety. It is also necessary to ensure that adequate drawings and commissioning procedures are ready when that stage of the contract commences.

Throughout the manufacturing or construction process, documents such as operating and maintenance procedures, spares lists and as-built drawings must be collated and indexed to enable them to be handed over in a complete state when the official handover takes place.

Commissioning and handover

Before a plant is handed over for operation, it must be checked and tested. This process is called commissioning and involves both the contractor and operator of the completed facility.

Careful planning is necessary, especially if the plant is part of an existing facility that has to be kept fully operational with minimal disruption during the commissioning process. Integration with existing systems (especially when computers are involved) has to be seamless, and this may require the existing and new systems to be operated concurrently until all the teething problems have been resolved.

In certain types of plants, equipment will have to be operated 'cold', that is, without the operating fluids (gases, liquids or even solids) being passed through the system. Only when this stage has been successfully completed can 'hot' commissioning commence with the required media being processed in the final operating condition (temperature, viscosity, pressure voltage, etc.) and at the specified rates of usage (flow, wattage, velocity, etc.).

It is often convenient to involve the operating personnel in the commissioning process so that they can become familiar with the new systems and operating procedures.

After the various tests and pilot runs have been completed and the operating criteria and KPIs have been met to the satisfaction of the client, the new facility can be formally handed over for operation. This involves handing over all the stipulated documents, such as built drawings, operating and maintenance instructions, lubrication schedules and spares lists, as well as commissioning records, test certificates of equipment, materials and operators, certificates of origins and compliance, guarantees and warranties.

A close-out report will have to be written, which records the major events and problems encountered during the manufacturing, construction and commissioning stages. This will be indexed and filed to enable future project managers to learn from past experience.

Types of contracts

Contracts consist of three main types:

1. Lump sum contracts
2. Remeasured contracts
3. Reimbursable contracts

It is possible to have a subcontract that is a combination of two or even all three types. For example, a mechanical erection subcontract could have design content that is a lump sum fixed fee, a remeasured piping erection portion and a reimbursable section for heavy lifts in which the client wants to be deeply involved. The main differences between the three types are as follows.

Lump sum contracts

A prerequisite to a lump sum (a fixed price) contract, with or without an escalation clause, is a complete set of specifications and construction drawings. These documents will enable the subcontractor to obtain a clear picture of the works, assess the scope and quality requirements and produce a tender that is not inflated by unnecessary risk allowances or hedged with numerous qualifications.

The trend in the United States is to have all the designs and drawings complete before tenders are invited and, provided such usual risk items as sub-soil, climatic and seismic conditions are clearly given, a good competitive price can be expected.

Most lump sum contract documents should, however, contain a schedule of rates, so that variations can be quickly and amicably costed and agreed. Clearly, these variations should be kept to a minimum and must not exceed reasonable limits, since the rates used by the subcontractor are based on the tender drawings and quantities and a major change could affect his man-hour distribution, supervision level and site organization. A common rough limit accepted as reasonable is a value of variations of 15% of the contract value.

It must be stated that a variation can be a decrease as well as an increase in scope and although the quantities may be reduced by the client, the reduction in price will not be proportional to that reduction. Indeed, when a subcontract includes a design element, a reduction in hardware (say the elimination of a small pump), may increase the contract value due to costly drawing changes and cancellation charges.

Remeasured contracts

Most civil engineering subcontracts in the United Kingdom are let as remeasured contracts. In other words, the work is measured and costed (usually monthly) as it is performed in accordance with a priced bill of quantities agreed between the purchaser and the subcontractor. The documents that are required for the tender are the following:

1. Specifications
2. General arrangement drawings
3. Bills of quantities

The bills of quantities are usually prepared by a quantity surveyor employed by the purchaser and are in fact only approximate, as the only drawings available for producing the bills are the general arrangement drawings and a few details or sketches prepared by the designers. Obviously, the more details available, the more accurate the bills of quantities are, but as one of the objectives of this type of contract is to invite tenders as quickly as possible after the basic design stage, full details are rarely available for the quantity surveyor. The subcontractor prices the items in the bills of quantities, taking into account the information given in the drawings, the specification, the preambles in the bills and, of course, the location and conditions of the proposed site.

Although the items in the bills of quantities are often described in great detail, they are included only for costing purposes and do not constitute a specification. In the same way, the quantities given in the bill are for costing purposes only and are no guarantee that they will be the actual quantities required.

As with lump sum contracts, variations, which inevitably occur, must not exceed reasonable limits in either direction, as these could invalidate the unit rates inserted by the subcontractor. For example, if the bills of quantities call for 10,000 m^3 of excavation of a depth of 2 m, and it subsequently transpires that only 1000 m^3 have to be excavated, the subcontractor is entitled to demand a rerating of this item on the grounds that a different excavator would have to be employed, which costs more per cubic metre than the machine envisaged at the time of tender.

In remeasured subcontracts, it is a common practice to carry out a monthly valuation on site as a basis for progress payments to the subcontractor. These valuations consist of three parts:

1. Value of materials on site, but not yet incorporated in the works
2. Value of work executed and measured in accordance with the method of measurement stated in the bills of quantities
3. Assessed value of preliminary items and provisional sums set out in the bills

The value of materials, which have been paid for in a previous month (when they were delivered, but not yet incorporated in the works), is deducted from the measured works in the subsequent month (by which time they were incorporated) since the billed rates will include the cost of materials as well as labour and plant.

At the end of the contract period, the final account will require a complete reconciliation of the cost and values, so that any overpayments or underpayments will be balanced out.

Reimbursable contracts

When a client (or purchaser) wishes to place a contract as early as possible, but is not in a position to supply adequate drawings or specifications, or when the scope has not been fully defined, a reimbursable contract is the most convenient vehicle.

In its simplest form, the contractor (or subcontractor) will supply all the materials, equipment, plant and labour as and when required, and will invoice the client at cost, plus an agreed percentage to cover overheads and profit. To ensure 'fair play', the client has the right to audit the contractor's books, check his invoices and labour returns, etc., but he has little control over his method of working or efficiency. Indeed, since the contractor will earn a percentage on every hour worked, he has little incentive in either minimizing his man-hour expenditure or finishing the job early.

To overcome the obvious deficiencies of a straight reimbursable (or cost-plus) contract, a number of variations have been devised over the years to give the contractor an incentive to be efficient and/or finish the job on time.

Most cost reimbursable contracts have two main components:

1. A fee component that can cover design costs, site and project management costs, overheads and profits
2. A prime cost component covering equipment, materials, consumables, plant, site labour, subcontracts and site establishment

In some cases, the site establishment may be in the fee component or the design costs may be in the prime cost portion. The very flexibility of a reimbursable contract permits the most convenient permutation to be adopted.

By agreeing to have the fee portion 'fixed', the contractor has an incentive to finish the job as quickly as possible, as his fee and profits are recoverable over a shorter period and he can then release his resources for another contract. Furthermore, as he only recovers his prime cost expenditure at cost, he has absolutely no advantage in extending the contract — indeed, his reputation will hardly be enhanced if he finishes late and costs the client more money.

If the scope of work is increased by the client, the contractor will usually be entitled to an increase of the fixed fee by an agreed percentage, but frequently such an increase only comes into effect if the scope charge exceeds by 10% of the original contract value.

The factors to be considered when deciding on the constitution of the fixed fee and prime cost components are as follows:

1. Time available for design
2. Extent of the client's involvement in planning and design
3. Extent of the client's involvement in site supervision and inspection
4. Need to permit operations of adjacent premises to continue during construction. This is particularly important in extensions for factories, hotels, hospitals or process plants
5. Financial interest of the client in the contractor's business or vice versa
6. Location of site in relation to the main area of equipment manufacture
7. Importance of finishing by a specified date, for example, weather windows for offshore operations or committed sales of the product
8. Method of sharing savings or other incentives; it can be seen that if both parties are in agreement a contract can be tailored to suit the specific requirements, but the client still has only an estimate of the final cost

Target contracts

To counter some of the disadvantages of straight reimbursable contracts, target contracts have been devised. In these contracts, an estimate of the works has been prepared by the employer (or his consultants) and agreed with a selected contractor. The prime cost is then frozen as a target cost the contractor must not exceed. The fee component is fixed, but it can also be calculated to be variable in such a way that the contractor has an incentive to complete the works below target or on programme or both.

Again, there are numerous variations of this theme, but the following are the more common methods of operation:

1. If the final measured prime costs are less than the target value, the difference is shared between the parties in a previously agreed proportion. If the final costs exceed the target value, the contractor pays the difference in full. The fee portion remains fixed.
2. As in 1, but the fee portion increases by an agreed percentage as the prime cost portion decreases. This gives the contractor a double incentive to complete the contract below the target value and thus increase the fee and ensure savings.
3. As in 1, but the fee portion increases by an agreed percentage for every week of completion before the contractual completion date.
4. The employer pays the final prime cost or the target cost, whichever is lower, but the difference, if any, is not shared with the contractor. The fee, however, is increased if there is a saving of prime cost or time or both.

If the contractor is responsible for purchasing equipment as part of the prime cost portion, the procurement costs such as purchasing, expediting and inspection may be reimbursable at an hourly rate (subject to an annual review), but all discounts, including bulk and prompt payment discounts, must be credited to the employer. The contractor still has an incentive to obtain the best possible prices, since the prime cost will be lower.

Design, build and operate contracts

These types of contracts have been developed since 1992 to reduce the financial burden on the public purse. This process is known as private public partnership (PPP). A subset of PPP is private finance initiative (PFI). These new types of contracts have a number of variations as described later, but their main purpose is for the private sector to take most, if not all, of the financial risk and employ their specialist knowledge and commercial experience, not always available in the public sector. Generally, a special service company, known as a special service vehicle is formed, consisting of the contractors (building, maintenance and operating) and the financiers (bank or other finance house) to construct, maintain and operate the asset for a contractually agreed period. This company then signs the contract with the public authority (central or local) and arranges any leases with the actual operating organization, such as a prison or hospital.

The big difference between PFI and more conventional contracts is that the contractor finances the whole project from his own resources, and then recoups the cost and profit from the operation revenues or from levies charged on the public sector organization that awarded the contract. The following three examples explain the different types.

Free standing

A typical example of this type of PFI contract is where the public authority drafts a performance specification for a new bridge and asks a number of large contractors to submit costed schemes. The successful contractor then designs, builds and maintains the bridge for a specified period. During this operating period, the cost and profit are recovered by the contractor/operator from tolls charged for the crossing. The contractor carries the risk that the volume of traffic will not reach the anticipated levels to yield the required revenues, possibly because the end user, the general public, considers the tolls to be too high.

Levies on the public sector

An example of such a contract is where the government requires a new prison. The successful contractor designs and builds the prison to the client's specification and operates it in accordance with the standards laid down by the prison service. As the end users, the prisoners, can hardly be expected to pay rent, an operating levy is charged on the prison service to cover building, financing and operating costs plus profit.

Joint ventures

Joint venture PFI projects are often used for road construction where there are no toll charges on the road users. Here, the design and construction costs are shared by an agreed amount by the public and private sector, and the contractor recovers his costs by a fee based on a benefit/revenue formula agreed in advance.

In all these types of contracts, the contractor is not chosen on the basis of the cheapest price, but on the viability of his business case, design concept, experience, technical expertise, track record and financial backing. In many cases, the contractor leads a consortium of specialist contractors, design consultants and financial institutions.

Basic requirements for success

Whichever formula is agreed upon, it is clear that in any reimbursable contract, the employer retains a measure of control and hence a considerable responsibility. If the employer is also involved in the design process, the release of process information, construction drawings and operating procedures must form part of the programme and should be marked as a series of key dates that must be kept if adherence to the programme is imperative.

The success of a contract (or subcontract) depends, however, on more factors than a well-drawn-up set of contract conditions, and the following points must not be overlooked:

1. Good cost control of prime cost items
2. Good site management
3. Careful planning and programming
4. Punctual release of information to site
5. Timely deliveries of equipment and materials
6. Elimination of late design changes
7. Good labour relations

8. Good relationship between contracting parties

The main types of contracts are summarized in Fig. 34.9.

Bonds

A bond is a guarantee given by third party, usually a bank or insurance company, that specified payments will be made by the supplier or contractor to the client if certain stipulated requirements have not been met. There are four main types of bonds a client may require to be lodged before a contract is signed. These are shown in order of submission:

1. Bid bond
2. Advance payment bond
3. Performance bond
4. Retention bond

Any of these bonds can be either conditional or on-demand. Conditional bonds, which are usually issued by an insurance company or similar financial institution, carry a single charge that is independent of the time the bond is in force and can only be called if certain predetermined conditions have been met. Such a condition might include that the supplier or contractor has to agree that the bond is called (i.e. that the money is paid to the client), or that the client or purchaser must prove loss to the satisfaction of the issuing house due to the default of the supplier. While such a bond may be very advantageous to the supplier, it is often regarded as unacceptable to a purchaser, as the collection and submission of evidence of default or proof of loss can be a time-consuming business.

The on-demand bond, on the other hand, has no such restrictions. As the name implies, it enables the purchaser to call in the bond as soon as and when he or she believes that a default by the supplier has occurred. Such a bond is normally issued by a recognized bank and will be paid without question and without the need for justification as soon as the demand for payment is made by the purchaser. Clearly, the main element of such a bond is trust. Both the bank and the supplier trust the purchaser to be reasonable and honourable not to call the bond until the contractual terms permit it. These bonds cost more than a conditional bond and are only for a fixed duration, usually a particular stage of the project. They can, however, be extended for a further period for an additional fee (see Fig. 34.4).

Apart from the benefit of speedier payment should there be a default by the supplier, another advantage to the purchaser of such a bond is that, as the cost of the bond depends on the bank's perception of the risk and the supplier's financial rating, a measure of the supplier's standing can be obtained. A low bond fee usually means that the supplier is regarded by the bank as reliable and financially stable.

Bid bond

In major contracts, many overseas clients require a bid bond to be submitted with the tender documents. The purpose of this bond, which is usually an on-demand bond, is to

Foster Wheeler Power Products Limited.,
P.O. Box 160,
Greater London House,
Hampstead Road,
LONDON   NW1 7QN

BANK GUARANTEE DRAFT

We understand that you have entered into contract number ....

with .... for ....

at an agreed price of ....

In this connection, we hereby give you our guarantee in the sum of ....

.... such guarantee being effective in the

event that .... fail in their contractual

obligations in respect of this contract.

Claims under this guarantee are to be received by ....

at .... no later than ....

Such claims will be payable by us upon your first demand accompanied by your

statement that .... have

failed in their contractual obligations in respect of the contract specified

above.

Authorised signatories of Bank.

Figure 34.4 Bank guarantee draft.

discourage the tenderer from withdrawing his bid after submission. This can be of considerable potential danger to a tenderer who discovers, after the bids have been dispatched, that there was an error in his tender price or that other contractual requirements have been overlooked. Unless his price was originally higher than that of his competitors, the unfortunate tenderer has to decide whether to proceed with a potentially loss-making contract or to forfeit his bond.

The client will undoubtedly argue that the main purpose of the bond is to eliminate frivolous bids and ensure that those bids submitted are not only serious but also firm.

However, there can be considerable financial disadvantages to a tenderer since a bid bond, if issued by his bank, is equivalent to an overdraft, so that the working capital can be greatly reduced for a considerable period. When one considers that it can take between 3 months and a year to know whether a large contract has been won or lost, the loss in financing facilities and interest charges for the bond can be so great as to deter all but the largest contractors from tendering.

Advance payment bond

There are circumstances when a seller requires payments to be made before the goods are delivered. This arrangement is frequently required to finance expensive raw materials. The purchaser may also wish to make advance payments to reserve a place in the manufacturing queue or, as in the case of public authorities, to meet an expenditure deadline.

Until the goods are delivered, however, the purchaser has little or no guarantee that the advance payments will not be completely lost, should the supplier go into liquidation or the directors disappear to South America. To eliminate this risk, the purchaser requires the supplier to deposit with him a bond, usually underwritten by a bank, which guarantees a refund should any of the above misfortunes of the above type occur. The bond usually has a time limit that is often geared to a physical stage of the contract, such as the receipt by the purchaser of preliminary drawings or the arrival of raw materials. The latter stage is often accompanied by a certificate of ownership which vests the proprietorial rights with the purchaser.

Such a certificate is often supplemented by labels, which are affixed to the equipment or materials, declaring that the items marked are the property of the purchaser. This enables the purchaser to recover his goods (for which he has after all paid) should the vendor go into liquidation. The wording of such a notice should be vetted by the purchaser's legal advisers to ensure that the goods can, indeed, be recovered without further court action.

Where bulk materials have to be protected in this way, it is usual to fence off a 'bonded' area and erect notice boards at a number of locations. A typical notice of transfer of ownership is shown in Fig. 34.5.

While an advance payment bond will usually be required for progress payments for work carried out off site, it is not normally required for work on site, as the completed works are the immediate property of the purchaser and could be finished by another subcontractor in the case of bankruptcy or default (see Fig. 34.6).

Performance bond

This type of bond is more usually associated with subcontracts and is an underwritten guarantee by a bank or other financial institution that the subcontractor will perform his contract and complete the work as specified.

To be typed on Company Headed Paper

TRANSFER OF OWNERSHIP CERTIFICATE

In pursuance of invoice(s)   , dated   presented
by (insert full Company Name & Address) in respect of Foster Wheeler Power
Products Limited Purchase Order No: dated   and in
accordance with the terms and conditions thereof it is hereby warranted that
the materials, components and equipment used and all employees, agents,
sub-contractors and sub-suppliers employed by (insert full Company Name) for
the purposes of manufacturing and delivering the Goods and/or rendering the
services have been properly and fully paid in respect of the Goods listed
below whether completed or otherwise and are marked, identified and set aside
to become the absolute property of Foster Wheeler Power Products Limited
wherever they may be situated. Transfer of ownership of the Goods set aside
as aforesaid under the above numbered Purchase Order shall become effective to
the extent payment(s) are made in respect of the Goods listed below.

(Insert full Company Name) hereby agrees that the transfer of ownership shall
be without prejudice to any other warranty provided in accordance with the
Purchase Order.

List of Goods

. . . .   (To be signed by a Director of
                                                            your Company).

                                                            _____

Figure 34.5 Transfer of ownership certificate.

Even if the subcontractor is paid by progress payments, the purchaser may still suffer considerable loss and frustration if the works are not completed due to the subcontractor withdrawing from the site.

The performance bond should be of sufficient value to cover the cost of finding and negotiating with a new subcontractor and paying for the additional costs that the new subcontractor may incur. There may, of course, be the additional costs of delays in completing the project, which are often far greater than the difference in the price of two subcontractors.

Usually, the value of a performance bond is between 2.5% and 5% of the contract value, which covers most contingencies.

TO BE ON BANK LETTERHEADED PAPER

ADVANCE PAYMENT BOND
BANK GUARANTEE
DRAFT

We understand that you have entered into a contract with PIPEFABRO LTD.

for ECONOMISER TUBES at an agreed price of £30,000.

and we are informed that in this connection a Bank Guarantee for £3000

being 10% of the contract value is required.

In consideration of ALBAN POWER CO. paying the sum of £3000.

as advanced payment for goods to be delivered, we hereby give you

our Guarantee in the sum of £3000, such Guarantee being

effective in the event that PIPEFABRO LTD. fail in their contractual

obligations in respect of this contract.

Claims under this Guarantee are to be received by NAT.CITY BANK

at 12 HOWE STREET, LEEDS, no later than 1 p.m. 19TH OCTOBER, 1988,

Such claims will be payable by us upon your first demand accompanied

by your statement that PIPEFABRO LTD. have failed in their

contractual obligations in respect of the contract specified above.

This Guarantee should be returned to us on expiry.

Our maximum liability under this Guarantee is limited to the sum of

£3000.

This Guarantee shall be construed and shall take effect in all respects

in accordance with English Law.

                         Authorised Signatories of Bank.

Figure 34.6 Advance payment bond draft.

Once the certificate of substantial completion has been issued, the performance bond is returned to the subcontractor. Alternatively, the bond can be extended to cover the maintenance period and thus takes the place of the retention bond (see the following section), provided, of course, that the percentage of the contract value is the same for both bonds (see Fig. 34.7).

Performance Bond

Bond No. ....   Amount £.............

Know all men by these presents

That we,

As Principal, and the .... (hereinafter called the "Principal")
a corporation duly organised under the laws of England, (hereinafter called the
"Surety"), as Surety, are held and firmly bound unto ....
(hereinafter called the "Obligee"), in the sum of .... Pounds
(£..............), for payment of which sum well and truly to be made, we,
the said Principal and the said Surety, bind ourselves, our heirs, executors,
administrators, successors and assigns, jointly snd several, firmly by these
presents.

THE CONDITION OF THIS OBLIGATION IS SUCH, that whereas the Principal
entered into a certain Contract with the Obligee, dated ....19... for

In accordance with the terms and conditions of said contract, which is hereby
referred to and made a part hereof as if fully set forth herein;

NOW THEREFORE, THE CONDITION OF THIS OBLIGATION SUCH, that if
the above bounden Principal shall well and truly keep, do and perform each
and every, all snd singular, the matters and things in said contract set forth
and specified to be by said Principal kept, done and performed, at the times
and in the manner in said contract specified, or shall pay over, make good
and reimburse to the above named Obligee, all loss and damage by which said
Obligee may sustain by reason of failure or default on the part of said Principal
so to do, then this obligation shall be null and void; otherwise shall remain in
full force and effect.

Sealed with our seals and dated this .... day of ....
A.D. nineteen hundred and ....

....
Principal
By..........................
By..........................
And..........................

Figure 34.7 Performance bond.

Retention (or maintenance) bond

Many purchase orders and most subcontracts require a retention fund to be established during the life of the manufacturing or construction stage. The purpose of a retention bond is to release the monies held by the purchaser at the end of the construction period and yet give the purchaser the available finance to effect any necessary repairs or replacements if the subcontractor or vendor fails to fulfil his contractual obligations during the maintenance period. The value of the bond is exactly equal to the value of the retention fund (usually between 2.5% and 10% of the contract value), and is issued by either a bank or an insurance company.

When the maintenance period has expired and the final certificate of acceptance has been issued, the retention bond is returned by the purchaser to the subcontractor, who in turn returns it to his bank (see Fig. 34.8).

Letter of intent

If protracted negotiations created a situation where it is vital to issue an order quickly to meet the overall project programme, it may be necessary to issue a letter or fax of intent. Formal purchase orders, especially if extensive amendments have to be incorporated, can take days if not weeks to type, copy and distribute. A device must thus be found to give the vendor a formal instruction to proceed to enable the agreed delivery period to be maintained.

The letter of intent fulfils this function, but unless properly drafted it can turn out to be a very dangerous document indeed. Invariably, the buyer tends to be brief, restricting the letter or fax to essentials only. The danger lies in the fact that by being too brief, he may underdefine the contract, leaving the position open for an unscrupulous or genuinely confused vendor to lodge claims for extras. To make matters worse, instructions to proceed may have to be given before a number of apparently minor contractual points have been fully agreed, and while the buyer may try to build a safeguard into the letter by a clause, such as: 'This authority is given subject to final agreement being reached on the outstanding matters already noted', he has not, in fact, protected anybody.

The following examples show how a letter or fax of intent should **not** and should be drafted.

Bad letter

Following our Invitation to Bid and your quotation No. 2687 of … …. together with all subsequent documentation, please accept this fax as your instruction to proceed with the works.
This Authority is given subject to final agreement being reached on the outstanding matters already noted.

MAINTENANCE RETENTION BOND

Bond No: ....   Amount £............

KNOW ALL MEN BY THESE PRESENTS,

That we,

                                        (hereinafter called the "Principal"),
as   Principal,   and   a
corporation duly organised under the laws of the   , and duly
licensed to transact business in the State of   (hereinafter
called the "Surety"), as Surety, are held and firmly bound unto

                                        (hereinafter called the "Obligee"),
in the sum of   Pounds
(£..............), for the payment of which sum well and truly to be made, we,
the said Principal and the said Surety, bind ourselves, our heirs, executors,
administrators, successors and assigns, jointly and severally, firmly by these
presents.

Sealed with our seals and dated this   day of   ,
A.D. nineteen hundred and   .

WHEREAS, the said Principal has heretofore entered into a contract with said
Obligee dated   , 19   , for

                                                                   and;

WHEREAS, the said Principal is required to guarantee the
installed under said contract, against defects in materials or workmanship,
which may develop during the period.

NOW, THEREFORE, THE CONDITION OF THIS OBLIGATION IS SUCH, that if said
Principal shall faithfully carry out and perform the said guarantee, and
shall, on due notice, repair and make good at its own expense any and all
defects in materials or workmanship in the said work which may develop during
the period

or shall pay over, make good and reimburse to the said Obligee all loss and
damage which said Obligee May sustain by reason of failure or default of said
Principal so to do, then this obligation shall be null and void; otherwise
shall remain in full force and effect.

                              ....
                              Principal

                              ....
                              BY

Figure 34.8 Retention bond.

Summary of main types of contract

	Lump sum	Remeasured	Reimbursable
Documents required	Schedule of rates for variations	Bill of quantities	Little definition
Design requirement	Full design	Almost full design	Basic design
	Full specification Full set of drawings	Full specification Almost full drawings	Part specification Basic drawings
Price	Fixed or subject to escalation	Fixed or subject to escalation	Preliminary, subject to escalation
Client's involvement	Minimum	Negotiations of star rates and extras	Monitoring of manhours. Auditing of costs of materials
Supervision	Quality only	Quality and variations to contract	Close quality control and variation
Advantages	Price known	Drawings need not be complete	Can start early on site
Disadvantages	Drawings must be complete	Costs could rise as design is changed	Final cost could be very high. Contractor has little incentive to reduce costs

Target contract
Fee for contractor is fixed
Prime cost (materials and labour) is frozen
If final prime cost is less than target, saving is shared
If final prime cost exceeds target, contractor pays the difference
Contactor's fee increases as prime cost decreases
Contractor's fee increases if overall project time is reduced

Figure 34.9 Summary of main types of contract.

Good letter

This fax gives the vendor the right to start work and incur costs which can be recovered by him even if the final negotiations breakdown and the formal contract is not issued. A fax of intent should be drafted on the following lines:
Following an Invitation to Bid of the … … and your quotation No. 2687 of … … together with Amendments Nos. 1, 2, 3 and 4, and Minutes of Meeting of … …, and … …, please proceed with the design portion of the works and the preparation of sub-order requisitions to a max. value of £2000 to maintain a contract completion of … … The firm order for the remainder of the contract of the agreed value of £59,090 (subject to adjustment) will be issued if the outstanding matters, i.e., amount of liquidated damages and cost of extended drive shafts are agreed by the … …

This fax of intent is undoubtedly longer, but it contains all the essential information and tells the vendor what his limits of expenditure are before the final order is placed. The vendor also knows the scope of supply (including all the agreed amendments) and the date by which the equipment has to be delivered. By releasing the vendor to commence the design and suborder preparation, the delivery date will not be jeopardized, provided, of course, that the stated outstanding issues are resolved.

The vendor realizes that he may, in fact, still lose the order if he does not come to terms with the purchaser, and this gives him an incentive to complete the deal.

The fax also states what the contract sum (subject to the negotiated adjustments) will be and what the items are that are subject to adjustment.

Clearly, the best procedure is to be in a position to issue the formal purchase order as soon as the negotiations have been completed. This can be done provided the buyer works up to the preparation of the purchase order during the negotiation phase. As clauses or specification details are amended and agreed, they are added to the draft purchase order document so that when the final meeting has taken place, any last-minute extra paragraphs can be added and the price and delivery boxes filled in. It should then be possible to send the final draft to the typing section within 24 hours.

A further advantage of following the above procedure is that the buyer is aware of, and can make quick reference to, the current status of the discussions with the vendor so that he can brief other members of the organization at short notice.

Subcontracts

Definition of subcontracts

The difference between a subcontract and a purchase order is that the subcontract has a site labour content. The extent of this content can vary from one operative to hundreds of men. The important point is that the presence of the man on site requires documents to be included in the enquiry and contract package that set out the site conditions for labour and advise the subcontractor of the limitations and restrictions on the site. While this distinction is undoubtedly true, there are numerous cases where the decision between issuing a relatively simple purchase order or a full set of subcontract documents is not quite as straightforward as it would appear.

For example, if an order is placed for a gas turbine and it is required that the manufacturers send a commissioning engineer to site to supervise setting up and commissioning, does this constitute a site labour content or not? Similarly, if a control panel vendor prefers to complete the wiring of a panel on site (possibly due to programming requirements) and has then to send two or three technicians to site, can this be classed as a subcontract?

There are undoubtedly good reasons why, if at all possible, the issuing of a full set of subcontract documents should be avoided. The cost of collating and issuing what is often a very thick set of contractual requirements, site conditions, specifications, safety regulations, etc., is obviously greater than the few pages that constitute a normal purchase order. Furthermore, the vendor has to read and digest all these instructions and warnings and may well be inclined to increase his price to cover for conditions that may not even relate to his type of work. On the other hand, if a vendor brings a man onto the site who performs similar work to other site operatives but is paid more, or belongs to an unacceptable trade union (or no union), or works longer hours, or enjoys unspecified conditions better than the other men, the effect on site labour relations may be catastrophic. The cost of even half a day's strike is infinitely greater than a bundle of contract documents.

It can be seen, therefore, that there is a grey area that can only be resolved in the light of actual site conditions known at the time, plus a knowledge of the scope of work to be carried out by the vendor's site personnel. The following guidelines may be of some assistance in deciding the demarcation between a purchase order and a subcontract, but the final decision must reflect the specific labour content and site conditions.

Typical subcontracts are as follows:

- Demolition
- Site clearance and fencing
- Civil engineering
- Steel erection
- Building work and decorating
- Mechanical erection and piping
- Electrical and instrumentation installation
- Insulation application
- Painting
- Specialist tray erection
- Specialist telecommunication installations
- Specialist tank erection
- Specialist boiler or heater erection
- Water treatment
- Effluent treatment
- Site refractory works
- Site cleaning (including office cleaning)
- Security and night watchmen
- Radiography and other non-destructive testing

Subcontract documents

The documentation required for a subcontract can be roughly classified into three main groups:

1. Commercial conditions
2. Technical specification
3. Site requirements

Although all three types of documents are interrelated, they cover very different aspects of the contract and are therefore prepared by different departments in the purchaser's organization.

The commercial conditions are usually standardized for a particular contract or industry, and if not actually written by the commercial or legal department, are certainly vetted and agreed by them.

The technical specification may be prepared by the relevant technical department and includes the necessary technical description, material and work specifications, standards, drawings, data sheets, etc.

The site requirements originate from the construction department or client and set out the site conditions, labour restrictions, safety and welfare requirements, and programme (sometimes called the schedule).

The subcontract manager's function is to pull these three sets of documents together and produce one combined set of papers that tell the subcontractor exactly what he must do, how, where and when.

Commercial conditions – general

The conditions of subcontract, like the general or main conditions of contract, are most effective if they follow a standardized and familiar form. Most civil engineers are conversant with the Institution of Civil Engineers' (ICE) General Conditions of Contract and NEC3, and every mechanical engineer should have at least the knowledge of MF/1/as published by the Institution of Mechanical Engineers (I.Mech.E.) and the Institution of Electrical Engineers (IEE). In 1993, the ICE published the New Engineering Contract (NEC), called the Engineering and Construction Contract. This has since been updated and is now known as NEC3. The NEC family of contracts now covers contract conditions for main contractors, subcontractors, professional services, supply and adjudicators. A table of the more important standard conditions of contract, which frequently form the basis of the subcontract conditions, is given in Fig. 34.10, but it is not imperative that any of these standard conditions be used. Many large companies, such as oil companies, chemical manufacturers or nationalized industries, have their own conditions of contract. In turn, many of the contractors, whether civil or mechanical, have their own conditions of subcontract. Generally, the terms and clauses of all these conditions are fairly similar, since if they were unreasonably onerous, contractors would either not quote or would load their tenders accordingly. However, there are differences in a number of clauses a prospective tenderer would be well advised to heed. Such differences are often incorporated by the purchaser in the light of actual unfortunate experiences he has no intention of repeating. One can well imagine the commercial officer writing these conditions and applying the adage that the difference between a wise man and a fool is that a wise man learns from his experience.

The alternative to using standard conditions, whether issued by established institutions or by the purchaser's organization, is to write tailor-made general conditions for a particular project. This is usually only viable when the project is very large and when a multitude of subcontracts is envisaged. There are considerable advantages for the purchaser or main contractor in tailoring the conditions to a particular project, as in this

STANDARD CONDITIONS OF CONTRACT

NEW ENGINEERING CONTRACT (NEC3) CONDITIONS OF CONTRACT AND FORMS OF TENDER, AGREEMENT AND BOND FOR USE IN CONNECTION WITH WORKS OF CIVIL ENGINEERING CONSTRUCTION:	Institution of Civil Engineering Association of Consulting Engineers Federation of Civil Engineering Contractors
JCT 80 STANDARD FORM OF BUILDING CONTRACT:	Joint Contracts Tribunal (JCT) Royal Institute of British Architects National Federation of Building Trades Employees Royal Institution of Chartered Surveyors
MODEL FORM OF GENERAL CONDITIONS OF CONTRACT (INCLUDING FORMS _ OF AGREEMENT AND GUARANTEE)	Institution of Mechanical Engineering Institution of Electrical Engineers Association of Consulting Engineers
GENERAL CONDITIONS OF CONTRACT FOR STRUCTURAL ENGINEERING WORKS:	Institution of Structural Engineers
MODEL CONDITIONS OF CONTRACT FOR PLANT (INCLUDING ERECTION)	EB (ELECTRICITY BOARD) B E A M A (British Electrical and Allied Manufacturers Association)
CONDITIONS OF CONTRACT (INTERNATIONAL) FOR WORKS OF CIVIL ENGINEERING CONSTRUCTION	F I D E C Fédération Internationale des Ingénieres - Conseils
MODEL FORM OF CONDITIONS OF CONTRACT FOR PROCESS PLANTS (SUITABLE FOR LUMP SUM CONTRACTS IN THE U.K.)	Institution of Chemical Engineers
MODEL CONDITIONS OF CONTRACT FOR REPAIR, MODIFICATION AND REHABILITATION OF BOILERS AND ASSOCIATED PLANT. (CONDITONS RMR)	GB(GENERATING BOARD) WTBA (WATERTUBE BOILERMAKERS ASSOCIATION)
GENERAL CONDITIONS OF GOVERNMENT CONTRACT FOR BUILDING AND CIVIL ENGINEERING WORKS (GC/WORKS 1)	H.M. GOVERNMENT

Figure 34.10 Standard conditions of contract.

way the same base documents can be used for every discipline. In other words, instead of the civil contractor being governed by the ICE conditions, the piping erection contractor by model form 'A', and the insulation contractor by the Thermal Insulation Contractors Association conditions, all the subcontractors must work to the same general conditions written especially for the project. To ensure that the various disciplines can work to one set of conditions, great care must be taken in their compilation. Since most of the clauses must be applicable to all the subcontracts, they should be of a general nature. Clauses specific to a particular discipline or trade are collected together in what are known as 'special conditions of subcontract'. These are described later.

Obviously, such a comprehensive set of conditions will contain clauses that are not relevant to some of the disciplines. This problem is overcome by either incorporating a list of non-relevant clauses in the accompanying special conditions, or relying on the common sense of all parties to ignore clauses that are not usually applicable by custom and practice. For example, a clause relating to underground hazards (usually in a civil contract) would be irrelevant in an insulation contract.

The advantages of a common set of general conditions are as follows:

1. There is no confusion at the issuing stage as to which conditions of the contract must be used for a specific subcontract.
2. The site subcontract administrator becomes conversant with the terms of the contract and will thus find it easier to administer them.
3. There is no risk of contradiction between certain terms that may have a different interpretation in different standard conditions. A typical example is Clause 24 in the I.Mech.E. model form 'A' of general conditions of contract. This clause lists industrial disputes as a reason for granting an extension of time. The corresponding clause in the ICE conditions (Clause 4A) does not list this particular occurrence as a valid claim for extension of time. Clearly, it is highly desirable that such an important factor as industrial disputes has the same implications for all contractors on a particular site.

Special conditions

As mentioned earlier, one way of advising the tenderer that certain clauses in the general conditions are not applicable to his particular contract is to list all those non-applicable conditions in special conditions of the contract that form part of the package.

Where the general conditions have *not* been tailor-made for a contract, the special conditions contain all those clauses peculiar to a particular site, especially the labour relations procedures. In theory, general conditions of contract apply to any site in the United Kingdom (overseas sites usually require separate conditions) so that particular items such as site establishment requirements, utility facilities, security, site car parking, site agreement notifications and other special clauses must be drawn to the attention of the tenderer in a separate document. Because of the specific nature of these clauses, special conditions of a contract usually precede the general conditions in the hierarchy of importance. In other words, a modification or qualification in the special conditions takes precedence over the unqualified clause in the general conditions. Other clauses in the special conditions are terms of payment and, of course, the form of agreement.

Technical specification

The technical portion of the subcontract document consists of six main sections:

1. Description of work
2. Specification and test requirements
3. Bills of quantities (if applicable)

4. List of drawings
5. List of reports to be submitted and details of cost codes
6. Payments schedule (if related to work packages)

Some organizations also include the planning schedule and insurance requirements in this section, but these two items are more logically part of the site requirements and will be dealt with later.

Description of work

This section is divided into two parts:

1. Description of the site and a general statement of the objectives relating to the project as a whole
2. Description of that portion of the work relating to the subcontract in question

Thus, part 1 states the purpose of the project (e.g. to produce 1000 tonnes of cement per day using the dry process, etc.). Part 2 describes (in the case of a civil subcontract) which structures are in concrete, which are steel with cladding, the extent of roads, pavings and sewers, and the soil conditions likely to be encountered.

Needless to say, more detailed technical descriptions will appear on the drawings, in the technical specifications and in the bills of quantities, giving, in effect, the scope of the subcontract.

Liquidated damages (or ascertainable liquidated damages)

Liquidated damages have been defined by Lord Dunedin in a court case in 1913 as 'a genuine covenanted pre-estimate of damages', and as such is the compensation payment by a vendor to a purchaser when the goods are not delivered by the contract date. In cases of subcontracts, liquidated damages can be imposed if the contract is not completed by the agreed date.

Liquidated damages are not penalties. They are primarily designed to cover the losses suffered by a purchaser because the goods or services were not available to him by the agreed date. As the amount of liquidated damages was agreed by both parties in advance, the purchaser does not have to prove he has lost money. The fact that the goods are late is sufficient reason for claiming the damages.

Over the years, however, liquidated damages have been assessed in quite an arbitrary way that bears no relationship to the losses suffered. Usually, they are calculated as a percentage of the contract value and vary with the number of days or weeks for which the goods have been delayed.

In most cases, the Courts will uphold such a clause, provided the actual amount of liquidated damages is less than the amount that could have been realistically shown as the loss. It is argued that both parties knew at the time of signing the contract that the loss would probably be greater, but agreed to the lower figure. If, on the other hand, the amount is greater than the real loss and the vendor could demonstrate to the Courts that the purchaser was, in fact, imposing a penalty, then the clause would not be enforceable.

A normal figure used for assessing liquidated damages is 0.5% per week of delay with a maximum of 2.5%. This means that the vendor's maximum liability becomes operative after a 5 weeks' delay and is limited to 2.5% of the contract value. If the purchaser does not really need the goods, even after 5 weeks' delay, he can still claim his 2.5%, which is, in fact, pure profit. On the other hand, if, because of the delay of one item of equipment, the whole plant remains inoperative, his losses could be enormous. The receipt of a miserable 2.5% of the value of one relatively small item is insignificant.

It can thus be seen that the real purpose of liquidated damages is to encourage the vendor to deliver on time, since a loss of 2.5% represents a large proportion of his profit. It is quite naïve to suggest that the vendor should pay the true value of a loss that could be suffered by a purchaser, which could be many times greater than the cost of the goods in question.

If no liquidated damages clause is included in the purchase order, the purchaser may claim damages at large, and may, indeed, recover the full, or a substantial proportion of the full amount of his loss, due to the goods being delayed. For this reason, many vendors actually request that a liquidated damages clause is inserted so that their liability is limited to the agreed amount.

For large subcontracts, it is prudent to produce some form of calculation for assessing the amount of liquidated damages, as if they are challenged, they must be shown to be reasonable. There are a number of ways, these can be assessed:

1. If the whole plant was prevented from producing the desired product, the loss of net profit per week of production can be used as a basis.
2. If the works are nonprofit earning, such as a road or reservoir, the additional weekly interest payment on the capital cost is a realistic starting point.
3. If the delayed items hold up work by another subcontractor, the waiting time for plant and additional site overheads are considered as real losses. To these could be added the standby time of labour, if it cannot be redirected to other work.

Liquidated damages may be imposed on the total contract or on sections. This means that the late delivery of layout or even final drawings could be subject to liquidated damages. The amount of these damages could easily be calculated as the man-hours of waiting time by engineers being held up for information.

After all these calculations have been produced, the total value of the damages must be compared with the contract value of the goods. If the amount is high in relation to the contract value, it must be reduced to a figure that a vendor can accept. At the end of the day, if the purchaser requires the goods, he must find a vendor who is prepared to supply them.

Insurance

Normally a purchaser—contractor requires his goods to be fully insured from the point of manufacture up to the stage when the client has taken over the whole project. In practice, this insurance is affected in a number of stages, which vary with the terms of the main contract between the contractor and his client. The more usual methods adopted are as follows:

1. The manufacturer insures the goods from the time they leave his works to the time they are off-loaded on site. The insurance cover for this stage ceases when the contractor's crane lifts the goods off the transport. The contractor's all-risk insurance policy now covers the goods until they are actually taken over by the client.
2. The manufacturer insures the goods as above − the *client's* overall site insurance policy covers the goods as soon as they are lifted off the transport. In such circumstances, the goods will be paid for at the next payment stage and will become the property of the client, although they may, in fact, not yet be erected or installed. Depending on the terms of the conditions of purchase, the goods will have become the property of the purchaser when delivered to the site or were paid for, whichever was earlier.

For large capital projects, the second method is the more common for the following reasons:

1. A large site may involve a number of contractors, all of whom have to insure their works. The cost of this insurance will, if provided by the different contractors, have to be paid eventually by the client as part of the contract sum. By taking out his own insurance for the total value of all the various contractors' works, the client will be able to negotiate far better terms with a large insurance company than if the different works were insured individually.
2. Most contractors require payments for materials delivered or erected in accordance with agreed terms of payment, which form part of the contract. When these payments are made, that portion of the finished works becomes the property of the client. It is reasonable, therefore, for the client to be responsible for the insurance also. It can be seen that if the contractor's insurance were to cover the goods from receipt on site to the date of payment, a whole series of insurance changeover dates would have to be agreed. The additional administrative problems would be both time consuming and costly.
3. In many cases, the new works will be constructed on a site close to, or even integrated in, an existing operational plant owned or run by the client. Any damage to the existing plant, due to an accident on the new plant, can be covered by the same insurance policy.
4. The project, though large in itself, may only be a part of an even bigger project, for example, an onshore oil terminal may be part of a major development of an offshore oil field involving a number of oil rigs. In such a situation, the client will negotiate a massive insurance policy, perhaps with a consortium of insurers, at a really attractive rate.

Needless to say, the goods will only be covered by the client's policy once they have arrived on the job site. If the goods have to be stored temporarily in an off-site warehouse, the contractor will have to arrange for insurance, even if the goods have been paid for in the form of advance payments.

The exact stage, at which the insurance risk passes from the seller to the buyer, depends on the conditions of the purchaser and the shipping terms. For a more detailed explanation, see the section on Incoterms, which discusses the shipping responsibilities used internationally by all the trading nations.

Discounts

During the pre-order discussions with the prospective supplier, the buyer must try to reduce the price as much as possible. This can be achieved by asking the supplier to give a price reduction in the form of a discount. These, often considerable, reductions can take the form of the following:

1. Negotiated and hidden discounts
2. Bulk purchase discounts
3. Annual order discounts
4. Prompt payment discounts
5. Discount for retention bond

Negotiated and idden discounts

There comes a stage during most negotiations when all the technical points have been resolved and all the commercial conditions agreed. However, the final price can still be unresolved as the very technical and commercial points discussed have probably affected the original bid. This is the time for the buyer to bring up the question of discounts. The arguments put forward could be the following:

1. The technical requirements are now to a different specification requiring less material, etc.
2. The commercial conditions are now less onerous.

Both these changes could warrant a price reduction. If, on the other hand, the opposite is the case, that is, the specification is higher or the conditions harsher, a 'hidden discount' can be obtained by insisting that the price remains as tendered. To clinch the deal, the vendor may well agree to this at this stage. A salesman would be very loath to return to his Head Office without an order, having got so far in the negotiations.

It must be remembered that there is no such thing as a fixed profit percentage. Most salesmen are allowed to negotiate between prescribed limits, and it is the buyer's job to take advantage of these margins. When the bid analysis is prepared, the discounts obtained should be shown separately so that the bid price can be checked against the original tender documents. This is especially important if the bid price is made up of a number of individual prices that have to be compared with those of competitors.

Bulk purchase discounts

When large quantities of a particular material have to be purchased, the vendor, in order to make the offer more attractive, may offer a bulk purchase discount on the basis that some of the economies of scale can be shared with the purchaser. If such a discount is not volunteered by the vendor, it can still be suggested by the buyer.

Annual order discounts

A vendor may offer (or be persuaded to offer) a discount if the purchaser buys goods whose total value over a year exceeds a predetermined amount. This will encourage a purchaser to order all similar items of equipment, say electric motors, from the same vendor. The items may be of different size or specification, but will still be obtained from the same supplier. At the end of the year, a percentage of the total value of all orders is paid back to the purchaser as a discount.

Prompt payment discount

Although the conditions of sale may stipulate payment within 30 days of the date of the invoice (assuming the item has been received by the purchaser in good condition), many companies tend not to pay their bills unless the vendor has issued repeated requests or even threatened legal action. To encourage the prompt payment of invoices, an additional discount is frequently offered. The value of this is usually only a few percent and reflects the financing charges the vendor may have to pay due to late receipt of cash.

Discount for retention bond

Most contracts or subcontracts contain a retention clause, which requires a percentage of the contract value to be retained by the purchaser for a period of 6–12 months. To improve the vendor's cash flow, a retention bond can often be accepted by the purchaser, which guarantees the retention value, but this will deprive the purchaser of the use of these monies during the retention period. To compensate the purchaser for this, a vendor may offer a discount, which in effect is a proportion of the interest charges the vendor would have to pay for borrowing the retention sum from a bank. A usual procedure is to split the interest charges 50:50 between the purchaser and the vendor. In this way, both parties gain from the transaction.

It can be seen that discounts can frequently be obtained from a supplier, especially if it is a buyer's market. In most reimbursable cost contracts, all discounts except prompt payment discounts must be passed on to the client for whom the goods or services have been purchased. For this reason, all negotiations including the discounts offered must be open and properly documented so that they can stand up to any subsequent audit.

Counter-trade

Despite the name, this is not meant to refer to trade carried out over a shop counter, although this use of the term is commonly applied to goods collected from a wholesaler's premises. In the case of international business, the term refers to the payment for goods or services by something other than money. In other words, it is akin to good old-fashioned barter.

The difference between barter and counter-trade is that in barter, one type of goods or services are exchanged for another without money being involved, while in counter-trade, the goods supplied by the buyer are delivered to a third party who sells them (usually at a profit) for the benefit of the seller who then receives cash.

A simple example illustrates how the system works: a potential client in a developing country may need to extend his production facilities. His business may be expanding and highly profitable, but because of government restrictions the company has no access to hard currency. It is in the country's national interest to encourage industrial growth at home, but not to increase its national debt by borrowing dollars or pounds. A new approach is needed and one solution is to resort to counter-trade. If, for example, the country is rich in some natural resources, such as coal, this may be the most convenient commodity to trade-off against the proposed factory extension. The expanding company will buy the coal from the mine in local currency. The UK supplier will provide the production facility expansion and receive an appropriate quantity of coal as payment.

Incoterms

World trade inevitably requires goods to be shipped from one country to another. Raw materials must be transported from the less developed countries to the developed ones, from which finished goods are sent in the opposite direction. Both movements have to be packed, insured, transported, cleared through customs and unloaded at their point of destination, and to standardize the different conditions required by the trading partners, Incoterms (Fig. 34.11) were developed. These trade terms cover 14 main variations and encompass the spectrum of cost and risk of shipments from 'ex works' where the buyer has all the risk and pays all the costs, to 'delivered duty paid' where the seller contracts to cover delivery costs and insurance.

Ex works

'Ex works' means that the seller's only responsibility is to make the goods available at his premises (i.e. works or factory). In particular, seller is not responsible for loading the goods in the vehicle provided by the buyer, unless otherwise agreed. The buyer bears the full cost and risk involved in bringing the goods from there to the desired destination. This term thus represents the minimum obligation for the seller.

Free carrier (named point)

This term has been designed to meet the requirements of modern transport, particularly such 'multi-modal' transport as container or 'roll-on roll-off' traffic by trailers and ferries. It is based on the same main principle as FOB, except that the seller fulfils his obligations when he delivers the goods into the custody of the carrier at the named point. If no precise point can be mentioned at the time of the contract of sale, the parties

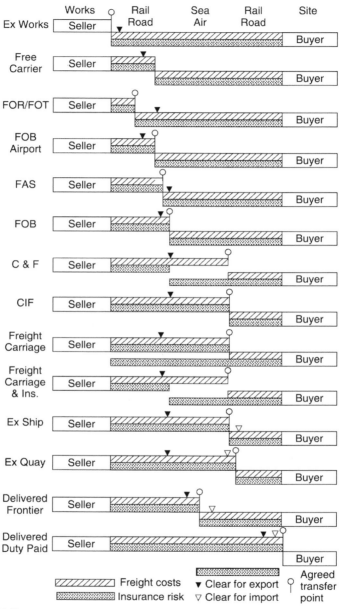

Figure 34.11 Incoterms.

should refer to the place or range where the carrier should take the goods into his charge. The risk of loss of or damage to the goods is transferred from seller to buyer at the time and not at the ship's rail. 'Carrier' means any person by whom or in whose name a contract of carriage by road, rail, air, sea or a combination of modes, has been made. When the seller has to furnish a bill of lading, waybill or carrier's receipt, he or she duly fulfils this obligation by presenting such a document issued by a person so defined.

FOR/FOT

FOR and FOT mean 'free on rail' and 'free on truck'. These terms are synonymous as the word 'truck' relates to the railway wagons. They should only be used when the goods are to be carried by rail.

FOB airport

FOB airport is based on the same main principle as the ordinary FOB term. The seller fulfils his obligations by delivering the goods to the air carrier at the airport of departure. The risk of loss of or damage to the goods is transferred from the seller to the buyer when the goods have been so delivered.

FAS

FAS means 'free alongside ship'. Under this term, the seller's obligations are fulfilled when the goods have been placed alongside the ship on the quay or in lighters. This means that the buyer has to bear all costs and risks of loss of or damage to the goods from that moment. It should be noted that, unlike FOB, this term requires the buyer to clear the goods for export.

FOB

FOB means 'free on board'. The goods are placed on board a ship by the seller at a port of shipment named in the sales contract. The risk of loss of or damage to the goods is transferred from the seller to the buyer when the goods pass the ship's rail.

C&F

C&F means 'cost and freight'. The seller must pay the cost and freight necessary to bring the goods to the named destination, but the risk of loss of or damage to the goods, as well as of any cost increases, is transferred from the seller to the buyer when the goods pass the ship's rail in the port of shipment.

CIF

CIF means 'cost, insurance and freight'. This term is basically the same as C&F but with the addition that the seller has to procure marine insurance against the risk of loss of or damage to the goods during carriage. The seller contracts with the insurer and pays the insurance premium.

Freight carriage — paid to ...

Like C&F, 'freight or carriage — paid to ... ' means that the seller pays the freight for the carriage of the goods to the named destination. However, the risk of loss of or damage to the goods, as well as of any cost increases, is transferred from the seller to the buyer when the goods have been delivered into the custody of the first carrier and not at the ship's rail. It can be used for all modes of transport including multi-modal operations and container, or roll-on or roll-off traffic by trailers and ferries. When the seller has to furnish a bill of lading, waybill or carrier's receipt, he or she duly fulfils this obligation by presenting such a document issued by the person with whom the contracted for carriage to the named destination was signed.

Freight carriage — and insurance paid to ...

This term is the same as 'freight or carriage paid to ... ' but with the addition that the seller has to procure transport insurance against the risk of loss of or damage to the goods during the carriage. The seller contracts with the insurer and pays the insurance premium.

Ex ship

'Ex ship' means that the seller shall make the goods available to the buyer on board the ship at the destination named in the sales contract. The seller has to bear the full cost and risk involved in bringing the goods there.

Ex quay

'Ex quay' means that the seller makes the goods available to the buyer on the quay (wharf) at the destination named in the sales contract. The seller has to bear the full cost and risk involved in bringing the goods there.

There are two 'ex quay' contracts in use, namely 'ex quay (duty paid)' and 'ex quay (duties on buyer's account)' in which the liability to clear the goods for import are to be met by the buyer instead of by the seller.

Parties are recommended to use the full description of these terms always, namely 'ex quay (duty paid)' and 'ex quay (duties on buyer's account)', or uncertainty may arise as to who is to be responsible for the liability to clear the goods for import.

Delivered at frontier

'Delivered at frontier' means that the seller's obligations are fulfilled when the goods have arrived at the frontier — but before 'the customs border' of the country named in the sales contract. The term is primarily intended to be used when goods are to be carried by rail or road, but it may be used irrespective of the mode of transport.

Delivered duty paid

While the term 'ex works' signifies the seller's minimum obligation, the term 'delivered duty paid', when followed by words naming the buyer's premises, denotes the other extreme — the seller's maximum obligation. The term 'delivered duty paid' may be used irrespective of the mode of transport.

If the parties wish the seller to clear the goods for import but that some of the costs payable upon the import of the goods should be excluded — such as VAT and/or other similar taxes — this should be made clear by adding words to this effect (e.g. 'exclusive of VAT and/or taxes').

Further reading

Boundy, C. (2010). *Business contracts handbook*. Gower.

Broome, J. C. (2002). *Procurement routes for partnering: A practical guide*. Thomas Telford.

El-Reedy, M. A. (2011). *Construction management for industrial projects*. Wiley.

Fisher, R., & Shapiro, D. (2007). *Building agreements*. Random House.

Grimsey, D., & Lewis, M. K. (2007). *Public private partnerships*. Edward Elgar.

Lewis, H. (2005). *Bids, tenders and proposals*. Kogan Page.

Thacker, N. (2012). *Winning your bid*. Gower.

Turner, R., & Wright, D. (2011). *The commercial management of projects*. Ashgate.

Ward, G. (2008). *The project manager's guide to purchasing*. Gower.

Yescombe, E. (2002). *Public−private partnerships*. Academic Press.

Value management

Chapter outline

In a constantly changing environment, methods and procedures must be constantly challenged and updated to meet the needs and aspirations of one or more of the stakeholders of a project. This need for constant improvement was succinctly expressed for the first time by Henry Ford when he said he could not afford to be without the latest improvement of a machine.

Value management and its subset, *value engineering*, aim to maximize the performance of an organization from the board room to the shop floor. Value management is mainly concerned with the strategic question of 'what' should or could be done to improve performance, while value engineering concentrates more on the tactical issues of 'how' these changes should be done.

Value can be defined as a ratio of function/cost, so in its simplest terms, the aim is to increase the functionality or usefulness of a product while reducing its overall cost. It is the constant search for reducing costs across all the discipline and management structures of an organization without sacrificing quality or performance that makes value management and value engineering such an essential and rewarding requirement.

The first hurdle to overcome in encouraging a value-management culture is inertia. The inherent conservatism of 'if it ain't broke, don't fix it' must be replaced with 'how can a good thing be made better?' New materials, better techniques, faster machines, more sophisticated programs and more effective methods are constantly being developed, and in a competitive global economy, it is the organization that can harness these developments and adapt them to its own products or services that will survive.

The search and questioning must therefore start at the top. Once the strategy has been established, the process can be delegated. The implementation, which could cover every department and may include prototyping, modelling and testing, must then be monitored and checked to ensure that the exercise has indeed increased the function/cost ratio. This process is called *value analysis*.

The objectives should be one or more of the following: eliminating waste, saving fuel, reducing harmful emissions, reducing costs, speeding production, improving deliveries, improving performance, improving design, streamlining procedures, cutting overheads, increasing functionality and increasing marketability. All this requires is to 'think value' and challenge past practices, even if they were successful.

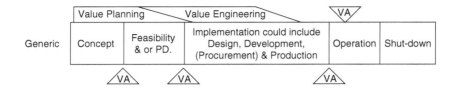

Figure 35.1 Value management and the project life cycle.

In an endeavour to discover what areas of the business should be subjected to value analysis, brainstorming sessions or regular review meetings can be organized, but while such meetings are fundamentally unstructured, they require a good facilitator to prevent them from straying too far off the intended route.

Value analysis can be carried out at any stage of the project as can be seen from the simplified life-cycle diagram of Fig. 35.1. For the first two phases, it is still at the 'What' stage and can be called value planning, while during the implementation phase it is now at the 'How' stage and is known as value engineering. The diagram has been drawn to show value management during the project phases, that is, before handover. However, value management can be equally useful when carried out during the operation and demolition phases in order to reduce the cost or manufacturing time of a product, or simplify the dismantling operations, especially when, as with nuclear power stations, the decommissioning phase can be a huge project in its own right.

In addition to brainstorming, a number of techniques have been developed to systemize or structure the value-engineering process of which one of the best-known ones is *function analysis system technique*. This technique has the following defined stages:

1. Collect and collate all the information available about the product to be studied from all the relevant departments, clients, customers and suppliers.
2. Carry out a functional analysis using the 'Verb and Noun' technique.

This breaks down the product into its components and the function (verb) of each component is defined. The appropriate noun can then be added to enable a cost value to be ascertained. This is explained in the following example.

It has been decided to analyse a prefabricated double glazed window unit. The functions in terms of verbs and nouns are as follows:

Verbs	Nouns
Transmit	Light, glass
Eliminate	Draughts, seals
Maintain	Heat, double glazing
Facilitate	Cleaning access, reversibility
Secure	Handles, locking catches

Each function and component can now be given a cost value and its percentage of the total cost calculated.

3. *Find alternative solutions.* For example, it may be possible to reduce the thickness of the glass but still maintain the heat-loss characteristics by increasing the air gap between the panes. It may also be cheaper to incorporate the lock in the handles instead as a separate fitting.
4. *Evaluation.* The suggested changes are now costed and analysed for a possible saving and the function/cost ratio compared with the original design.
5. *Acceptance.* The proposed changes must now be approved by management in terms of additional capital expenditure, marketability, sales potential, customer response, etc.
6. *Implementation.* This is the production and distribution stage.
7. *Audit.* This is carried out after the product has been on the market for a predetermined time and will confirm (or otherwise) that the exercise has indeed given the perceived additional value or function/cost ratio. If the results were negative, the process may have to be repeated.

Value management is not only meeting the established success criteria or key performance indexes but improving them by periodic reviews. Having previously carried out a stakeholder analysis and identifying their needs, it should be possible to meet these requirements even if the costs have been reduced. Indeed, customer satisfaction may well be improved and environmental damage reduced, resulting in a win—win situation for all the parties.

Further reading

Davies, R. H., & Davies, A. J. (2011). *Value management*. Gower.
European Committee for Standardisation. (2012). *CEN. FprEN 1627:2012 (E)value management*.
Schwartz, M. (2016). *The art of business value*. Portland, Oregon: IT Revolution.

Health, safety and environment

36

Chapter outline

In the light of some spectacular company collapses following serious lapses and short-comings in safety, health and safety is now on the very top of the project-management agenda. Apart from the pain and suffering caused to employees and the public by accidents attributable to lax maintenance of safety standards, the inability to provide high standards of safety and a healthy environment is just bad business. Good reputations built up over years can be destroyed in a day due to one serious accident caused by negligence or lack of attention to safety standards. In addition, under the Corporate Manslaughter Act 2007, directors of companies can now be held responsible for fatalities caused by contraventions of Health and Safety (H&S) regulations.

It is for this reason that the British Standards Institution's 'Guide to project management in the construction industry' BS 6079 Part 4: 2006 has placed the 'S' for safety in the centre of the project-management triangle, indicating that a project manager can juggle the priorities between cost, time and performance, but he must never compromise safety.

Unfortunately, safety is not emphasized nearly enough in current project-management standards and manuals. In BS ISO 21500: 2012 'Guidance on project management', the word 'safety' does not even appear!

In the latest BS 6079: 2010, 'Principles and guidelines for the management of projects', the only reference to safety is in clause 4.6.2 'Health & Safety & Environment' but relates to safety mainly in the context of safety in the workplace. While clause 5.1.1 shows the accepted project-management triangle with the essential criteria: schedule, cost and specification (variously termed performance or quality) at the corners, it misses the opportunity of placing the even greater criterion of safety in the centre of the triangle as shown in the aforementioned BS 6079 Part4: 2006.

In the latest (7th) edition of the *APM Body of Knowledge*, the word 'safety' is not even mentioned in the index, and is only used in a reference to the 'Health and Safety Executive' in relation to the definition of stress.

The PMI *Body of Knowledge* does not even mention the word 'Safety'.

Project Management, Planning and Control. https://doi.org/10.1016/B978-0-12-824339-8.00036-5

One of the characteristics of a project manager should be to learn from experience and from the disastrous results of previous project failures. The following list of some of the more spectacular safety failures (a number of which have been mentioned in Chapter 1) are given here in chronological order and should be a constant reminder to the project manager to place safety at the top of the list of project criteria:

Hindenburg 6 May 1937, Lakehurst N.J. Hydrogen was used instead of Helium, 36 dead.

Aberfan colliery spoil tip, NCB, 21 October 1966 Water build up not monitored, 116 dead.

Challenger space shuttle, NASA, O-ring failure due to cold weather, 28 January 1986, 7 dead.

Nuclear power station, Chernobyl, 26 April 1986, safety and power regulating system intentionally disabled, 54 dead confirmed, estimated total 4000 dead.

Zeebrugge Ferry, Herald of Free Enterprise, Townsend Thorensen, 6 March 1987, 193 dead.

Piper Alpha Oil platform, Occidental, 6 July 1988, 167 dead.

Potters Bar Rail disaster, Jarvis and Railtrack, 10 May 2002, 7 dead.

Texas City oil refinery, BP, 23 March 2005, 15 dead.

Buncefield oil depot, Total UK, 11 December 2005, 43 injured.

Copiapo mine, Chile, 5 August 2008, Long list of safety violations. Fortunately the 33 trapped miners could be rescued.

Deepwater Horizon, Macondo Well, Gulf of Mexico, BP, 20 April 2010, blowout of well, 11 dead.

In every case, there were serious failures of the safety systems and safety management. Health and Safety was given a legal standing with the British Health and Safety at Work Act 1974. This creates a legal framework for employers to ensure that a working environment is maintained in which accidents and unhealthy and hazardous practices are kept to a minimum.

Subsequent legislation included:

- Management of Health and Safety Regulations 1992
- Control of Asbestos at Work Regulations 1987
- Noise at Work 1989
- Workplace (H, S & Welfare) Regulation 1992
- Personal Protective Equipment Regulation 1992
- Manual Handling Operations Regulations 1992
- Fatal Accidents Act 1976
- Corporate Manslaughter Act 2007
- As well as a raft of European Community directives such as EC Directive 90/270/EEC.

The Act set up a Health and Safety Executive that has wide powers to allow its inspectors to enter premises and issue improvement or prohibition notices as well as instigating prosecutions where an unsafe environment has been identified. The Act also gives legal responsibilities to employers, employees, self-employed persons, designers, manufacturers, suppliers and persons generally in control of premises where work is performed.

Each of these groups has been identified as to their responsibilities for health, safety and welfare, which broadly are as follows.

Employer

- Provide and maintain safe equipment
- Provide safe and healthy systems
- Provide a safe and healthy workplace
- Ensure safe handling and storage of chemicals and toxic substances
- Draw up a health and safety policy statement
- Provide information, training, instruction, and supervision relating to safety issues

Employee

- Cooperate with employer
- Take care of one's own health and safety
- Look after the health and safety of others
- Do not misuse safety equipment
- Do not interfere with safety devices

Self-employed

- Must not put other people at risk by their method of working

Designers, manufacturers, suppliers and installers

- Must use safe substances
- Must ensure designs are safe
- Must ensure testing and construction operations are safe
- Must provide information, instructions and procedures for safe operation and use

People in control of premises

- Ensure the premises are safe and healthy

To ensure that the requirements of the *Health and Safety at Work Regulations* are met, employers are required to manage the introduction and operation of safety measures by following the procedures:

- Setting up planning, control and monitoring procedures
- Training and appointing competent persons
- Establishing emergency procedures
- Carrying out regular risk assessments
- Auditing and reviewing procedures
- Disseminating health and safety information

The Act is given teeth by the formation of an enforcement authority called the *Health and Safety Executive* (*HSE*), which appoints inspectors with wide powers to conduct investigations, enter premises and sites, take photographs and samples, issue, where necessary, improvement or prohibition notices and even initiate prosecution.

However, a well-run organization will make sure that visits from the HSE are not required. The watchword should always be: prevention is better than cure. Accidents do not just happen. They are caused by poor maintenance, inappropriate equipment, unsafe practices, negligence, carelessness, ignorance and any number of human frailties.

Because accidents are caused, they can be prevented, but this requires a conscious effort to identify and assess the risks that can occur, and then make sure that any possibility of accident or hazard is avoided. Such risk assessment is a legal requirement and does assist in increasing the awareness of health and safety, and reducing the high costs of accidents.

The most common forms of accidents in commercial, manufacturing and construction or even domestic premises are caused by the following factors:

- Equipment failure
- Fire
- Electricity
- Hazardous substances
- Unhealthy conditions
- Poor design
- Unsafe operating practices
- Noise and lighting

Each of the previously mentioned factors can be examined to find what hazards or shortcomings could cause an accident.

Equipment failure

- Poor maintenance
- Sharp edges to components
- Points of entrapment or entanglement
- Ejection of finished products
- High temperatures of exposed surfaces
- Ill-fitting or insecure guards
- No safety features (overload or pressure relief devices)
- Badly sited emergency stop buttons
- Lack of operating manuals or procedures
- Lack of operator training
- Tiredness of operator

Fire

- Overheated equipment
- Sparks from electrical equipment
- Naked flames
- Hot surfaces
- Hot liquids
- Combustible liquids
- Combustible rubbish
- Explosive gases
- Smoking
- Blocked vents
- Deliberate sabotage (arson)
- Lack of fire extinguishers

Electricity

- Poor insulation
- Bad earthing
- Overrated fuses
- Lack of overload protectors
- Underrated cables or switches
- Unprotected circuits
- No automatic circuit breakers
- No warning signs
- Trailing cables
- Unqualified operators or installers
- Poor maintenance
- Lack of testing facilities
- Dirty equipment

Hazardous substances

- Badly sealed containers
- Corroded containers
- Poor storage
- Unlockable enclosures
- Lack of sign-out procedures
- Poor ventilation
- Bad housekeeping, dirt, spillage
- Inadequate protective clothing
- Lack of emergency neutralizing stations
- Badly designed handling equipment
- Lack of staff training

Unhealthy conditions

- Dirty work areas
- Dusty work areas
- Lack of ventilation
- Fumes or dusty atmosphere
- Smoke
- Noxious smells
- Poor lighting
- Lack of protective equipment
- Excessive heat or cold
- Vibration of handles
- Slippery floors, etc.

Poor design

- No safety features
- Awkward operating position
- Lack of guards
- Poor ergonomic design
- Poor sight lines

- Vibration
- Poor maintenance points
- Poor operating instructions
- Awkward filling points

Unsafe operating practices

- No permit system
- Inadequate lifting equipment
- Inadequate handling equipment
- Poor protective clothing
- Untied ladders
- No obligatory rules for hard hats, boots, etc.
- Inadequate fencing
- Poor warning notices
- Poor supervision
- Poor reporting procedures
- No hazard warning lights, etc.
- Blocked or inadequate emergency exits
- No emergency procedures
- No safety officer
- Poor evacuation notices
- Long working hours

Noise and lighting

- Excessive noise
- High-pitched noise
- Vibration and reverberation
- Inadequate noise enclosures
- Inadequate silencers
- No ear protectors
- Poorly designed baffles
- Poor lighting
- Glare
- Intermittent light flashes
- Poor visibility
- Haze or mist

All the previously mentioned hazards can be identified and either eliminated or mitigated. Clearly, such a risk assessment must be carried out at regular intervals, say every 6 months, as conditions change, and new practices may be incorporated as the project develops.

Apart from the direct effect of accidents and health-related illness on the individual who may suffer great physical and mental pain, the consequences are far reaching. The following list gives some indication of the implications:

- Cost of medical care
- Cost of repair or replacement
- Absence of injured party
- Cost of fines and penalties

- Cost of compensation claims
- Loss of customer confidence
- Loss of public image
- Loss of market due to disruption of supply
- Loss of production
- Damages for delays
- Loss of morale
- Higher insurance premiums
- Legal costs
- Possible loss of liberty (imprisonment)
- Closure costs

Construction, design and management regulations

A special set of regulations came into force in 1994 to cover work in the construction industry, which has a poor safety record. These regulations are called the *Construction (Design and Management) Regulation 1994* (CDM). These are concerned with the management of health and safety, and apply to construction projects including not only the client and contractors but also the designers, associated professional advisers and, of course, the site workers. The duties of the five main parties covered by these regulations are the following:

1. **Client**
   a. Ensure adequate resources are available so that the project can be carried out safely.
   b. Appoint only competent designers, contractors and planning supervisors.
   c. Provide planning supervisor with relevant health and safety information.
   d. Ensure health and safety plan has been prepared before start of construction.
   e. Ensure health and safety file is available for inspection at the end of project.

   These duties do not apply to domestic work where the client is the householder.

2. **Designer**
   a. Design structures that are safe and incorporate safe construction methods.
   b. Minimize the risk to health and safety while structures are being built and maintained.
   c. Provide adequate information on possible risks.
   d. Safe designs to be inherent in drawings, specifications and other documents.
   e. Reduce risks at source, and avoid risks to health and safety where practicable.
   f. Cooperate with planning supervisor and other designers.
3. **Planning supervisor**
   a. Coordinate the health and safety aspects of the design and planning phases.
   b. Help draw up the health and safety plan.
   c. Keep the health and safety file.
   d. Ensure designers cooperate with each other and comply with health and safety needs.
   e. Notify the project to the HSE.
   f. Give advice on health and safety to clients, designers and contractors.
4. **Main contractor**
   a. Take into account health and safety issues during tender preparation.
   b. Develop and implement site health and safety plan.
   c. Coordinate activities of subcontractors to comply with health and safety legislation.

   d. Provide information and training for health and safety.
   e. Consult with employees and self-employed persons on health and safety.
   f. Ensure subcontractors are adequately resourced for the work in their domain.
   g. Ensure workers on site are adequately trained.
   h. Ensure workers are informed and consulted on health and safety.
   i. Monitor health and safety performance.
   j. Ensure only authorized persons are allowed on site.
   k. Display the HSE notification of the project.
   l. Exchange information on health and safety with the planning supervisor.
   m. Ensure subcontractors are aware of risks on site.

5. Subcontractors and self-employed
   a. Cooperate with main contractor on health and safety issues.
   b. Provide information on health and safety to main contractor and employees.
   c. Provide information on health and safety risks and mitigation methods.

The CDM regulations apply to the following factors:

a. *Notifiable* construction work, that is, if it lasts for more than 30 days.
b. Work that will involve more than 500 person days of work (approx. 4000 man-hours).
c. Non-notifiable work which involves five or more persons being on site at any one time.
d. Demolition work regardless of the time taken or the number of workers.
e. Design work regardless of the time taken or the number of workers on site.
f. Residential property where business is also carried out.

The regulations do not apply to the following factors:

a. Domestic dwellings
b. Very minor works

However, the requirement on designers still applies and the project must be submitted to the HSE.

Health and safety plan

The *health and safety plan* consists of two stages:

1. Pre-tender health and safety plan
2. Construction phase health and safety plan

Pre-tender health and safety plan

This is drawn up by the employer under the direction of the planning supervisor. Its main purpose is to set a pattern for the construction phase plan and should include the following items:

a. General description of the work to be carried out
b. Programme and key milestones for the project
c. Table of risks envisaged at this stage and their effects on workers and staff

d. Information to be submitted by the contractor to demonstrate his capabilities regarding resources and management

e. Information to be submitted by the contractor regarding the preparation of the health and safety plan for the construction phase and welfare arrangements

Construction phase health and safety plan

This plan is prepared by the main contractor and has to be submitted to the planning supervisor for approval before work can start on site. Its main constituents are as follows:

a. Health and safety arrangements for all persons on site or who may be affected by the construction work

b. Managing and monitoring the health and safety of construction work

c. Detailed arrangements of the site welfare facilities

d. Evidence of arrangements for keeping the health and safety file

Health and safety file

The *health and safety file* is a record of events on site relating to health and safety, and in particular, the risks encountered and their mitigations as well as possible risks still to be anticipated. The file must be handed to the client at the end of the contract to enable him to manage and deal with possible risks during the carrying out of subsequent renovations or repairs.

The planning supervisor is responsible for ensuring that the file is compiled properly as the project proceeds and that it is handed to the client at the end of the project.

The client must make the file available to all persons involved in future designs of similar structures or those concerned with alterations, additions, maintenance or demolition of the structure.

Warning signs

Standard warning signs have been developed to draw attention to prohibit certain actions, take certain safety precautions or warn of particular environmental hazards.

These signs are colour- and shape-coded to indicate quickly what the type of warning is. The following samples of signs, taken from BS 5499-10-2006, are the most common ones in use, but the selection is not exhaustive:

* Hazard signs have a yellow background in a triangle (Fig. 36.1).
* Prohibition signs are a red circle with a red diagonal (Fig. 36.2).
* Mandatory signs have a blue background, usually in a round disc (Fig. 36.3).
* Safe condition signs have a green background in a square (Fig. 36.4).

Figure 36.1 Hazard signs [*yellow background* (grey in print versions)].

Figure 36.2 Prohibition signs [*red circles* (light grey in print versions)].

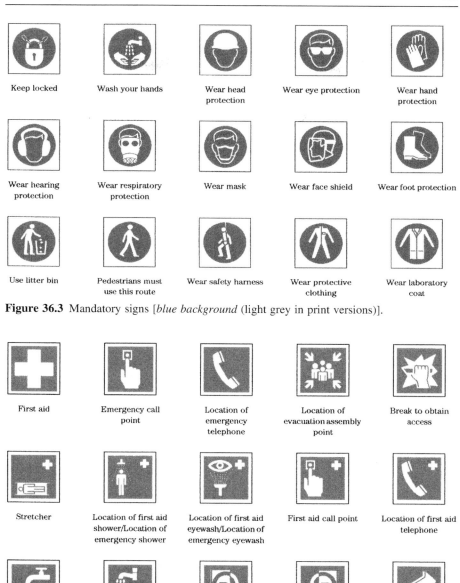

Figure 36.3 Mandatory signs [*blue background* (light grey in print versions)].

Figure 36.4 Safe condition signs [*green background* (light grey in print versions)].

Information management

37

Chapter outline

Information, together with communication, is the very lifeblood of project management. From the very beginning of a project, information is required to enable someone to prepare a cost and time estimate, and it is the accuracy and ease of acquisition of this information that determines the quality of the estimate.

The success of a project depends greatly on the smooth and timely acquisition, preparation, exchange, dissemination, storage and retrieval of information, and to enable all these functions to be carried out efficiently, an information system enshrined in an *information policy plan* is an essential ingredient of the project-management plan.

As with many procedures, a policy document issued by management is the starting point of an information system. If issued at a corporate level, such an information policy document ensures not only that certain defined procedures are followed for a particular project, but that every project carried out by the organization follows the same procedures and uses the same systems.

The following list indicates the most important topics to be set out in an information policy plan:

1. Objectives and purpose for having an information-management plan
2. Types of documents to be covered by the plan

Project Management, Planning and Control. https://doi.org/10.1016/B978-0-12-824339-8.00037-7

3. Authority for producing certain documents
4. Methods of distribution of information
5. Methods for storing information and virus protection
6. Methods for retrieving information and acquisition/modification permits
7. Methods for acknowledging receipt of information
8. Security arrangements for information, especially classified documents
9. Disaster recovery systems
10. Configuration control for different types of information
11. Distribution schedule for different documents
12. Standards to be followed
13. Legal requirements regarding time period for information retention
14. Foreseeable risks associated with information

This *information-management plan* sets out the basic principles, but the actual details of some of the topics must then be tailored to a particular project. For example, the document distribution schedule, which sets out which document is produced by whom and who receives it, must clearly be project-specific. Similarly, the method of distribution depends on the availability of an IT system compatible with the types of information to be disseminated and the types and styles of documents produced.

Objectives and purpose

The purpose of explaining the objectives of the policy document is to convince the readers that it is not bureaucratic red tape, but an essential aid to a smooth-running project. There is no doubt that there are numerous projects being run that do not have such a document, but in these cases the procedures are either part of another document or are well known and understood by all parties due to company custom and practice.

Types of documents

The types of documents covered include correspondence with clients and suppliers, specifications, data sheets, drawings, technical and financial reports, minutes of meetings, records of telephone conversations and other selected data. All these data will be subject to configuration management to ensure correct distribution and storage.

Authority

Certain types of documents can only be issued by specified personnel. This covers mainly financial and commercial documents such as purchase orders, invoices and cheques.

Distribution of information

Distribution of information can be done electronically or by hard copy. While in most cases, the sender has the option of which one to use, certain documents may only be sent by a specified method. For example, in some organizations, legal documents may be e-mailed or faxed but must also be followed up with hard copies by mail. Generally nowadays, most data are available electronically and can be accessed by selected stakeholders using appropriate access codes or passwords.

Storing information and virus protection

Most data can be stored electronically either by the sender or the receiver, but where special measures have to be taken, suitable instructions must be given. Data must be protected against viruses and hackers, and must be arranged and filed in a structured manner according to a predetermined hierarchy based on departmental or operational structures such as work breakdown structures or work areas. In some cases, hard copies of documents and drawings will have to be filed physically where electronic means are not available, as, for example, on remote construction sites. Important documents such as building leases, official purchase orders and contracts require storage space that must be both easily accessible and safe from natural disasters and theft. Special fire and waterproof storage facilities may have to be installed.

Retrieving information and acquisition/modification permits

Retrieving data electronically will generally not be a problem provided a good configuration-management system and effective indexing and identification methods are in place. In many cases, only certain personnel will be able to access the file and of these only a proportion will have the authority to make changes. Appropriate software will have to be installed to convert data from external sources, in the form of e-mails, faxes, spread sheets or other computerized data transmission processes, into a format compatible with the database in use. When handling hard copies, the method of filing documents in a central filing system must be firmly established and an order of search agreed upon. For example, the filing could be by suppliers' names (alphabetically), by product type, by order number, by requisition number or by date. Again, some organizations have a corporate system, while others file by project. Whatever system is used, an enormous amount of time will be wasted searching for documents if the filing system is badly designed and, equally importantly, not kept up to date regularly.

Acknowledging receipt of information

In many cases, it may be necessary to ensure that the information sent is received and read. The policy of acknowledging certain documents must therefore be set down. A recipient may glance at a letter or scan an e-mail without appreciating its importance. To ensure that the receiver has understood the message, a request must be made to reply either electronically or, in the case of a hard copy, by asking for the return of an attached 'confirmation of receipt' slip. This is particularly important when documents such as drawings go to a number of different recipients. Only by counting the return slips can the sender be sure that all the documents are received.

Security arrangements

Classified or commercially sensitive documents usually have a restricted circulation. Special measures must therefore be put in place to ensure the documents do not fall into the wrong hands. In some government departments, all desks must be cleared every evening and all documents locked away. Electronic data in this category requires special passwords that may have to be changed periodically. Documents no longer required must be shredded and where necessary incinerated. It is known that some private investigators have retrieved the wastepaper bags from the outside of offices and reassembled the paper shreds.

Disaster recovery systems

In light of both major natural disasters, such as earthquakes and hurricanes, and terrorist attacks, disaster recovery, also known as business continuity, is now a real necessity. Arrangements must be made to download important data regularly (usually daily) on disk, tape or even film and store them in a location far enough removed from the office to ensure that they can be retrieved if the base data has been destroyed. It will also be necessary to make arrangements for the replacement of any hardware and software systems that may have been destroyed or corrupted.

Configuration control

Configuration management is an integral part of information management. Version control, change control and distribution control are vital to ensure that everybody works to the latest issue and is aware of the latest decisions, instructions or actions. The subject is discussed in more detail in the chapter on configuration management, but the information policy plan must draw attention to the configuration-management procedures and systems being employed.

Distribution schedule

The distribution of documents can be controlled either by a central computerized data distribution system activated by the originator of the document or by a special department charged with operating the agreed configuration-management system. One of the key sections of the project-management plan is the document distribution schedule, which sets out who originates a document and who receives it in tabular form. If hard copies have to be sent, especially in the case of drawings, the number of copies for each recipient must also be stated. If this schedule is lodged with the project office or the distribution clerk, the right persons will receive the right number of copies of latest version of a document at the right time. While most document distribution will be done electronically, hard copies may still be required for remote locations or unsophisticated contractors, suppliers or even clients.

Standards to be followed

Standard procedures relating to information management must be followed wherever possible. These standards may be company standards or guidelines, codes of practice or recommendations issued by national or international institutions. The British Standards Institution has issued guides or codes of practice for configuration management, project management, risk management and design management, all of which impact on information management. An International Standard BS ISO 15489-I-2001 'Information and Documentation— Records Management — Part 1 — General' is of particular importance.

Legal requirements

Every project is constrained by the laws, statutory instruments and other legal requirements of the host country. It is important that stakeholders are aware of these constraints and it is a part of the information-management process to disseminate these standards to all the appropriate personnel. The Freedom of Information Act and Data Protection Act are just two such legal statutes that have to be observed. There may also be legal requirements for storing documents (usually for about 7 years) before they can be destroyed.

Foreseeable risks

All projects carry a certain amount of risk, and it is vital that warnings of these risks are disseminated in a timely fashion to all the relevant stakeholders. In addition to issuing the usual risk register, which lists the perceived risks and the appropriate mitigation strategies, warnings of unexpected or serious risks, imminent political upheaval or potential climatic disasters must be issued immediately to enable effective countermeasures to be taken. This requires a preplanned and rehearsed set of procedures to be set up that can be implemented very rapidly. The appointment of an information risk manager will be part of such a procedure.

An important part of information management is issuing reports. A project manager has to receive reports from other members of the organization to enable him to assess the status of the project and in turn produce his own reports to higher management. Systems must be set up to ensure that these reports are produced accurately, regularly and timely. The usual reports required by a project manager cover progress, cost, quality, exceptions, risks, earned value, trends, variances and procurement or production status. These data will then be condensed into the regular (usually monthly) progress report to the programme manager, sponsor or client as the case may be. Templates and standard formats greatly assist in the production of these reports and modern technology enables much of these data to be converted into graphs, charts and diagrams, but presentation can never be a substitute for accuracy.

Data collection

As the technical advances of IT expand, more and more data will be collected and screened by more and more organizations and companies. These data are then used for a great number of reasons ranging from security and defence to stock control, marketing, design and planned maintenance, as well as populating expert systems.

The way the data is collected also varies from direct face to face or telephone interviews and questionnaires, through data-logging techniques as used by utility companies to read domestic or commercial metre usage, to the sophisticated methods used by the security services to monitor phone conversations and internet traffic.

However, while the quantity of the data may be impressive, it will be useless, if not dangerous, if the quality is suspect. Equally important is the timing, as late data can seriously affect the making of a decision or the issue of a report.

A case in point is the need for vigorous expediting of data required by procurement departments and design engineers from manufacturers or subcontractors. For this reason, the methods described in Chapter 34 (expediting, monitoring and inspection) should form part of the standard procedures of any design organization. Fig. 34.2 shows a typical questionnaire sent out to prospective bidders to enable their capability, competence and financial stability to be assessed.

Another example is given in Fig. 27.5 which shows what information is required from different suppliers to enable the designers to incorporate the information into their overall scheme. Accompanying such a table would be the organization's conditions of purchase and the dates for which the different data are required. For example, in the case of pumps, setting plans and information on flanges are required before impeller details.

There are two main types of data collection: qualitative (depth) and quantitative (breadth).

Qualitative

This method includes techniques such as structured interviews, direct observation and discussion groups. Much depends on the skills of the interviewer or discussion leader who must retain an open mind and not ask leading questions. Advantages include high-response rates and the ability to explore in greater depth after an initial analysis.

Quantitative

This covers questionnaires sent by mail or e-mail, telephone surveys, sample surveys based on a large number of participants, etc. Such surveys, which take into account age, ethnicity, income bracket and geographical location, are often used by governments, academic researchers, retail businesses and manufacturers. The results of such surveys can affect policies and marketing strategies.

There should be a corporate strategy on data management which covers the operation of central data base, access to data, security, methods of quality verification, evaluation techniques, dissemination and any time or cost constraints.

Further reading

Laudon, K. C. (2008). *Management of information systems* (11th ed.). Prentice-Hall.

Marchewska, J. T. (2012). *Information technology project management with CD-rom* (4th ed.). Wiley.

Communication

38

Chapter outline

While it is vital that a project manager has a good information system, without an equally good communication system such information would not be available when it is needed.

Generally, all external communications of a contractual nature, especially changes in scope or costs, must be channelled via the project manager or his/her office. This applies particularly to communications with the client and suppliers/subcontractors. The danger of not doing this is that an apparent small change agreed between technical experts could have considerable financial or program (or even political) repercussions due to the experts not being aware of the whole picture.

'An agreement to this policy should be reached between the parties'.

Unless a project manager decides to do everything him/herself, which should certainly not be the case, he/she has to communicate his/her ideas, plans and instructions to others. This requires communication, whether verbally, in writing, by mail, electronically or by carrier pigeon.

Communications can be formal and informal, and while contractual, organizational and technical information should always follow the formal route, communication between team members is often most effective when carried out informally. There are many occasions when a project manager has the opportunity to meet his/her team members, client or other stakeholders, all of which will enable him/her to discuss problems, obtain information, elicit opinions and build up trusting relationships which are essential for good project management.

Management by walkabout is an accepted method of informal communication, which not only enables an exchange of information and ideas to take place, often in a relaxed atmosphere, but also has the advantage of seeing what is actually going on as well as setting the framework for establishing personal relationships.

Project Management, Planning and Control. https://doi.org/10.1016/B978-0-12-824339-8.00038-9

Probably more errors occur in a project due to bad communications than any other cause. Ideas and instructions are often misunderstood, misinterpreted, misheard or just plainly ignored for one reason or another; in other words the communication system has broken down. Every communication involves a sender and a receiver. The sender has a responsibility to ensure that the message is clear and unambiguous, and the receiver has to make sure that it is correctly understood, interpreted, confirmed and acted upon.

There are a number of reasons why and how failures in communication can occur. The most common of these, generally known as communication barriers, are the following:

- Cultural differences
- Language differences
- Pronunciation
- Translation errors
- Technical jargon
- Geographical separation of locations
- Equipment or transmission failure
- Misunderstanding
- Attitude due to personality clash
- Perception problems due to distrust
- Selective listening due to dislike of sender
- Assumptions and prejudice
- Hidden agendas
- Poor leadership causing unclear instructions
- Unclear objectives
- Poor document distribution system
- Poor document retention system or archiving
- Poor working environment, such as background noise
- Unnecessarily long messages
- Information overload, such as too many e-mails
- Withholding of information
- Poor memory or knowledge retention

Clearly, some of these barriers are closely related. Some of these have been collected and will be discussed in more detail in the following sections, together with the techniques which can be used to overcome these communication problems.

Cultural differences, language differences, pronunciation, translation and technical jargon

Problems may arise because different cultures have different customs, etiquettes and trading practices. In some instances, where two countries use the same language, a particular word may have a totally different meaning. This occurs not only between England and America but, for example, also between Germany and Austria, who are, as some cynics might say, all 'divided by a common language'.

For example, a lift in England would be an elevator in the United States and a water tap would be a faucet. Most project managers will be familiar with the English term *planned* being called *scheduled* in the United States. In addition, regional accents and variations in pronunciation can cause misunderstandings in verbal communications. The solution is simple. Always speak clearly and confirm the salient points in writing.

Confusion can occur with dates. In the United States the month precedes the day, while in the United Kingdom the day precedes the month. It is best therefore to always write the month in words; that is, 3 August 2016 instead of 3.8.2016.

Forms of address may be fairly informal in some countries like United Kingdom or the United States, but unless one knows the other party well, the formal personal pronoun *sie* or *vous* must be used in Germany or France, respectively. The incorrect form of address could easily cause offence. It is advisable, therefore, to seek guidance or attend a short course before visiting a country where such rules apply.

Incorrect translations are not only a source of amusement but can be a real danger. To overcome such errors, the translator should always be a native speaker of the language the text is translated *into*. This will enable the correct word for a particular context to be chosen and the right nuances to be expressed.

Most disciplines or industries have their own technical jargon which can cause difficulties or misunderstandings when the recipient is from a different environment or culture. There may be reluctance of the receiver to admit to his/her ignorance of the terms used, which can cause errors or delays in the execution of an instruction. The sender should, therefore, refrain from using jargon or colloquialisms, but by the same token, it is up to the receiver to request that any unfamiliar term be explained as it is mentioned.

Geographical separation, location equipment or transmission failure

Where stakeholders of a project are located in different offices or sites, good electronic transmission equipment is essential. The necessary equipment must be correctly installed, regularly checked and properly maintained. Generally, it is worthwhile to install the latest updates, especially if these increase the speed of transmission, even if they do not reduce the often high-operating and high-transmission costs. Where persons in countries with different time zones have to be contacted, care must be taken to take these into consideration. A person, from whom one wants a favour, will not be very cooperative if woken up at four o'clock in the morning!

Misunderstanding, attitude, perception, selective listening, assumptions and hidden agendas

Senders and receivers of communications are human beings and are therefore prone to prejudice, bias, tiredness and other failings, often related to their mood or health at the time. Misunderstandings can occur due to bad hearing or eyesight, or because there was not sufficient time to properly read and digest the message. Cases are known where, because the receiver did not like or trust the sender, the transmitted information was perceived as being unimportant or not relevant and was therefore not been acted upon with the urgency it actually required. The receiver may believe the sender to have a hidden agenda or indeed have his/her own agenda, and may therefore deliberately not cooperate with a request. To avoid these pitfalls, all parties must be told in no uncertain manner that the project has priority over their personal opinions. It also helps to arrange for occasional face-to-face meetings to take place.

It is not unusual for the receiver to make assumptions which were not intended. For example, the sender may request a colleague to book some seats to a theatre. The receiver may assume the sender wants the best seats when the opposite may be true. The fault here lies with the sender who was not specific in his/her request.

Poor leadership, unclear instructions, unclear objectives, unnecessarily long messages and withholding of information

Instructions, whether verbal or written, must be clear and unambiguous. They should also be as short as possible as the receiver's as well as the sender's time is often costly. Winston Churchill required all important documents to be condensed onto 'one sheet of foolscap paper' (approximately the size of an A4 sheet). Time is money, and the higher one is in the hierarchy, the more expensive time becomes. As with instructions, objectives must also be set out clearly and unequivocally. It is often advantageous to add simple sketches to written communications. These are often more explicit than long descriptions.

When information has to be communicated to a number of recipients, it may not be advisable to tell everything to everybody. For example, an instruction to a technical department may not include the cost of certain quoted components. Some information is often only disseminated on a 'need to know' basis. The sender therefore has the responsibility to decide which parts of the documentation are required by each receiver. Clearly, particular care has to be taken with sensitive or classified information, which may be subject to commercial distribution restrictions or even the Official Secrets Act.

It can be seen that while there are many potential communication barriers, they can all be overcome by good communication planning and sensitive project management.

An example of how ones attitude can be affected by receiving good or bad communications is clearly shown by the following scenario.

You are standing on a railway platform waiting for a late train, and you hear the usual bland announcement which simply says:

'This train will be 40 minutes late. We are sorry for the inconvenience this has caused.'

This will probably make you angry and blame the train operators for incompetence.

If, on the other hand, the announcement says: 'This train will be delayed by 40 minutes due to a young girl falling onto the line at the last station', you will be mollified and probably quite sympathetic to the operators who now have the problem of dealing with a very unhappy circumstance.

The difference is that in the second announcement you were given an explanation.

Meetings

Meetings are an essential part of project management, as they enable two or more stakeholders to communicate and discuss issues in such a way that quick decisions can be made and implemented. The meeting may be face to face or via video link, tele-conference or other virtual system.

The main types of meetings are as follows:

Board meetings at director level

Pre-bid meetings with prospective suppliers or subcontractors

Kick off meetings at the start of the project; these may be in the office or on site

Progress meetings at regular (usually monthly) intervals

Site meetings which may be ad hoc or related to progress

Ad hoc meetings to discuss unforeseen issues

Technical meetings between experts

Meetings with other stakeholders (clients, public authorities, contractors, etc.)

Team or staff meetings

Whatever the type or purpose, meetings should always be structured, starting with apologies for absence, approval of the last meeting's minutes and discussions arising from points relating to the previous meeting. A previously drawn-up agenda and attendance list should be distributed in advance to all attendees. The chairman/chairwoman of the meeting should ensure that the subject matter being discussed does not stray from the topic in question and must resist the temptation of spending too much time on their pet subject or on areas which are his/her base discipline or expertise. Similarly, unless the issue is urgent or crucial, discussions of details by experts should be cut short and deferred to a follow-up meeting at which the subject can be discussed and examined in greater depth.

If the project manager is not present at such a meeting, one of the attendees must issue a short report on the results of the discussions and the recommendations made.

For all meetings, minutes should be kept of all the topics discussed and circulated to all attendees within a few days of the meeting. All minutes of meetings should include an action column for each topic, which contains the name of the person designated to perform the action and the date by which it is to be completed.

Meetings are costly as they are usually attended by people of managerial rank, experts or specialists, all of whom command high salaries. For this reason, meetings should be attended only by persons related to the issues discussed and only for as long as their topic is being discussed. Once the discussion to which they contributed has been concluded, departmental managers and specialists should be allowed to return to their departments.

There is some truth in the old cynical saying which states that 'The success of a meeting is inversely proportional to the number of people present'.

Guidelines for managing meetings should be a part of the company standards, which should include a standardized format for minutes of meetings.

Further reading

De Vito, J. A. (2012). *The interpersonal communications book*. Pearson.

Team building and motivation

Chapter outline

Large or complex projects usually require many different skills that cannot be found in one person. For this reason, teams have to be formed whose members are able to bring their various areas of expertise and experience together to fulfil the needs of the project and meet the set criteria. The project manager is usually the team leader and it may be his or her responsibility to select the members of the team, although in many instances he or she may be told by the senior management or HR department who will be allocated to the team. If the project is run as a matrix-type organization, the different specialist team members will almost certainly be selected by the relevant functional department manager so that the project manager has to accept whoever has been allocated.

There are considerable advantages in operating as a team, which need not require all the members to be fully allocated to the project all the time. Nevertheless, the project manager must create an atmosphere of cooperation and enthusiasm whether the members are permanent or not.

Project Management, Planning and Control. https://doi.org/10.1016/B978-0-12-824339-8.00039-0

The main advantages of teams are as follows:

- Teams engender a spirit that encourages motivation and cooperation.
- Different but complementary skills and expertise can be brought to bear on the project.
- Problems can be resolved by utilizing the combined experience of the team members.
- New ideas can be 'bounced' between team members to create a working hypothesis.
- Members gain an insight into the workings of other disciplines within the team.
- Working together forms close relationships which encourage mutual assistance.
- Lines of communications are short.
- The team leader is often able to make decisions without external interference.

The following characteristics are some of the manifestations of a successful team:

- Mutual trust
- A sense of belonging
- Good team spirit
- Firm but fair leadership
- Mutual support
- Loyalty to the project
- Open communications
- Cooperation and participation
- Pride in belonging to the team
- Good mix of talents and skills
- Confidence in success
- Willingness to overcome problems
- Clear goals and objectives
- Enthusiasm to get the job done
- Good teams tend to receive good support from top management and sponsors. They are often held up as examples of good project management during discussions with existing and potential clients.

Clearly, if too many of the above characteristics are absent, the team will be ineffective. Merely bringing a number of people together with the object of meeting a common objective does not make a team. The difference between a group and a team is that the team has a common set of objectives, and is able to cooperate and perform as a unified entity throughout the period of the project. However, to create such a team requires a conscious effort by the project manager to integrate and motivate them, and instil an *esprit de corps* to create an efficient unit, whether they are in industry, in the armed services or on the playing field.

Team development

Building a team takes time, and its size and constituency may change over the life of the project to reflect the different phases. Team development was researched by Tuckman who found that a team has to undergo four stages before it can be said to operate as a successful entity. These stages are as follows:

1. Forming
2. Storming

3. Norming
4. Performing

To these could be added a fifth stage termed mourning, which occurs when the project is completed and the team is being disbanded.

Forming

As the word implies, this is the stage when the different team members first come together. While some of them may know each other from previous projects, others will be new and unsure, not only of themselves but also of what they will be required to do. There will be an inevitable conflict between the self-interest of the team member and the requirements of the project, which may impose pressures caused by deadlines and cost restraints.

Clearly, at this stage the project manager will have to 'sell' the project to the team and explain what role each member will play. There may well be objections from some people who feel that their skills are not being given full rein, or conversely that they do not consider themselves to be well suited for a particular position. The project manager must listen to and discuss such problems, bearing in mind that the final decision rests with him or her and once decided, must be adhered to. There is no virtue in forcing a square peg into a round hole.

Storming

Once the team has been nominally formed and the main roles allocated, the storming stage will start. Here the personalities and aspirations of the individuals will become apparent. The more dominant types may wish to increase their sphere of influence or their limits of authority, while the less aggressive types may feel they are being sidelined. There will be some jockeying for position and some attempts to write their own terms of reference, and it is at this stage that the conflict-management skills of the project manager are most needed. It is vital that the project manager asserts his or her authority and ensures that the self-interests of the individual become subservient to the needs of the project.

Norming

When the storming is over, the project should run smoothly into the norming stage. Here, all the team members have settled down and accepted their roles and responsibilities, although the project manager may use a more participative approach and do some 'fine tuning'. The important thing is to ensure that the team is happy to work together, fully aware of the project objectives and the required regulations and standards, and is motivated to succeed.

Performing

At this stage, the team can now be considered a properly integrated working entity with every member confident of his or her role. All the energy will be focused on the well-being of the project rather than the individual. Communications are well established and morale is high. The project manager can now concentrate on the work at hand, but must still exercise a degree of maintenance on the team. The organization should now run as 'on well-oiled casters' with everyone being fully aware of the three main project criteria: cost, time and quality/performance.

Mourning

There is an inevitable anticlimax when a project has come to an end. Members of a project team probably feel what soldiers feel at the end of a war. There is a mixture of relief, satisfaction and apprehension of what is to follow. Unless there is another similar project ready to be started, the team will probably be disbanded. Some people will return to their base discipline departments, some will leave on their own accord and some will be made redundant. There is a sense of sadness when friendships break up and relationships built up over many months, based on respect and mutual cooperation, suddenly cease.

The project manager now has to take on the mantle of a personnel officer and keep the team spirit alive right up to the end. There is always a risk, on large long-running projects, that as the end of the project approaches some people will leave before final completion to ensure further employment without a break. It may then be necessary for the organization to offer termination bonuses to key staff to persuade them to stay on, to ensure that there are sufficient resources to finish the job.

The Belbin team types

While the main requirement of a team member must be his or her expertise or experience in his or her particular field, in the ideal team, not only the technical skills but also the characteristics of the team members should complement each other. A study of team characteristics was carried out by Meredith Belbin after 9 years of research at the Industrial Training Research Unit in Cambridge. At the end of the study, Belbin identified nine main types that are needed to a greater or lesser extent to make up the ideal team.

Unfortunately, in practice, it is highly unlikely that the persons with the right skills and ideal personal characteristics will be sitting on a bench waiting to be chosen. More often than not, the project manager has to take whatever staff is assigned by top management or functional managers. However, the benefit of the Belbin characteristics can still be obtained by recognizing what Bebin 'type' each team member is, and then exploiting his or her strengths (and recognizing the weaknesses) to the benefit of the project. In any case, most people are a mix of Belbin characteristics, but some will no doubt be more dominant than others.

The nine Belbin characteristics are as follows:

- Plant
- Resource investigator
- Coordinator
- Shaper
- Monitor/evaluator
- Team worker
- Implementor
- Completer/finisher
- Specialist

The strengths and weaknesses of each of these characteristics are as follows.

Plant

Such persons are creative, innovative, imaginative, self-sufficient and relish solving difficult problems often using new ideas and fresh approaches. Their unorthodox behaviour may make them awkward to work with and their dislike of criticism, discipline and protocol may make them difficult to control.

Resource investigator

These persons are very communicative, probably extroverted, show curiosity in new ideas and are enthusiastic in responding to new challenges. Once the initial challenge or fascination is over, their interest tends to wane.

Coordinator

Coordinators are self-controlled, stable, calm, self-confident, can clarify goals and objectives and are good at delegating and maximizing people's potentials. When given the opportunity they tend to hold the stage.

Shaper

These persons are outgoing, dynamic and thrive on pressure. Drive and courage to shape events, overcome difficulties and a desire to challenge inertia or complacency are part of their character. They may therefore be anxious, impatient and easily irritated by delays and blockages.

Monitor/evaluator

These people are sober, prudent and are able to evaluate the options. They have a good sense of judgement, are analytical and can make critical and accurate appraisals. They could be easily judgemental and their tactless criticism may be destructive.

Team worker

Such persons are cooperative, sensitive, socially orientated and help to build a good team. They are often only noticed when they are absent. They may have difficulties in making decisions and tend to follow the crowd.

Implementer

Disciplined and reliable, conservative and practical, such persons turn ideas into actions systematically and efficiently. They could be inflexible and averse to new unconventional methods.

Completer/finisher

Such people are painstaking, conscientious and self-controlled perfectionists with a strong sense of urgency. They are good at checking and seeking out errors and omissions. They tend to be over-concerned with minor faults and find it hard to give in.

Specialist

Specialists supply skills that are in short supply. They tend to be single-minded, self-reliant and dedicated to their profession. Their independence is not easily controlled, especially if they know they are difficult to replace. Being absorbed in their speciality, they may at times have difficulty in seeing the larger picture.

Motivation

The simplest dictionary definition of motivation is 'the desire to do'. The strength or degree depends on the individual's character and the reason or cause of the desire. In many cases, the individual may be self-motivated, due to an inner conviction that a particular action or behaviour is necessary for personal, political or religious reasons, but in a project context, it may be necessary for an external stimulant to be applied. It is undoubtedly the function of a project manager to motivate all the members of the project team and convince them that the project is important and worthwhile. The *raison d'etre* and perceived benefits, be they political, economic, social or commercial, must be explained in simple but clear terms so that each team member appreciates the importance of his or her role in the project. In a wartime scenario, motivation can well be a question of survival and is often the result of national pride or convincing propaganda, but such clear objectives are seldom the case in a normal peacetime project, which means that the project manager has to provide the necessary motivation, encouragement and enthusiasm.

There is little doubt that a large part of the success of the 2012 London Olympics was due to the collaborative approach, the team spirit of the design/construction team and the motivation to complete the project on time, for what was regarded by everybody as a project of national pride and international importance.

Apart from the initial indoctrination and subsequent pep talks, a project manager can reinforce the message by the conventional management practices of giving credit where it is due, showing appreciation of good performance and offering help where an individual shows signs of stress or appears to be struggling, mentally or physically.

A good example of the effect of motivating people by explaining the objective of a project or even a work package is shown in the following little story.

A man walking along a street notices that a bricklayer building a wall is very lethargic, clearly not enthusiastic and looks generally unhappy about his work.

'Why are you so unhappy?' he asks the workman.

'I have just been told to build this wall. Just placing one brick on another is monotonous and boring' was the reply.

The man walked further up the street and met another bricklayer clearly building the other end of the same wall. This workman, on the other hand, worked quickly, was clearly interested in the work and whistled a happy tune while he laid the bricks.

'Why are you so happy?' he asked the man.

The man looked up with shining eyes and said proudly: 'I am building a cathedral'.

Maslow's hierarchy of needs

A.H. Maslow carried out research on why people work and why some are more enthusiastic than others. He discovered that in general there was, what he called, a *hierarchy of human needs*, which had to be satisfied in an ascending order. These can be conveniently demonstrated as a series of steps in a flight of stairs where a person has to climb one step before proceeding to the next (see Fig. 39.1).

The five levels on Maslow's needs are: *physiological, security and safety, social, esteem and self-actualization*. Maslow argued that the first needs are the ones that enable the human body to perform its functions, that is, air for the lungs, food and water for the digestive system, exercise for the muscles and, of course, sex for the continuation of the species. Once these needs have been met, the next requirement is *shelter, security* in employment and a *safe* environment. This is then followed by *social acceptance* in the society one frequents, such as at work, clubs or pubs and, of course, the family. The next step is *self-esteem*, which is the need to be appreciated and respected. Praise, attention, recognition and a general sense of being wanted, all generate self-confidence and well-being. The last aspiration is *self-actualization*. This is the need to maximize all of one's potentials, utilize one's abilities fully and be able to meet new challenges.

As in all theories, there are exceptions. The proverbial starving artist in his garret is more concerned about his esteem and self-actualization than his security or even social acceptance. Similarly, the ideals of missionaries take precedence over the desire for physical comfort. However, for the majority of wage or salary earners, the theory is valid and must be of benefit to those wishing to understand and endeavouring to fulfil the needs of people in their charge.

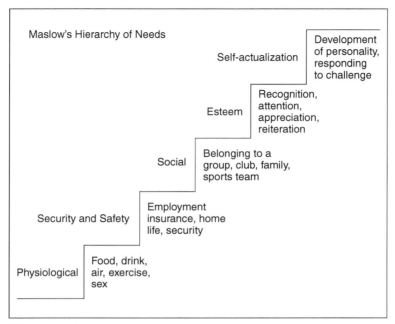

Figure 39.1 Maslow's hierarchy of needs.

Herzberg's motivational hygiene theory

Herzberg has tried to simplify the motivational factors by suggesting two types:

- Hygiene factors
- Motivators

Hygiene factors

- Physiological needs
- Security
- Safety
- Social

Motivators

- Recognition of achievement
- Interesting work
- Responsibility
- Job freedom
- Pleasant working conditions
- Advancement and growth prospects

The hygiene factors represent the first three steps of Maslow's needs, that is, physiological needs, security and safety, and social. The motivators are then esteem and self-actualization. From a management point of view, the first three can almost be taken for granted, as without reasonable pay or security, staff will not stay. To obtain the maximum commitment from an employee (or even oneself), motivators such as recognition of achievement, interesting or challenging work, responsibility, job freedom, pleasant working conditions and possibility of advancement and growth must be present.

In general, people like doing what they are good at and what gives them satisfaction. At the same time, they tend to shun what they are less able at or what bores them. It is of benefit to the organization therefore to reinforce these behaviours, once they have been identified.

Further reading

Katzenbach, J. R., & Smith, D. K. (2005). *The wisdom of teams*. Harper Business.

Turner, J. R. (2003). *People in project management*. Gower.

Leadership

Chapter outline

Leadership can be defined as the ability to inspire, persuade or influence others to follow a course of action or behaviour towards a defined goal. In a political context, this can be for good or evil, but in a project environment it can generally be assumed that good leadership is a highly desirable attribute of a project manager.

Leadership is not the same as management. Leadership is about motivating, influencing and setting examples to teams and individuals, whilst management is concerned with the administrative and organizational facets of a project or company. Therefore, it can be seen that a good project manager should be able to combine his leadership and management skills for the benefit of the project.

Whether leadership is attributed to birth, environment or training is still a subject for debate, but the attributes required by a leader are the same. The following list gives some of the most essential characteristics to be expected from a good leader. To dispel the impression that there is a priority of qualities, they are given in alphabetical order.

Adaptability	Ability to change to new environment or client's needs
Attitude	Positive can-do outlook, optimism despite setbacks
Charisma	Presence and power to attract attention and influence people
Cognitive ability	Ability to weigh up options, give clear instructions

Continued

Project Management, Planning and Control. https://doi.org/10.1016/B978-0-12-824339-8.00040-7

Commitment	Will to succeed and achieve set goals
Common sense	Ability not to be hoodwinked by irrational suggestions or solutions
Creativity	Able to do some innovative or lateral thinking
Drive	Energy, willpower and determination to push forward
Fairness	Fair and considerate attitude to human needs and staff problems
Flexibility	Willingness to modify ideas and procedures to new circumstances
Honesty	Trustworthy, reliable, will not tolerate cover-ups
Integrity	Ability to make sound moral judgements, approachable, principled
Intelligence	Clear thinking and ability to understand conflicting arguments
Open-mindedness	Open to new ideas and suggestions even if unconventional
Prudence	Ability to weigh up and take risks without being reckless
Self-confidence	Trust in own decisions and abilities without being self-righteous
Technical knowledge	Understanding of technical needs of the project and deliverables

Whilst these 'paragonial' attributes (apart from being charismatic) sadly do not seem to be necessary in a politician, they are desirable in a project leader and in fact many good project managers do possess these qualities which, in practice, result in the following abilities:

- Good communication skills, such as giving clear, unambiguous instructions and listening to others before making decisions.
- Inspiring the team by clearly setting out the aims and objectives, and stressing the importance of the project to the organization or indeed, where this is the case, the country.
- Fostering a climate in which new suggestions and ideas are encouraged and giving due credit when and where these can be implemented.
- Allocating the roles and tasks to the selected members of the team to suit the skills, abilities and personal characteristics of each member irrespective of race, creed, colour, sex or orientation.
- Gaining the confidence and respect of the team members by resolving personnel issues fairly, promptly and sympathetically.

Situational leadership

Situational leadership simply means that the management style has to be adapted to suit the actual situation the leader finds himself or herself in.

According to Hersey and Blanchard, who made a study of this subject as far back as 1960, managers or leaders must change their management style according to the level of maturity of the individual or group. Maturity can be defined as an amalgam of education, ability, confidence and willingness to take responsibility. Depending on this level of maturity, a leader must then decide, when allocating a specific task, whether to give firm, clear instructions without inviting questions or delegating the performance of the task, giving the follower a virtual freehand. These are the extreme outer (opposite) points of a behavioural curve. In between these two extremes lies the bulk of management behaviour. For convenience, the level of maturity can be split into four categories:

Category 1

Low skill, low confidence, low motivation

Category 2

Medium skill, fair confidence, fair ability, good motivation

Category 3

Good skill, fair confidence, good ability, high motivation

Category 4

High skill, high confidence, high ability, high motivation

The degree of direction or support given to the follower will depend on the leader's perception of the follower's maturity, but always in relation to a specific task. Clearly, a person can be more confident about one task or another, depending largely on his or her level of experience of that task, but situational leadership theory can only be applied to the situation (task) to be performed at this particular time.

The simplest way to illustrate situational behaviour is to look at the way tasks are allocated in the army.

High task, low support

A sergeant will give clear direction to a category 1 recruit, which he or she will not expect to be questioned on. There will be little technical or emotional support − just plain orders to perform the task.

High task, high support

A captain will give an order to the sergeant (category 2) but will listen to any questions or even suggestions the sergeant may make, as this follower may have considerable experience.

Low task, high support

A colonel will suggest a course of action to a major but will also discuss any fears or problems that may arise before deciding on the exact tactics.

Low task, low support

A commander in chief will outline his strategy to his general staff, listen to their views and will then let them get on with implementing the tasks without further interference.

Clearly, in every case, the leader must continue to monitor the performance of any follower or group, but this will vary with the degree of confidence the leader has in the follower. At the lowest level, it could be a check every half an hour. At the highest level, it could be a monthly report.

It is not possible to apply mathematical models to managing people who are not only diverse from one another, but can also change themselves day by day depending on their emotional or physical situation at the time.

Fig. 40.1 shows the four maturity categories set against the behaviour grid. It also superimposes a development curve that indicates the progression of behavioural changes from the lowest to the highest, assuming that the follower's maturity develops over the period of the project.

The leader can help to develop the maturity of the follower by gradually reducing the task behaviour, which means explaining the reasons for instructions and increasing the support by praising or rewarding achievements as soon as they occur. There should be a high degree of encouragement by openly discussing mistakes without direct criticism or apportioning blame. Phased monitoring and a well-structured feedback mechanism will highlight a problem before it gets out of hand, but probably the most important point to be hammered home is the conviction that the leader and follower are on the same side, have the same common interest to reduce the effect of errors and must therefore work together to resolve problems as soon as they become apparent.

It is the fear of criticism that inhibits the early disclosure of problems or mistakes, which tend to get worse unless confronted and rectified as soon as possible. Even senior managers risk instant dismissal if they deliberately submit incorrect information or unduly withhold an unpalatable financial position from the board of directors.

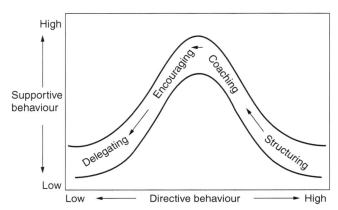

Figure 40.1 Situational leadership.
Modified from Hersey, P.H., & Blanchard, K.H. (2012). Management of organisational behaviour. Prentice-Hall

Leaders who are confident in their abilities and able to practise the low-task, low-support style will be able to delegate without completely abdicating their own accountability.

Delegation means transferring both the responsibility and authority to another person, but still retaining the right to monitor the performance as and when required. Clearly, if this monitoring takes place too frequently or too obtrusively, the confidence of the follower is soon undermined. Generally speaking, a monthly report is a reasonable method of retaining overall control, provided of course that the report is up to date, honest and technically correct.

Professionalism and ethics

All the major professional institutions expect their members to observe rules of conduct that have been designed to ensure that the standards of ethical behaviour set out in the charter and by laws of the institution are adhered to.

The rules set out the duties of professional members towards their employers and clients, their profession and its institution, their fellow members, the general public and the environment. These rules apply equally to professionals in full- or part-time employment and to those acting as professional consultants or advisers on a fee-basis for and on behalf of private or public clients. Contravention of this code could result in disciplinary proceedings and possible suspension, or even expulsion from the institution.

Project management is a relatively new profession when compared with the established professions such as law, medicine, architecture, accountancy, civil engineering and surveying.

A standard code of conduct was therefore produced by the Association for Project Management (APM), which sets out the required standards of professional behaviour expected from a project manager. These duties and responsibilities can be divided into three main categories as follows.

Further Discussion on Ethics Is Given in Chapter 43 (Governance).

Responsibilities to clients and employers

- Ensure that the terms of engagement and scope are agreed by both parties.
- Act responsibly and honestly in all matters.
- Accept responsibility for own actions.
- Declare possible conflicts of interest.
- Treat all data and information as confidential.
- Act in the best interest of the client or employer.
- Where required, provide adequate professional indemnity insurance.
- Desist from subcontracting work without the client's consent.

Responsibilities to the project

- Neither give nor accept gifts or inducements, other than of nominal value, from individuals or organizations associated with the project.
- Forecast and report realistic values in terms of cost, time and performance, and quality.
- Ensure the relevant health and safety regulations are enforced.
- Monitor and control all tasks.
- Ensure sufficiency and efficient use of resources as and when required.
- Take steps to anticipate and prevent contractual disputes.
- Act fairly and equitably in resolving disputes if called upon to do so.

Responsibilities to the profession of project management

- Only accept assignments for which he or she considers himself or herself to be competent.
- Participate in continual professional development.
- Encourage further education and professional development of staff.
- Refuse to act as project manager in place of another professional member without instructions from the client and prior notification to the other project manager.
- Always act in a manner that will not damage the standing and reputation of the profession, or the relevant professional institution.

Normally, one of the conditions of membership of a professional institution is that one accepts the rules of conduct without question and that any decision by the disciplinary committee is final.

Competence

The competence of a project manager can best be described as an amalgam of technical knowledge, experience, ability to handle people and the ability to work in a sometimes stressful environment.

A sponsor or senior director of an organization must be reasonably sure that a person designated as project manager or project director has the attributes described above and is therefore qualified for the particular project to be undertaken. As projects vary in size and complexity, the competence of the project manager will also vary, and it is the job of the sponsor to ensure that the project manager's competence profile matches the requirements of the project.

While some organizations have developed their own competence framework for their own staff, the need for a more universal set of metrics was clearly a step in the right direction so that smaller organizations with limited resources and less developed managerial maturity could benefit from an accepted standard.

For this reason, the APM designed and published the user-friendly APM Competence Framework in 2015, and with their kind permission some of their major suggestions for measuring competence are described below.

The framework applies equally to project management, programme management, portfolio management, project context and the project office. A manager in any of these five key concepts must satisfy the three relevant domains of technical competence, behavioural competence and contextual competence. Of these, the technical domain and the behavioural domain have been fleshed out by 30 and 9 competence elements respectively, which can be measured relatively easily to enable an assessment to be made of the manager's competence for the specific project to be undertaken.

The table in Fig. 40.2 shows the elements for the two domains based on the APM Body of Knowledge 6th edition.

The APM then defines four levels of competence A—D, which describe the experience required for each level ranging from project director, through two levels of project manager to associate project manager. A scoring matrix for knowledge and experience can then be drawn up either by a senior manager or by self-assessment.

Technical Competence	Behavioural Competence
Concept	Communication
Project success & benefit management	Teamwork
Stakeholder management	Leadership
Requirements management	Conflict management
Project risk management	Negotiation
Estimating	Human resource management
Business case	Behavioural characteristics
Marketing and sales	Learning and development
Project review	Professionalism and ethics
Definition	
Scope management	
Modelling and testing	
Methods and procedures	
Project quality management	
Scheduling	
Resource management	
Information management & reporting	
Project management plan	
Configuration management	
Change control	
Implementation	
Technology management	
Budgeting & Cost management	
Procurement	
Issue manageent	
Development	
Value management	
Earned value management	
Value engineering	
Handover & close-out	

Figure 40.2 The elements for the two domains based on the APM Body of Knowledge 6th edition.

In practice, many assessments on competence are made (often informally) at the job interview. An experienced senior manager can, by asking the right questions and studying the applicant's CV, obtain a fairly good idea of the technical and behavioural competence of the interviewee. Experience on a similar project in terms of size and complexity, which was successfully completed, is probably the best measure of competence.

Further reading

Hersey, P. H., & Blanchard, K. H. (2012). *Management of organisational behaviour*. Prentice-Hall.

Lewis, J. P. (2003). *Project leadership*. McGraw-Hill.

Negotiation

Chapter outline

However well a project is managed, it is inevitable that sooner or later a disagreement will arise between two persons or parties, be they different stakeholders or members of the same project team. If this disagreement escalates to become a formal dispute, a number of dispute resolutions exist (see Chapter 42) that have been designed to resolve the problem. However, it is far better, and certainly cheaper, if the disagreement, which may be financial, technical or organizational, can be resolved by negotiation.

Negotiation can be defined as an attempt to reach a result by discussion acceptable to both parties. This does not mean that either one or both parties are particularly happy with the outcome, but whatever compromise has been agreed, business or relationships between the parties can continue.

The ideal negotiation will end in a win—win situation, where both parties are satisfied that their main goals have been met, even at the expense of some minor concessions. More often, however, one party is not able to achieve the desired result and may well leave the negotiating table aggrieved, but the fact that the work can continue and the dispute does not escalate to a higher level, or that a commercial deal is struck rather than a complete breakdown of a business relationship, indicates that the negotiation has been successful.

Once it has been agreed by both parties to enter into negotiation, both parties should follow a series of phases or stages to achieve maximum benefit from the negotiation.

Phase 1: preparation

As with claims or legal proceedings, negotiations will have a greater chance of succeeding if the arguments are backed by good documentation. The preparation phase consists largely of collecting and collating these documents, and distilling them into a concise set of data suitable for discussion. These data could be technical data, test

Project Management, Planning and Control. https://doi.org/10.1016/B978-0-12-824339-8.00041-9

results and commercial forecasts, and could include precedents of previous discussions. There is really no limit as to what this back-up documentation should be, but the very act of reading these data and condensing them into a few pages will give the negotiator a clear picture of what the issues are. (See Fig. 41.1 for stages of negotiation.)

Phase 2: planning

It is pointless even considering a negotiating process if there is no intention to compromise. The degree of compromise and the limits of concessions that can be accepted have to be established in this phase. There is usually a threshold, below (or above), that must be respected, and the upper and lower limits in terms of time, delivery, money and payment arrangements as well as the different levels of compromise for each area must be established in advance. There must be a clear appreciation of what concessions can be accepted and at what stage one must either concede or walk away.

Generally, the party that has the most to gain from a negotiated settlement is automatically in the weaker position. In addition, factors such as financial strength, future business relationships, possible publicity (good or bad), time pressures and legal restraints must all be taken into consideration.

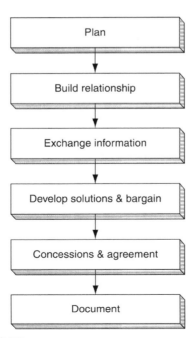

Figure 41.1 Negotiation stages.

The location of the negotiations must be given some consideration as it may be necessary to call in advisers or experts at some stage. There is some psychological advantage in having the negotiation on one's home ground and for this reason the other party may insist on a neutral venue such as a hotel or conference centre.

Phase 3: introductions

Negotiations are carried out by people and the establishment of a good relationship and rapport can be very beneficial. Knowledge of the other party's cultural background and business norms can help to put the other side at ease, especially where social rituals are important to them. Past cooperative ventures should be mentioned, and a discussion of common acquaintances, alliances and interests all help to break the ice and tend to put all parties at their ease. A quick overview of the common goals as well as the differences may enable the parties to focus on the important issues, which can then be categorized for the subsequent stages.

Phase 4: opening proposal

One of the parties must make an initial offer that sets out their case and requirements. The wording of this opening would give some indication of the flexibility as an inducement to reaching a mutually acceptable settlement. Often the requirements of the opening gambit are inflated to increase the negotiation margin, but the other party will probably adopt the same tactics. It is at this opening stage that the other party's body language such as hand gestures, posture, eye movements and facial expressions can give clues as to the acceptance or non-acceptance of particular suggestions or offers. The common identification of important points will help to lead the discussion into the next phase.

Phase 5: bargaining

The purpose of bargaining is to reach an agreement that lies somewhere between the initial extreme positions taken by the parties. Both parties may employ well-known tactics such as veiled threats, artificial explosions of anger or outrage, threats of walking out or other devices, but this is all part of the process. Often a concession on one aspect can be balanced by an enhancement on another. For example, a supplier may reduce his price to the level required by the buyer, provided his production (and hence his delivery) period can be increased by a few weeks or months. The buyer has to decide which aspect takes priority: money or time.

Concessions should always be traded for a gain in another area, which may not be necessarily in the same units or terms. For example, a reduction in price can be balanced by an increase in the number of units ordered or a later delivery. There should always be a number of issues on the table for discussion, so that quid pro quo deals can be struck between them.

Phase 6: agreement

Negotiations are only successful if they end with an agreement. If both parties walk away without an agreement, one or other (or possibly both) of the negotiators have not done their job and the case will probably end up in adjudication, arbitration or litigation. Concessions, which are not just given away, should not be regarded as a sign of weakness, but a realization that the other party has a valid point of view that merits some consideration. Both parties should be satisfied enough to wish to continue working or trading together and both are probably aware that there is always the risk that the legal costs of an action can exceed the amount in dispute. This realization often concentrates the mind to agree on a settlement. It may even be prudent, if there is no great time pressure, to leave the door open for a further discussion at a later date, or allow the future discussions to take place at a higher level of management.

Phase 7: finalizing

When an agreement has been reached, the deal has to be formalized by a written statement setting out the terms of the agreement. This must be signed by both parties attending the negotiations. In some cases, the agreement reached will be subject to ratification by senior management, but if the settlement is reasonable, such confirmation is usually given without question (Fig. 41.2).

It is a fact that the further a person is removed from the 'coal face' of the dispute, the more likely he or she is to ratify a settlement.

It must be pointed out that the negotiations involving labour disputes are best carried out by specialist negotiators with experience in industrial relations and national agreements, local working practices and labour laws. The procedures for such negotiations, which often end up with applications to conciliation boards or tribunals, are outside the scope of this book.

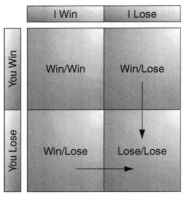

Figure 41.2 Negotiation outcomes.

However, differences of opinion can sometimes be reconciled by resorting to mediation involving the help of an independent third party.

When an agreement between the parties appears to be impossible, but neither party relishes the idea of potentially expensive and drawn-out arbitration or litigation, a practical next step would be for both parties to consider resorting to the relatively inexpensive and quick process of mediation. If this procedure fails, there is still the option of adjudication. Both these dispute resolution procedures are described more fully in Chapter 42.

Conflict management and dispute resolution

42

Chapter outline

Conciliation   395
Mediation   395
Adjudication   396

Conflict management covers a wide range of areas of disagreement from smoothing out a simple difference of opinion to settling a major industrial dispute.

Projects, as life in general, tend to have conflicts. Wherever there is a wide variety of individuals with different aspirations, attitudes, views and opinions there is a possibility that what may start out as a misunderstanding escalates into a conflict. It is one of the functions of a project manager to sense where such a conflict may occur and, once it has developed, to resolve it as early as possible to prevent a full-blown confrontation that may end in a strike, mass resignations or a complete stoppage of operations.

Conflicts can be caused by differences in opinions, cultural background or customs, project objectives, political aspirations or personal attitudes. Other factors that tend to cause conflicts are poor communications, weak management, competition for available resources, unclear objectives and arguments over methods and procedures.

Conflict between organizations can often be traced back to loose contractual arrangements, sloppy or ambiguous documentation and non-confirmation in writing of statements or instructions.

Thomas and Kilman published a study on conflict management and suggested five techniques that can be employed for resolving conflicts. These are as follows:

- Forcing
- Confronting the problem
- Compromising
- Smoothing
- Withdrawing

Forcing involves one party using its authority acquired by virtue of position in the organization, rank or technical knowledge to force through its point of view. While such a situation is not uncommon in the armed forces where it is backed up by strict discipline, it should only be used in a project environment where there is a health and safety issue, or in an emergency posing a risk of serious physical damage. In most of the situations where forcing has been used to solve a conflict, one party almost certainly feels aggrieved with a consequent adverse effect on morale and future cooperation.

Project Management, Planning and Control. https://doi.org/10.1016/B978-0-12-824339-8.00042-0

Confronting the problem is, by contrast, a more positive method. In this situation, both parties will try to examine what the actual issue is and will make a concerted effort to resolve it by reasoning and showing mutual respect for each other's point of view. The most likely situation where this method will succeed is when both the parties realize that failure to agree will be disastrous for everybody, and when success will enhance both their positions, especially when it is understood that future cooperation is vital for the success of the project. Often there are useful by-products such as innovative solutions or a better understanding of the wider picture.

Compromising is probably the most common method to resolve disputes, but generally both parties have to give up something or part with something, whether it is a point of principle, a financial claim, pension rights or an improvement in conditions. This means that the settlement may only be temporary and the dispute may well flare up again when one of the parties believes itself to be in a better bargaining position. No one really wins, yet both lose something and it may well be the subject of regret later when the effects of the compromise become apparent. Often commercial or time (programme) pressures make it necessary to reach a quick compromise solution, which means that if these pressures had not existed a more rational discussion could have produced a more lasting result.

Smoothing is basically one party acceding to the other party's demands because a more robust stance would not be in their best interest. This could occur where one party has more authority or power (financial, political or organizational) or where the arguments put forward are more cogent. Smoothing does not mean complete surrender, as it may just not be opportune or politically wise at this particular time to be more assertive.

Withdrawing in effect means avoiding the issue or ignoring it. While this may appear to be a sign of weakness, there may be good reasons for taking this stance. One may be aware that the dispute will blow over when the other party's anger has cooled down or a confrontation is likely to inflame the situation even more. One may also feel that the possibility of winning the argument is small, so that by making what may be considered a small concession, good relations are maintained. In practice, this procedure is only suitable for minor issues since by ignoring important ones, the problem is only shelved and will have to be resolved at a later date. If the issue is a major one and unlikely to be resolved by the other four options, it may still be correct for one or both parties to withdraw and agree to take the dispute to adjudication, arbitration or litigation as described later.

Whatever techniques are adopted in resolving disputes, the personality of the project manager or facilitator plays a major role. Patience, tact, politeness and cool-headedness are essential irrespective of the strength or weakness of the technical case. Any agreement or decision made by a human being is to a large extent subjective, and human attributes (or even failings) such as honour, pride, status or face-saving must be taken into account. It is good politics to allow the losing party to keep their self-respect and self-esteem. Team members may or may not like each other, but any such feelings must not be allowed to detract from the professionalism required to do their job.

In general, confrontation is preferable to withdrawal, but to follow such a course, project managers should practise the following:

- Be a role model and set an example to the team members in showing empathy with the conflicting parties.
- Keep an open-door policy and encourage early discussion before it festers into a more serious issue.
- Hear people out and allow them to open up before making comments.
- Look for a hidden agenda and try to find out what is really going on as the conflict may have different (very often personal) roots.

When a dispute involves organizations outside the project team, such as suppliers, subcontractors or labour unions, professional specialized assistance is essential in the form of commercial lawyers or industrial (labour) relations officers.

When the conflict is between two organizations and no agreement can be reached by either discussions or negotiations between the parties, it may be necessary to resort to one of the following five established methods of dispute resolution available to all parties to a contract. These, roughly in order of cost and speed, are as follows:

- Conciliation
- Mediation
- Adjudication
- Arbitration
- Litigation

Conciliation

The main purpose of conciliation, which is not used very often in commercial disputes, is to establish communications between the parties so that negotiations can be resumed. Conciliators should not try to apportion blame, but to focus on the common interests of the parties and the systemic reasons for the breakdown of relationships.

Mediation

In mediation, the parties in dispute contact and engage a third party either directly or via the mediation service of one of the established professional institutions. Although the parties retain control over the final outcome, which is not enforceable, the mediator, who is impartial and often experienced in such disputes, has control over the proceedings and pace of the mediation process. The mediator must on no account show him or herself to be judgemental or give advice or opinions, even if requested to do so. His or her main function is to clarify and explore all the common interests and issues as well as possible options, which may lead to a mutually beneficial and acceptable settlement. Once an agreement has been reached, it must be recorded in writing.

If mediation is started early enough before the differences become entrenched, the possibility of an amicable settlement is high. Provided legal advisers are not employed, the only costs are the fees of the mediator, which makes the procedure much cheaper and certainly quicker than any of the three more formal and legally binding dispute resolution procedures described later.

Adjudication

Although adjudication has always been an option in resolving disputes, it required the agreement of both parties. This also meant that both parties had to agree who would be the adjudicator. As this in itself could be a source of disagreement, it was not a common method of dispute resolution until the 1996 Construction Act, more accurately called 'Housing Grants, Construction and Regeneration Act 1996' was passed. This Act allowed one party to apply to one of a number of registered institutions called the *Adjudicator Nominating Body* (ANB) to appoint an independent adjudicator. The other party is then obliged by law to accept both the adjudication process and the nominated adjudicator.

Certain types of contract such as mining, extraction of coal, oil and gas and power generation are not covered

The process of adjudication has to follow strict procedures which can vary slightly depending on the ANB, most of whom have produced their own set of adjudication procedures. In the absence of such an ANB procedure being available, the act requires that the procedure to be followed is that set out in the 'The Scheme for Construction Contracts (England and Wales) Regulations 1998' known as the 'Scheme'. A slightly different Scheme applies for Scotland, but most procedures drawn up by the various ANBs follow the principles of the Scheme. The initiating party, called the *referring party* can choose which ANB to use and can also decide whether to use the Scheme or the chosen ANB's procedure.

The procedure to be followed is as follows:

(a) The referring party serves a Notice of Adjudication to the other party of the dispute, known as the *responding party* and to the adjudicator
(b) Once appointed, the adjudicator must follow the terms specified in the Contract.
(c) The adjudicator invites the *referring party, to* submit details of the dispute, called the *referral*, which must also be sent to the responding party.
(d) The adjudicator issues a programme giving the dates by which specified documents must be submitted
(e) After about 7 days the responding party must issue a Response in which they to put forward their case.
(f) The adjudicator then reviews these submissions together with any other papers or evidence he may request, and is obliged to give a ruling (called a *decision*) within 28 days after the Referral. However the Referring party can grant an extension of a further 14 days. Further

extension of time is only possible with the agreement of both parties. Although in the early days of adjudication, the adjudicator had to be requested to give *reasons* for his decision, in most cases giving reasons is now the norm.

In practice project management principles should be used, for example, only one person is nominated from each party to communicate with the other parties including the adjudicator. Confirmation of receipt for all letters, e-mails and faxes should be requested and filed with the original. As with all documents, it is vital to read the small print.

Originally the adjudicator dealt direct with the two disputing parties. Now both parties appoint lawyers or claims consultants to prepare their case. This has inevitably increased the cost and led to an emphasis on procedure, often resulting in subjecting the adjudicator to intimidatory tactics and a challenge to the adjudicator's jurisdiction. As the adjudicator cannot confirm his own jurisdiction this causes some confusion which may have to be decided by a court.

At times, the decision by the adjudicator is challenged by consultants or lawyers based purely on alleged non-compliance of procedure and this too will have to be decided by a court.

While a court will generally uphold an adjudicator's decision, a Court of Appeal ruling in 2012 (PC Harrington Contractors Ltd vs Systech International Ltd [2012] EWCA Civ 1371) stipulates that if the Adjudicator's decision has been revoked as a result of default or misconduct, the Adjudicator loses his fees unless a special clause is inserted in the Adjudicators contract with the parties, which provides for payment even in the event of an unenforceable decision. Fees charged by the adjudicator are levied jointly and severally so that if one party refuses to pay, the other party is liable for payment including interest for late payment.

The original Housing Grants, Construction and Regeneration Ac came into force in May 1998. This was replaced by the Local Democracy Economic Development and Construction Act in 2011 and a number of important changes were incorporated in Part 8, which covers adjudication.

(1) 'Pay when paid' clauses will become ineffective.
(2) Contracts need not be in writing. In other words, oral contracts are equally binding.
(3) A Payment Notice will have to be issued within 5 days of the payment due.
(4) The paying party must pay the notified sum.
(5) If the notified sum has not been paid, work can be suspended.
(6) Performance obligations which are conditional on related conditions are invalid.

Clearly even with these amended clauses, Contracts, Notices and Instructions should preferably still be in writing.

To reduce the cost of the Adjudication process, The Construction Industries Council (CIC) introduced a Low Value Disputes (LVD) Model Adjudication Procedure in

May 2020, for disputes under £50,000. This restricts the documentation to four lever arch files per party and limits the Adjudicator's fee as follows:

Value of claim (excluding VAT)	Max. adjudicator's fee
Up to £10,000	£2000
£10.001–£25,000	£3500
£25,001–£50,000	£6000
Over £50,000	To be negotiated by the parties

It does not however limit the fees of legal or other advisors to the parties, which can be substantially more than the Adjudicator's fees.

Governance of project management

Chapter outline

Introduction

The 37 Country Organisation for Economic Co-operation and Development (OECD) describes corporate governance as involving 'a set of relationships between a company's management, its board, its shareholders and other stakeholders.' It states that corporate governance 'provides the structure through which the objectives of the company are set, and the means of attaining those objectives and monitoring performance are determined'.

Within the United Kingdom, the Financial Reporting Council[1] (FRC), in order to promote transparency and integrity in financial markets, publishes the 'The UK Corporate Governance Code 2018'.[2] The FRC defines corporate governance simply as 'the system by which companies are directed and controlled'.

These definitions require that Governance applies to both the once-off activities of an organization as well as to ongoing operations. Once-off activities are managed through project management, however immature this discipline maybe in any particular organization.

---

[1] As of 2020 due to be superseded by The Audit, Reporting and Governance Authority.
[2] 'The UK Corporate Governance Code 2018', Financial Reporting Council.

Competence in directing and managing complex change is increasingly recognized as a competitive advantage in the business world and a requirement in the public sector. Hence, the importance of governance of project management. For UK companies, this significance is reinforced by the requirement to issue Strategic Reports which should, where material, refer to their project management performance.[3]

Governance of project management is the framework within which projects are managed. The governance of any one particular project being a specific subset of the governance of project management, that is, the governance of the wider portfolio, programme and project management in an organization. This chapter is concerned with this wider scope.

Within a company, the board of directors is responsible for governance. However, as implied by the OECD definition, governance achievement is not by any one body but involves relationships between interested parties. For corporations, these interested parties include the board, management and shareholders as well as the legislature and other stakeholders. For public sector organizations, social enterprises and other non-corporate organization, the parties responsible for and providing governance will differ. The following sections will be of special benefit to corporations, public sector and non-corporate private sector organizations including charities.

The delivery of governance, of once-off as of ongoing activities, requires Annual General Meetings, Board meetings, Benchmarking, Standard procedures, Assurance, Marketing and Public Relations. Another key governance competence is responding to changes in context in any of the social, technical, economic, environmental, political, legal and regulatory and ethical external factors influencing the organization. Good practice is evidenced by documents such as of Memorandum of Understanding, adopted policies, job descriptions, delegated authority schedules, management reports and external communications, supplemented by observed behaviour and results.

After identifying the necessary principles, it is important to concentrate on the components of the required systems, on the main roles involved and on the complicating factors in assessing performance.

Principles

To assist directors and equivalents to ensure that governance requirements are applied to project management throughout their enterprise, 16 principles can be identified. These are published by the UK chartered Association for Project Management (APM) in 'Directing Change, A guide to governance of project management'.[4] With the permission of APM, these 16 principles are repeated in the following abbreviated form:

(1) The organization differentiates between projects and business as usual.
(2) The Board has overall responsibility for the governance of project management.

---

[3] 'Guidance on the Strategic Report', 2018, Financial Reporting Council.
[4] 'Directing Change, A guide to the governance of project management', 3rd Edition 2018.

(3) There is a demonstrably alignment and coherence between the project portfolio and the business strategy.

(4) The Board reviews regularly its portfolio of projects

(5) Projects are formally started and are formally closed when they are completed or no longer justified.

(6) Roles and responsibilities for the governance of project management are defined clearly and applied.

(7) Each change initiative has a competent and engaged sponsor.

(8) Members of the board and delegated authorization bodies have appropriate competence, representation, authority and resources.

(9) Disciplined governance arrangements, supported by appropriate ethics, cultures, methods, resources and controls are applied to all change initiatives throughout the project life cycle.

(10) The organization fosters a culture of improvement and of frank internal disclosure of information.

(11) The board or its delegated agents decide when independent scrutiny is required.

(12) Project business cases are supported by relevant and realistic information that provides a reliable basis for making authorization decisions. Business cases are used as control documents.

(13) All programmes and projects have an approved plan containing authorization points at which the business case, risks, viability and strategic alignment are reviewed. Decisions made at authorization points are recorded and communicated.

(14) There are clearly defined processes and criteria for reporting status and for the escalation of risks and issues to the appropriate levels for action.

(15) Stakeholders are engaged in a manner that fosters understanding and trust.

(16) Lessons from other change initiatives are consciously embedded into new initiatives.

Components and roles

To fulfil these principles, five components of portfolio, programme and project direction and management have been identified. These are the following:

(a) Portfolio direction and alignment
(b) Programme and project sponsorship
(c) Project management capability
(d) Transparency and assurance
(e) Culture and ethics

Typical roles effecting governance are the following:

1. The Board
2. Programme or Project Sponsor
3. Programme or Project Manager
4. Business Change Manager
5. Independent Reviewer

Appendix 2 of 'Directing Change, A guide to governance of project management' provides useful checklists for each of these role's contribution. Other organizational units sometimes present include Project Management Offices, Centres of Excellence, Organization Planning Departments and Internal Auditors. To these, elements of governance responsibility may be delegated along with relevant authority and account-ability. The sections below note key governance requirements under the headings of the five components.

Part of good governance is ensuring that appropriate management methods are applied across the range of work. Where the increasingly popular Agile[5] techniques are used for project management, 12 additional governance principles have been identified in the publication 'Directing Agile Change'.[6] This guide includes a section on the importance of appropriate behaviour in the governance of agile methods. Also, the UK National Audit Office's 2012 review 'Governance for Agile Delivery'[7] identified four specific principles.

Portfolio direction and alignment

The Board should ensure that the organization's portfolio of complex change projects is within its capacity and aligned with its strategy. Priorities of projects within the portfolio should reflect key performance objectives.
The organization should discriminate effectively between activities requiring project man-agement and the remaining ongoing operations.
The Board should recognize its responsibility to apply governance to project management, devoting sufficient time to this.
The impact of the various projects must be compatible with the ongoing operations of the organization.
Financial planning and control must be applied specifically to individual projects as also to the entire portfolio.
Project risks should be regularly assessed and integrated so as to determine their combined impact on the organization as a whole.
Sponsoring organizations together with external stakeholders such as suppliers, customers, finance providers and regulators must be sufficiently engaged to ensure portfolio success.

Sponsorship

Whilst this subject is discussed in Chapter 5, the following aspects of this role are particularly relevant in relation to governance:

Sponsors link the Board to the governance and management of every programme or project.
The Board should ensure formal appointment of competent well motivated sponsors.
Sponsors should devote sufficient time to their programme or project from concept to final hand-over and closure.

---

[5] See https://agilemanifesto.org/2001.
[6] 'Directing Agile Change', APM, 2016.
[7] See https://www.nao.org.uk/wp-content/uploads/2012/07/governance_agile_delivery.pdf.

Sponsors should be held accountable for the business case, right through to benefit realization. Transfers of sponsorship duties should be formalized so that responsibility is not obscured.

Sponsors should ensure that appropriate management methods are used.

Sponsors should keep up to date with status by convening regular meetings with their managers and be prepared to seek independent advice to assist them with appraisal.

Sponsors should ensure that sufficient resources are available. These include appropriate skilled personnel.

Project management capability

Where delegated, the key roles and responsibilities for programme and project governance should be clearly specified and applied.

The success of programmes and projects depends on the experience and leadership qualities of the managers, the skills of team members, the timely availability of the necessary resources and on the processes and procedures employed. It is necessary to ensure therefore that:

- Business cases fully reflect the strategic objectives of the organization. Projects have clear objectives, scope definitions, envisaged business outcomes and critical success criteria which will enable realistic decisions to be taken.
- The project manager, team members and operatives are competent, aware of their roles and responsibilities and motivated to improving the performance and delivery.
- Management authority is delegated to the right levels, balancing efficiency and control.
- Risks and contingencies are identified, assessed and controlled in accordance with delegated powers. Consequential allowances are made for contingencies which are controlled by authorized persons.
- The organization's project management processes, procedures and management tools are appropriate for its range of projects.
- Dynamic stakeholder management procedures are applied with particular emphasis on communications to, from and between stakeholders.
- The suppliers, contractors and providers of services (internal and external) have competent staff and sufficient resources to meet the project's requirements.

The Board should ensure that the organization's change management processes specify the need for approved plans and review points, are subject to continual improvement.

Transparency and assurance

Project management information flows up, down and across as well as to and from organizations. Regular, timely and reliable disclosure and reporting is an essential part of good project management.

The Board must ensure that it, and others involved in governance, receive relevant communications that are timely and reliable. Thresholds should be established to escalate significant issues, risks and opportunities through the organization to the Board.

Checks should be carried out to ensure that the reporting procedures comply with company policies.

Top-level reports on any project should cover the following:

Progress

Financial forecasts, including commitments

Completion forecasts
Major risks
Major quality issues
Performance
Compliance with or deviations from key performance indicators
When appropriate Boards should also seek independent verification of reported project and portfolio information.
Boards should reflect their portfolio status in communications with key stakeholders, commensurate with their interests. Thus the Principle Risks and Uncertainties section of an Annual Report should reflect the impact of the Project Portfolio. Where projects are important, assurance should be provided that appropriate governance is in place.
To ensure that effort is only expended to produce necessary information, periodic appraisal is recommended by peer and/or external groups of the cost and value of information flows. Reporting should be commensurate with each initiative's complexity and significance.

Culture and ethics

Performance depends as much on an organization's culture and ethics as on systems and procedures. The Board should develop and promote policies on the desired culture, including on ethics. Governance of project management should require that project management complies with the sponsoring organization's policies.
It is the duty of project management practitioners to be fully informed about their organization's and their profession's policies on ethics. As a good example, the UK National Audit Office requires that 'all those who are members or students of professional bodies must uphold the codes of ethics of those bodies in addition to their obligations under this Code'.
The culture of the organization should encourage honest and open reports with arrangements in place to support 'whistle blowing' and to avoid fraudulent and illegal activity. Nevertheless, targeted independent reviews may be required.
The culture may also promote creativity and innovation. In projects this translates to risk management focussing on and exploiting opportunities as well as on responding to threats. Both external and internal stakeholders should be engaged proportionately and honestly in a manner that fosters understanding and trust.
Continual improvement should apply to cultural and ethical performance as to other areas of project management.[8] Good and bad examples of behaviour related to project management should be provided to ensure staff understanding.
There are however contexts where compatible ethical standards differ or do not exist. Indeed, it may be difficult if not impossible to carry out business operations in certain countries unless 'on-costs', often called euphemistically mobilization costs, introduction fees or facilitation payments, are added to the contract sum. Such allowances are often added to the fees paid to the agent representing the organization locally and that is the end of the matter as far as the company is concerned. Care must be taken in all cases to comply with the requirements of the UK 'Bribery Act 2010' or any other applicable Corruption Practices legislation.[9]
Another difficulty is where corporate control is confused due to complex arrangements made through tax havens. These often obscure sources of finance, ruling regulations, effective

[8] See 'Exit, Voice and Loyalty' by Albert Hirschman, 1970.
[9] See US 'Foreign Corrupt Practices Act', 1977.

ownership and the amount and destination of fees, interest and dividend payments. Such cases are increasingly common. Professionals should take care that they are operating within their codes of conduct, particularly with regard to the public interest. It is not sufficient to rely on the stewardship performance of institutional investors or on regulators. Governance should ensure compliance with relevant regulations.[10]

On a more personal level, there is the issue of how far one can go when giving or receiving seasonal gifts, entertaining clients or being entertained by suppliers. This should be addressed in the organization's policies and codes. Where this is not available, perhaps the clearest guidance was given by a senior manager of a construction company to a project manager 'You can accept anything − provided you can carry it home in your stomach'. The National Audit Office provides an excellent example[11] of guidance on this topic.

Evidence

With the increased focus on governance, second and third parties such as clients, investors, suppliers and regulators are looking for evidence of good governance. Some have developed scoring systems for the maturity of governance, inclusive of the governance of change. Difficulties must be recognized in this external scoring.

Governance arrangements for an organization should be tailored to its own needs. Significantly different interests in an organization's governance apply different weights to the same aspect of governance. As found by the UK Institute of Directors[12] it is unlikely that any standard scheme of measurement will usefully apply across organizations at different stages of development or in differing sectors and cultures. They state 'We can reject the naïve hypothesis that all components of corporate governance have an equal impact on the perception of corporate governance'. The difficulties arising in applying standard governance models are especially complicated where one organization does not have sole control of one or more projects.[13]

Particular governance issues arise in the public sector, such as political direction, confidentiality constraints and conflicts in applying professional codes of conduct. As successfully demonstrated by the National Audit Office, these require different treatment to that in the private sector.

The quality of applied governance is difficult to measure. The reality is that dynamic governance decisions can be made under stress and subject to group dynamics that are not recorded. Assurance schemes too often rely entirely on verifiable documented evidence. These capture only a part of the significant reality.

However measured, there is wide agreement about the significance of governance of project management. Research by APM[14] showed that, with clarity of goals and good planning and review, effective governance is amongst three factors with the highest bearing on project management success.

---

[10] 'The Money Laundering and Terrorist Financing (Amendment) Regulations 2019', UK statutory instrument.
[11] 'Code of Conduct', 2020, National Audit Office.
[12] 'The 2016 Good Governance Report', Institute of Directors.
[13] See 'Governance of Co-Owned Projects', APM, 2016.
[14] 'Conditions for Project Success', APM, 2015.

Conclusion

Too many projects fail not because of poor project management effort, but because the project's governance regime is poorly developed and not integrated into the wider governance of the organization. Implementing the principles above through the five components described will go a long way to ensure the reliable delivery of projects contributing to sustainable organizational success.

Project close-out and handover

Chapter outline

Close-out

Most projects involving construction or installation work include a *commissioning* stage during which the specified performance tests and operating trials are carried out with the objective of proving to the client that the deliverables are as specified and conform to the required performance criteria. The *snagging process*, which should have taken place immediately prior to the start of commissioning, often overlaps the commissioning stage so that adjustments and even minor modifications may be necessary. Commissioning is often carried out with the assistance of the client's operatives, to ensure that the person who runs the plant or system learns how to operate the controls and make necessary adjustments. This is as true for a computer installation as a power station.

On more complex projects, it may be necessary to run special training and familiarization programmes for clients' staff and operatives, in both workplace and classrooms.

When the project is complete and all the deliverables are tested and approved, the project must be officially closed out. This involves a number of checks to be made and documents to be completed to ensure that there is no 'drip' of man-hours being booked against the project. Unless an official, dated close-out instruction is issued to all members of the project team, there is always a risk of time and money being expended on additional work not originally envisaged. Even where the work was envisaged, there is the possibility of work being dragged out because no firm cut-off date has been imposed.

All contracts (and subcontracts) must be properly closed out and (if possible) all claims and back charges (including liquidated damages) agreed and settled.

A few unpopular, but necessary, tasks prior to commissioning are collation, indexing and binding of all the operating and maintenance manuals, drawings, test certificates, lubrication schedules, guarantees and priced spare lists that should have been collected and stored during the course of the project. Whether this documentation is in electronic format or hard copy, the process is the same. Indeed some

Project Management, Planning and Control. https://doi.org/10.1016/B978-0-12-824339-8.00044-4

client organizations require both, and the cost of preparing this documentation is often underestimated.

Many of these documents obtained and collated during various phases of the project have to be bound and handed over to the client enabling the plant or systems to be operated and maintained. It goes without saying that all these documents have to be checked and updated to reflect the latest version and as-built condition.

The following list gives some of the documents that fall into this category:

- Stage acceptance certificates.
- Final handover certificate.
- Operating instructions in electronic or hard copy format or both.
- Maintenance instructions or manuals.
- A list of operational and strategic spares with current price lists and anticipated delivery periods as obtained from the individual suppliers. These are divided into operating and strategic spares.
- Lubrication schedules.
- Quality-control records and audit trails.
- Material test certificates including confirmation of successful testing of operatives' (especially welders') test certificates and performance test results.
- Radiography and other non-destructive testing records.
- A dossier of the various equipment, material and system guarantees and warranties.
- Equipment test and performance certificates.

On completion, the site must be cleared; all temporary buildings, structures and fences have to be removed and access roads must be made good.

Arrangements should be made to dispose unused equipment or surplus materials. These may be sold to the client at a discounted rate or stored for use on another project. However, certain materials, such as valves, instruments and even certain piping and cables, cannot be used on other jobs unless the specified test certificates and certificates of origin are literally wired to the item being stored. Materials that do not fall into these categories will have to be sold for scrap and the proceeds credited to the project.

Project managers who want to appreciate their team may decide to use this money for a closing-down party. The team will now have to be disbanded, a process that is the 'mourning' stage of the Tuckman team phases. On large projects that required the team to work together for many months or years, the close-out can be a terrible anticlimax and the human aspect must be handled diplomatically and sympathetically.

Handover

The formal handover involves an exchange of documents, which confirm that the project has been completed by the contractor or supplier and accepted by the client. These documents, which include the signed acceptance certificate, will enable the contractor to submit his final payment certificate, subject to agreed retentions. If a retention bond has been accepted by the client, payment has to be made in full.

Project close-out report and review

Chapter outline

Close-out report

Currently, most organizations require the project manager to produce a close-out report at the end of the project. This is often regarded by some project managers as a time-consuming chore, as in many cases the project manager would already have been earmarked for a new project which he or she is keen to start as soon as possible.

Provided a reasonably detailed project diary has been kept by the project manager throughout the various stages of the project, the task of producing a close-out report is not as onerous as it would appear. Certainly, if the project included a site construction stage, the site manager's diary, which is in most companies an obligatory document, will yield a mass of useful data for incorporation in the close-out report. The information given in the report should cover not only what went wrong and why, but also the successes and achievements in overcoming any particularly interesting problem.

The following is a list of some of the topics that should be included in a close-out report:

- Degree to which the original objectives have been met
- Degree of compliance with the project brief (business case)
- Degree to which the original KPIs have been achieved
- Level of satisfaction expressed by client or sponsor
- Comparison between original (budgeted) cost and actual final cost
- Reasons for cost overruns (if any)
- Major changes incorporated due to:
  - Client's approved requirements
  - Internal modifications caused by errors or omissions
  - Other possible reasons (statutory, environmental, legal, health and safety, etc.)
- Comparison between original project time and actual total time expended
- Reasons for time overruns or underruns
- Major delays and the causes of these delays
- Special actions taken to reduce or mitigate particular delays
- Important or interesting or novel methods adopted to improve performance
- Performance and attitude of project team members in general and some in particular

Project Management, Planning and Control. https://doi.org/10.1016/B978-0-12-824339-8.00045-6

- Performance of consultants and special advisers
- Performance of contractors, subcontractors, and suppliers
- Attitude and behaviour of client's project manager (if there was one)
- Attitude and behaviour of client's staff and employees
- Comments on the effectiveness of the contract documents
- Comments on the clarity or otherwise of specifications, datasheets, or other documents
- Recommendations for actions on future similar projects
- Recommendations for future documentation to close loopholes
- Comments on the preparation and application/operation of major project management tools such as CPA, EVA, and data gathering/processing

The report will be sent to the relevant stakeholders and discussed at a formal close-out meeting at which the stakeholders will be able to express their views on the success (or otherwise) of the project. At the end of this meeting, the project can be considered to be formally closed.

Close-out review

Using the close-out report as a basis, the final task of the project manager is to carry out a post-project review (or a post-implementation review), which should cover a short history of the project and an analysis of the successes and failures together with a description of how these failures were handled.

The review will also discuss the performance of the project team and the contributions (positive and negative) of the other stakeholders. All this information can then be examined by future project managers employed on similar projects or working with the same client/stakeholders so that they can be made aware of the difficulties and issues encountered and ensure (as far as is practicable) that the same problems do not arise. Learning from previous mistakes is a natural process developed from childhood. Even more beneficial and certainly wider reaching, is learning from other people's mistakes. For example, where a new project manager finds that he has to deal with people, either in the client's or contractor's camps, who were described as 'difficult' in a previous close-out report, he or she should contact the previous project manager and find out the best ways of 'handling' these people.

For this reason, the close-out review, together with the more formal close-out report, has to be properly indexed and archived in hard copy or electronic format for easy retrieval.

The motto is: 'Forewarned is Forearmed'.

Stages and sequence

46

Summary of project stages and sequence

The following are the stages and sequences in diagrammatic and tabular format:

1. Fig. 46.1 shows the normal sequence of controls of a project from business case to close-out.
2. Fig. 46.2 provides a diagrammatic version of the control techniques for the different project stages.
3. Fig. 46.3 is a hierarchical version of the project sequence, which also shows the chapter numbers in the book where the relevant stage or technique is discussed.
4. Table 46.1 is a detailed tabular breakdown of the sequence for a project control system, again from business case to project close-out.

While the diagrams cover most types of projects, it must be understood that projects vary enormously in scope, size and complexity. The sequences and techniques given may therefore have to be changed to suit any particular project. Indeed certain techniques may not be applicable in their entirety or may have to be modified to suit different requirements. The principles are, however, fundamentally the same.

Project Management, Planning and Control. https://doi.org/10.1016/B978-0-12-824339-8.00046-8

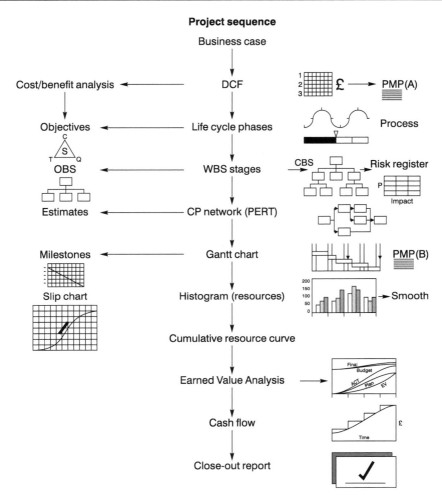

Figure 46.1 Project sequence.

Project

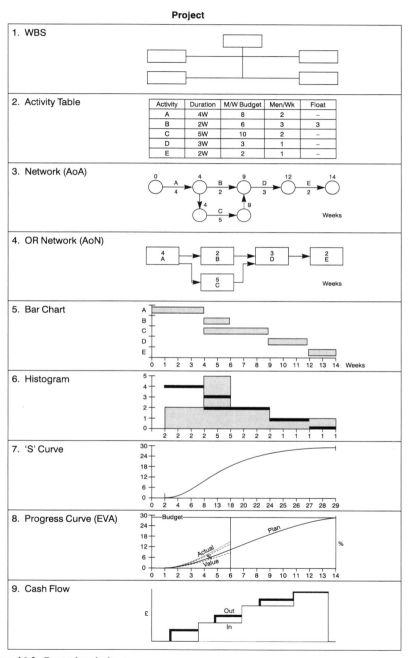

Figure 46.2 Control techniques.

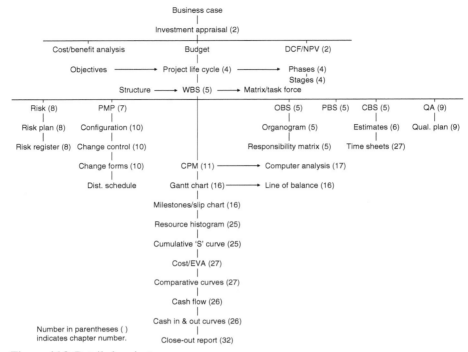

Figure 46.3 Detailed project sequence.

Table 46.1 Sequence for project control system.

Business case
Cost/benefit analysis
Set objectives
DCF calculations
Establish project life cycle
Establish project phases
Produce project management plan
Produce budget (labour, plant, materials, overheads, etc.)
Draw work breakdown structure
Draw product breakdown structure
Draw organization breakdown structure
Draw responsibility matrix
List all possible risks

Table 46.1 Sequence for project control system.—cont'd

Carry out risk analysis
Draw up risk management plan
Produce risk register
Draw up activity list
Draw network logic (freehand)
Add activity durations
Calculate forward pass
Revise logic (maximize parallel activities)
Calculate second forward pass
Revise activity durations
Calculate third forward pass
Calculate backward pass
Mark critical path (zero float)
Draw final network on grid system
Add activity numbers
Draw bar chart (Gantt chart)
Draw milestone slip chart
Produce resource table
Add resources to bar chart
Aggregate resources
Draw histogram
Smooth resources (utilize float)
Draw cumulative 'S' curve (to be used for EVA)
List activities in numerical order
Add budget values (person hours)
Record weekly actual hours (direct and indirect)
Record weekly % complete (in 5% steps)
Calculate value hours weekly
Calculate overall % complete weekly
Calculate overall efficiency weekly
Calculate anticipated final hours weekly

Continued

Table 46.1 Sequence for project control system.—cont'd

Draw time/person hour curves (budget, planned, actual, value, anticipated final)

Draw time/% curves (% planned, % complete, % efficiency)

Analyse curves

Take appropriate management action

Calculate cost per activity (labour, plant, materials)

Add costs to bar chart activities

Aggregate costs

Draw curve for plant and material costs (outflow)

Draw curve for total cash OUT (this includes labour costs)

Draw curve for total cash IN

Analyse curves

Calculate overdraft requirements

Set up information distribution system

Set up weekly monitoring and recording system

Set up system for recording and assessing changes and extra work

Set up reporting system

Manage risks

Set up regular progress meetings

Write close-out report

Close-out review

Worked example 1: Bungalow

47

Chapter outline

The previous chapters described various methods and techniques developed to produce meaningful and practical network programmes. In this chapter, most of these techniques are combined in two fully worked examples. One is mainly of a civil engineering and building nature and the other is concerned with mechanical erection — both are practical and could be applied to real situations.

The first example covers the planning, man-hour control and cost control of a construction project of a bungalow. Before any planning work is started, it is advantageous to write down the salient parameters of the design and construction, or what is grandly called the 'design and construction philosophy'. This ensures that everyone who participates in the project knows not only what has to be done, but why it is being done in a particular way. Indeed, if the design and construction philosophy is circulated *before* the programme, time- and cost-saving suggestions may well be volunteered by some recipients which, if acceptable, can be incorporated into the final plan.

Design and construction philosophy

1. The bungalow is constructed on strip footings.
2. External walls are in two skins of brick with a cavity. Internal partitions are in plasterboard on timber studding.
3. The floor is suspended on brick piers on an oversite concrete slab. Floorboards are T&G pine.
4. The roof is tiled on timber-trussed rafters with external gutters.
5. Internal finish is plaster on brick finished with emulsion paint.
6. Construction is by direct labour specially hired for the purpose. This includes specialist trades such as electrics and plumbing.
7. The work is financed by a bank loan, which is paid four-weekly on the basis of a regular site measure.
8. Labour is paid weekly. Suppliers and plant hires are paid 4 weeks after delivery. Materials and plant must be ordered 2 weeks before site requirement.
9. The *average* labour rate is £5 per hour or £250 per week for a 50-hour working week. This covers labourers and tradesmen.
10. The cross-section of the bungalow is shown in Fig. 47.1 and the sequence of activities is set out in Table 47.1, which shows the dependencies of each activity. All durations are in

Project Management, Planning and Control. https://doi.org/10.1016/B978-0-12-824339-8.00047-X

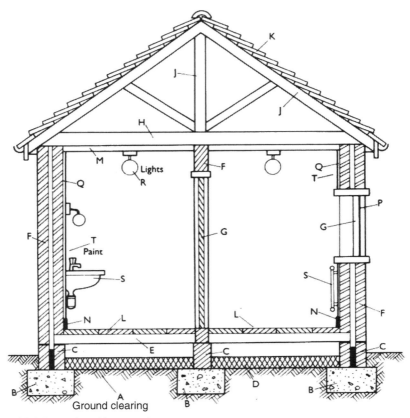

Figure 47.1 Bungalow (six rooms).

weeks. The network in Fig. 47.2 is in activity on arrow (AoA) format and the equivalent network in activity on node (AoN) format is shown in Fig. 47.3.

The activity letters refer to the activities shown on the cross-section diagram of Fig. 47.1, and on subsequent tables only these activity letters will be used. The total float column can, of course, only be completed when the network shown in Fig. 47.2 has been analysed (see Table 47.1).

Table 47.2 shows the complete analysis of the network including TL_e (latest time end event), TE_e (earliest time beginning event), total float and free float. It will be noted that none of the activities have free float. As mentioned in Chapter 21, free float is often confined to the dummy activities, which have been omitted from the table.

To enable the resource loading bar chart in Fig. 47.4 to be drawn, it helps to prepare a table of resources for each activity (Table 47.3). The resources are divided into two categories:

1. Labourers
2. Tradesmen

Table 47.1 List of activities.

Activity letter	Activity–Description	Duration (Weeks)	Dependency	Total float
A	Clear ground	2	Start	0
B	Lay foundations	3	A	0
C	Build dwarf walls	2	B	0
D	Oversite concrete	1	B	1
E	Floor joists	2	C and D	0
F	Main walls	5	E	0
G	Door and window frames	3	E	2
H	Ceiling joists	2	F and G	4
J	Roof timbers	6	F and G	0
K	Tiles	2	H and J	1
L	Floorboards	3	H and J	0
M	Ceiling boards	2	K and L	0
N	Skirtings	1	K and L	1
P	Glazing	2	M and N	0
Q	Plastering	2	P	2
R	Electrics	3	P	1
S	Plumbing and heating	4	P	0
T	Painting	3	Q, R and S	0

0 = Critical.

This is because tradesmen are more likely to be in short supply and could affect the programme.

The total labour histogram can now be drawn, together with the total labour curve (Fig. 47.5). It will be seen that the histogram has been hatched to differentiate between labourers and tradesmen, and shows that the maximum demand for tradesmen is eight men in weeks 27 and 28. Unfortunately, it is possible to employ only six tradesmen due to possible site congestion. What is to be done?

The advantage of network analysis with its float calculation is now apparent. Examination of the network shows that in weeks 27 and 28 the following operations (or activities) have to be carried out:

Activity Q	Plastering	3 men for 2 weeks
Activity R	Electrics	2 men for 3 weeks
Activity S	Plumbing and heating	3 men for 4 weeks

The first step is to check which activities have floats. Consulting Table 47.2 reveals that Q (plastering) has 2 weeks float and R (electrics) has 1 week float. By delaying Q (plastering) by 2 weeks and accelerating R (electrics) to be carried out in 2 weeks by 3 men per week, the maximum total in any week is reduced to 6. Alternatively, it may be possible to extend Q (plumbing) to 4 weeks using 2 men per week for the

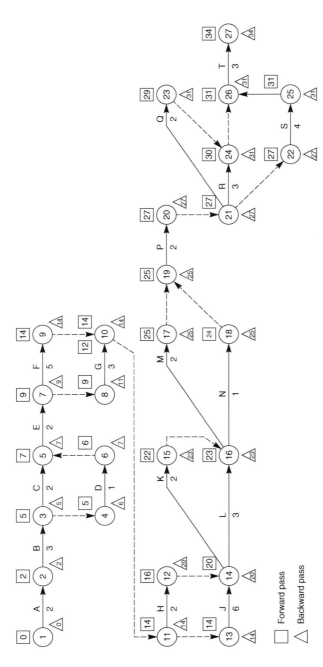

Figure 47.2 Network of bungalow (duration in weeks).

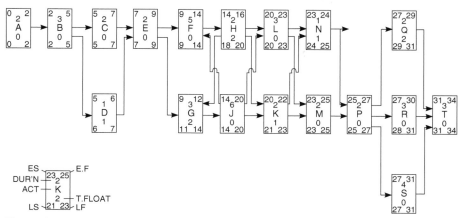

Figure 47.3 Network diagram of bungalow AoN format.

Table 47.2

a	b	C	d	e	F	g	H
Activity letter	Node no.	Duration	TL_e	TE_e	TE_b	d-f-c total float	e-f-c free float
A	1–2	2	2	2	0	0	0
B	2–3	3	5	5	2	0	0
C	3–5	2	7	7	5	0	0
D	4–6	1	7	6	5	1	0
E	5–7	2	9	9	7	0	0
F	7–9	5	14	14	9	0	0
G	8–10	3	14	12	9	2	0
H	11–12	2	20	16	14	4	0
J	13–14	6	20	20	14	0	0
K	14–15	2	23	22	20	1	0
L	14–16	3	23	23	20	0	0
M	16–17	2	25	25	23	0	0
N	16–18	1	25	24	23	1	0
P	19–20	2	27	27	25	0	0
Q	21–23	2	31	29	27	2	0
R	21–24	3	31	30	27	1	0
S	22–25	4	31	31	27	0	0
T	26–27	3	34	34	31	0	0

first 2 weeks and 1 man per week for the next 2 weeks. At the same time, R (electrics) can be extended by 1 week by employing 1 man per week for the first 2 weeks, and 2 men per week for the next 2 weeks. Again, the maximum total for weeks 27–31 is 6 tradesmen.

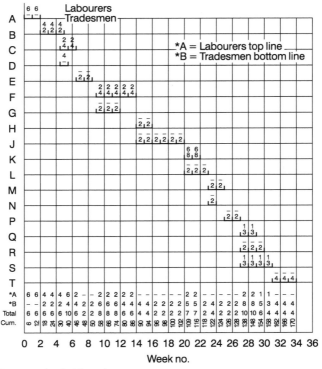

Figure 47.4 Resource loaded bar chat.

Table 47.3 Labour resources per week.

Activity letter	Resource A labourers	Resource B tradesmen	Total
A	6		6
B	4	2	6
C	2	4	6
D	4	—	4
E	—	2	2
F	2	4	6
G	—	2	2
H	—	2	2
J	—	2	2
K	2	3	5
L	—	2	2
M	—	2	2
N	—	2	2
P	—	2	2
Q	1	3	4
R	—	2	2
S	1	3	4
T	—	4	4

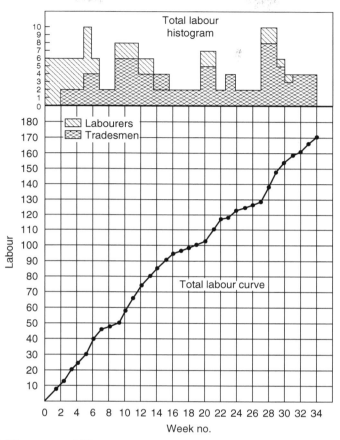

Figure 47.5 Histogram and 'S' curve.

The new partial disposition of resources and revised histograms after the two alternative smoothing operations are shown in Figs. 47.6 and 47.7. It will be noted that:

1. The overall programme duration has not been exceeded because the extra durations have been absorbed by the float.
2. The total number of man-weeks of any trade has not changed, that is, Q (plastering) still has 6 man-weeks and R (electrics) still has 6 man-weeks.

If it is not possible to obtain the necessary smoothing by utilizing and absorbing floats, the network logic may be amended, but this requires a careful reconsideration of the whole construction process.

The next operation is to use the EVA system to control the work on site. Multiplying for each activity, the number of weeks required to do the work by the number of men employed yields the number of man-weeks. If this is multiplied by 50 (the average number of working hours in a week), the man-hours per activity can be

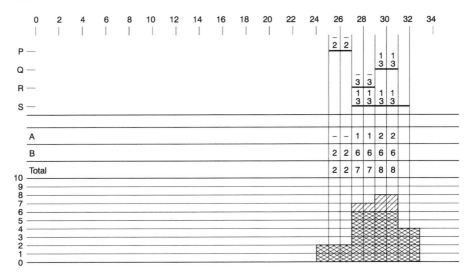

Figure 47.6 Resource smoothing 'A'.

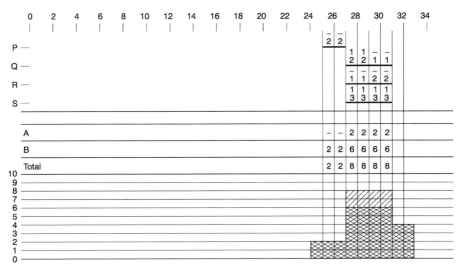

Figure 47.7 Resource smoothing 'B'.

obtained. A table can now be drawn up listing the activities, durations, number of men and budget hours (Table 47.4).

As the bank will advance the money to pay for the construction in four-weekly tranches, the measurement and control system will have to be set up to monitor the work every 4 weeks. The anticipated completion date is week 34, so that a measure in weeks 4, 8, 12, 16, 20, 24, 28, 32 and 36 will be required. By recording the *actual* hours

Table 47.4

a	B	C	D
Activity letter	**Duration (Weeks)**	**No. of men**	**b × c × 50 budget hours**
A	2	6	600
B	3	6	900
C	2	6	600
D	1	4	200
E	2	2	200
F	5	6	1500
G	3	2	300
H	2	2	200
J	6	2	600
K	2	5	500
L	3	2	300
M	2	2	200
N	1	2	100
P	2	2	200
Q	2	4	400
R	3	2	300
S	4	4	800
T	3	4	600
Total			8500

worked each week and assessing the percentage complete for each activity each week the value hours for each activity can be quickly calculated. As described in Chapter 32, the overall percent complete, efficiency and predicted final hours can then be calculated. Table 47.5 shows a manual EVA analysis for four sample weeks (8, 16, 24 and 32).

In practice, this calculation will have to be carried out every week, either manually as shown or by computer using a simple spreadsheet. It must be remembered that only the activities actually worked on during the week in question have to be computed. The remaining activities are entered as shown in the previous week's analysis.

For purposes of progress payments, the *value* hours for every 4-week period must be multiplied by the average labour rate (£5 per hour) and when added to the material and plant costs, the total value for payment purposes is obtained. This is shown later in this chapter.

At this stage, it is more important to control the job, and for this to be done effectively, a set of curves must be drawn on a time base to enable the various parameters to be compared. The relationship between the actual hours and value hours gives a measure of the efficiency of the work, while that between the value hours and planned hours gives a measure of progress. The actual and value hours are plotted straight from the EVA analysis, but the planned hours must be obtained from the labour expenditure curve (Fig. 47.5) and multiplying the labour value (in men) by 50 (the number of working hours per week). For example, in week 16, the total labour used to date is 94 man-weeks, giving 94 × 50 = 4700 man-hours.

Table 47.5

Period	Budget	Week 8			Week 16			Week 24			Week 32		
		Actual cum.	%	V	Actual cum.	%	V	Actual cum.	%	V	Actual cum.	%	V
A	600	600	100	600	600	100	600	600	100	600	600	100	600
B	900	800	100	900	800	100	900	800	100	900	800	100	900
C	600	550	100	600	550	100	600	550	100	600	550	100	600
D	200	220	90	180	240	100	200	240	100	200	240	100	200
E	200	110	40	80	180	100	200	180	100	200	180	100	200
F	1500	—	—	—	1200	80	1200	1550	100	1500	1550	100	1500
G	300	—	—	—	300	100	300	300	100	300	300	100	300
H	200	—	—	—	180	60	120	240	100	200	240	100	200
J	600	—	—	—	400	50	300	750	100	600	750	100	600
K	500	—	—	—	—	—	—	500	100	500	550	100	500
L	300	—	—	—	—	—	—	250	80	240	310	100	300
M	200	—	—	—	—	—	—	100	60	120	180	100	200
N	100	—	—	—	—	—	—	50	40	40	110	100	100
P	200	—	—	—	—	—	—	—	—	—	220	100	200
Q	400	—	—	—	—	—	—	—	—	—	480	100	400
R	300	—	—	—	—	—	—	—	—	—	160	60	180
S	800	—	—	—	—	—	—	—	—	—	600	80	640
T	600	—	—	—	—	—	—	—	—	—	100	10	60
Total	8500	2280	27.8	2360	4450	52	4420	6110	70.6	6000	7920	90.4	7680
Efficiency			103			99			98			96	
Estimated final hours			8201			8557			8654			8761	

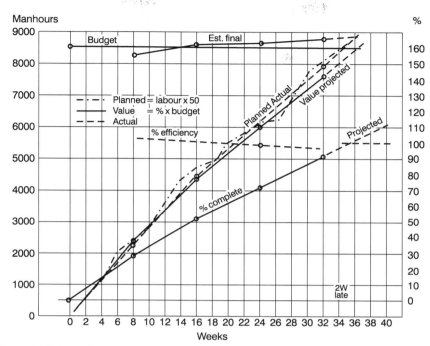

Figure 47.8 Control curves.

The complete set of curves (including the efficiency and percent complete curves) is shown in Fig. 47.8. In practice, it may be more convenient to draw the last two curves on a separate sheet, but provided the percentage scale is drawn on the opposite side to the man-hour scale; no confusion should arise. Again, a computer program can be written to plot these curves on a weekly basis as shown in Chapter 32.

Once the control system has been set up, it is essential to draw up the cash flow curve to ascertain what additional funding arrangements are required over the life of the project. In most cases where project financing is required, the cash flow curve will give an indication of how much will have to be obtained from the finance house or bank and when. In the case of this example, where the construction is financed by bank advances related to site progress, it is still necessary to check that the payments will, in fact, cover the outgoings. It can be seen from the curve in Fig. 47.10 that virtually permanent overdraft arrangements will have to be made to enable the men and suppliers to be paid regularly.

When considering cash flow, it is useful to produce a table showing the relationship between the usage of a resource, the payment date and the receipt of cash from the bank to pay for it — even retrospectively. It can be seen in Table 47.6 that

1. Materials have to be ordered 4 weeks before use.
2. Materials have to be delivered 1 week before use.
3. Materials are paid for 4 weeks after delivery.

Table 47.6

Week intervals	1	2	3	4	5	6	7	8
Order date			X	X			O	
Material delivery				X				
Labour use				X				
Material use								
Labour payments								
Pay suppliers								
Measurement							M	
Receipt from bank								R
Every 4 weeks								
Starting week no. 5								
First week no.	−3	−2	−1	1	2	3	4	5

Table 47.7

Activity	No. of weeks	Labour cost per week	Material and plant per week	Material cost and plant
A	2	1500	100	200
B	3	1500	1200	3600
C	2	1500	700	1400
D	1	1000	800	800
E	2	500	500	1000
F	5	1500	1400	7000
G	3	500	600	1800
H	2	500	600	1200
J	6	500	600	3600
K	2	1300	1200	2400
L	3	500	700	2100
M	2	500	300	600
N	1	500	200	200
P	2	500	400	800
Q	2	1000	300	600
R	3	500	600	1800
S	4	1000	900	3600
T	3	1000	300	900
Material total				33,600

4. Labour is paid in the same week of use.
5. Measurements are made 3 weeks after use.
6. Payment is made 1 week after measurement.

The next step is to tabulate the labour costs and material and plant costs on a weekly basis (Table 47.7). The last column in the table shows the total material and plant cost for every activity because all the materials and plant for an activity are being delivered

Figure 17.9 Resource bar chart

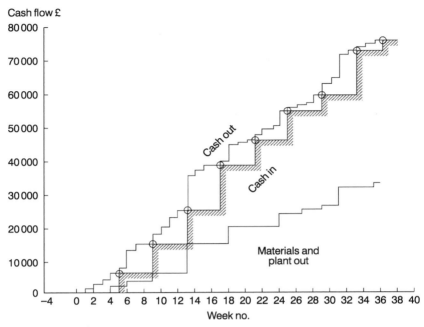

Figure 47.10 Cash flow curves.

1 week before use and have to be paid for in one payment. For simplicity, no retentions are withheld (i.e. 100% payment is made to all suppliers when due).

A bar chart (Fig. 47.9) can now be produced, which is similar to that shown in Fig. 47.4. The main difference is that instead of drawing bars, the length of the activity is represented by the weekly resource. As there are two types of resources – men and materials and plant – each activity is represented by two lines. The top line represents the labour cost in £100 units and the lower line the material and plant cost in £100 units. When the chart is completed, the resources are added vertically for each week to give a weekly total of labour out (i.e. men being paid, line 1) and material and plant out (line 2). The total cash out and the cumulative outflow values can now be added in lines 3 and 4, respectively.

The chart also shows the measurements every 4 weeks, starting in week 4 (line 5), and the payments 1 week later. The cumulative total cash is shown in line 6. To enable the outflow of materials and plant to be shown separately on the graph in Fig. 47.10, it was necessary to enter the cumulative outflow for material and plant in row 7. This figure shows the cash flow curves (i.e. cash in and cash out). The need for a more-or-less permanent overdraft of approximately £10,000 is apparent.

Worked example 2: Pumping installation

48

Chapter outline

Design and construction philosophy

1. A 3-tonne vessel arrives on-site complete with nozzles and manhole doors in place.
2. Pipe gantry and vessel support steel arrive in small pieces.
3. Pumps, motors and bedplates arrive as separate units.
4. Stairs arrive in sections with treads fitted to a pair of stringers.
5. Suction and discharge headers are partially fabricated with weldolet tees in place. Slip-on flanges to be welded on site for valves, vessel connection and blanked-off ends.
6. Suction and discharge lines from pumps to have slip-on flanges welded on site after trimming to length.
7. Drive, couplings to be fitted before fitting of pipes to pumps, but not aligned.
8. Hydro test to be carried out in one stage, hydro pump connection at discharge header end and vent at the top of vessel. Pumps have drain points.
9. Resource restraints require Sections A and B of suction, and discharge headers to be erected in series.
10. Suction to pumps is prefabricated on site from slip-on flange at valve to field weld at high-level bend.
11. Discharge from pumps is prefabricated on site from slip-on flange at valve to field weld on high-level horizontal run.
12. Final motor coupling alignment to be carried out after hydro test in case pipes have to be re-welded and aligned after test.
13. Only pumps no. 1 and 2 will be installed.

In this example, it is necessary to produce a material take-off from the layout drawings so that the erection man-hours can be calculated. The man-hours can then be translated into man-days and, by assessing the number of men required per activity, into activity durations. The man-hour assessment is, of course, made in the conventional manner by multiplying the operational units, such as numbers of welds or tonnes of steel, by the man-hour norms used by the construction organization. In this exercise, the norms used are those published by the OCPCA (Oil and Chemical Plant Contractors Association). These are the base norms that may or may not be factorized to take account of market, environmental, geographical or political conditions of the area in

Project Management, Planning and Control. https://doi.org/10.1016/B978-0-12-824339-8.00048-1

which the work is carried out. It is obvious that the rate for erecting a tonne of steel in the United Kingdom is different from erecting it in the wilds of Alaska.

The sequence of operations for producing a network programme and EVA analysis is as follows:

1. Study layout drawing or piping isometric drawings (Fig. 48.1).
2. Draw a construction network. Note that at this stage, it is only possible to draw the logic sequences (Fig. 48.2) and allocate activity numbers.
3. From the layout drawing, prepare a take-off of all the erection elements, such as number of welds, number of flanges, weight of steel and number of pumps.
4. Tabulate these quantities on an estimate sheet (Fig. 48.3) and multiply these by the OCPCA norms given in Table 48.1 to give the man-hours per operation.
5. Decide which operations are required to make up an activity on a network and list these in a table. This enables the man-hours per activity to be obtained.
6. Assess the number of men required to perform any activity. By dividing the activity man-hours by the number of men the actual working hours and consequently working days (durations) can be calculated.
7. Enter these durations in the network programme.
8. Carry out the network analysis, giving floats and the critical path (Table 48.2).
9. Draw up the EVA analysis sheet (Table 48.3) listing activities, activity (SMAC) numbers and durations.
10. Carry out EVA analysis at weekly intervals. The basis calculations for value hours, efficiency, etc. are shown in Table 48.4.
11. Draw a bar chart using the network as a basis for start and finish of activities (Fig. 48.4).

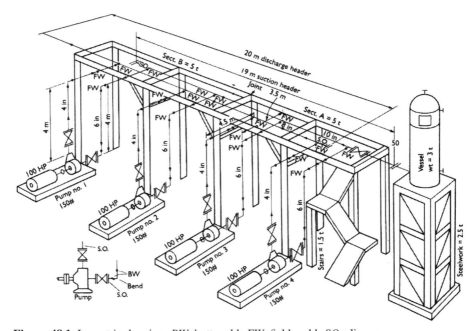

Figure 48.1 Isometric drawing. *BW*, butt weld; *FW*, field weld; *SO*, slip-on.

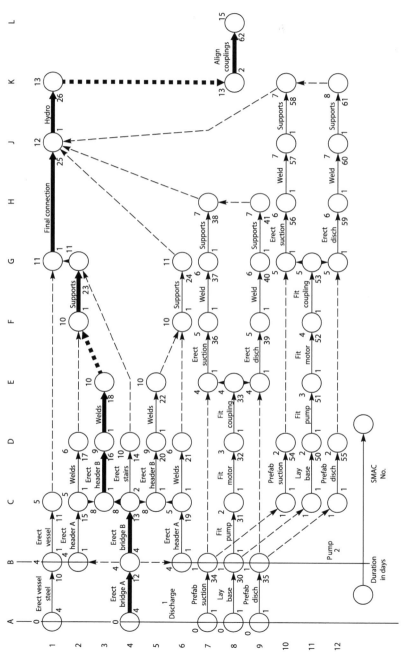

Figure 48.2 Network (using grid system).

ESTIMATE SHEET / SMAC ALLOCATION

A Item	B Unit	C Quant 1 set	D Hours rate	E =C+D man hours 1 set	F Pump man hours 2 sets	SMAC no. 1 set	SMAC man hours 1 set	SMAC no. pump no. 2	SMAC man hours pump no. 2	Duration days 1 set / 2 man/act
Erect vessel steelwork	Tonne	2.5	24.7	61.75		10	62			4
Erect vessel 3 T.	No. + Tonne	1	6.5 + 3.9	10.40		11	11			1
Erect bridge sect A	Tonne	5	12.3	61.50		12	62			4
Erect bridge sect B	Tonne	5	12.3	61.50		13	62			4
Erect stairs	Tonne	1.5	19.7	29.55		14	30			2
10" Suct. head erect sect A	Metre	10	0.90	9.00		15	9			1
10" Suct. head erect sect B	Metre	9	0.90	8.10		16	8			1
10" Suct. head slip-on (valve)	No	2	2.92	5.84		17.1	15			1
10" Suct. head butt joint	No	1	3.25	5.25		17.2	-			-
10" Suct. head fit valve	No	1	3.41	2.41		17.3	-			-
10" Suct. head slip-on (vessel)	No	1	2.92	2.92		17.4	-			-
10" Suct. head slip-on (end)	No	1	2.92	2.92		18.1	4			-
10" Suct. head fit blank	No	1	0.90	0.90		18.2	-			-
10" Suct. head fit supports	No	4	1.44	5.76		23	6			1
10" Suct. head final conn.	No	1	0.90	0.90		25	1			1
8" Disch. head erect sect A	Metre	8	0.80	6.40		19	6			1
8" Disch. head erect sect B	Metre	12	0.80	9.60		20	10			1
8" Disch. head butt joint	No	1	2.77	2.77		22	3			1
8" Disch. head slip-on (end)	No	1	2.49	2.49		21.1	3			-
8" Disch. head fit blank	No	1	0.50	0.50		21.2	-			-
8" Disch. head fitt supports	No	4	1.44	5.76		24	6			1
Erect base plate	No	1	4.00	4.00	8.00	30	4	50	4	1
Fit pump 100 HP	No	1	14.00	14.00	28.00	31	14	51	14	1
Fit motor	No	1	14.00	14.00	28.00	32	14	52	14	1
Fit coupling	No	1	10.00	10.00	20.00	33	10	53	10	1
Fit 2 valves 6" & 4"	No	2	0.77	1.54	3.08	36.1	7	56.1	7	-
6" Suction erect	Metre	7.5	0.70	5.25	10.50	36.2	-	56.2	-	-
6" Suction make joint	No	1	0.44	0.44	0.88	36.3	-	56.3	-	-
6" Suction bend	No	2	2.30	4.60	9.20	37.1	7	57.1	7	1
6" Suction butt header	No	1	2.30	2.30	4.60	37.2	-	57.2	-	-
6" Suction fit supports	No	3	1.44	4.32	8.64	38	4	58	4	1
6" Suction 2 butts bend	* No	2	2.41	4.82	9.64	34.1	6	54.1	6	1
6" Suction slip-on	* No	1	1.44	1.44	2.88	34.2	-	54.2	-	-
4" Disch. erect	Metre	8.5	0.59	5.01	10.03	39	6	59	6	1
4" Disch. make joint	No	1	0.37	0.37	0.74	60.1	4	40.1	4	1
4" Disch. butt joint	No	1	1.82	1.82	3.64	40.2	-	60.2	-	-
4" Disch. butt header	No	1	1.82	1.82	3.64	40.3	-	60.3	-	-
4" Disch. fit supports	No	3	1.44	4.32	8.64	41	4	61	4	1
4" Disch. 2 butts bend	* No	2	1.89	3.78	7.56	35.1	5	55.1	5	1
4" Disch. slip-on	* No	1	1.14	1.14	2.28	35.2	-	55.2	-	-
Hydro-test 54 m	No	1	12.00	12.00		26	12			1
Align couplings	No	2	25.00	50.00		62	50			2
Total					†		445	+	85	41 12
										= 530

* Pre-fabricate on site

† Item 62 is performed in 1 day due to overtime working

No. of man days = (41 + 12)/2 = 53 × 2 = 106

Average hours/man day = 530/106 = 5

Figure 48.3 EVA analysis.

Table 48.1 Applicable rates from OCPCA norms.

Steel erection		Hours
Pipe gantries	—	12.3/tonne
Stairs	—	19.7/tonne
Vessel support	—	24.7/tonne
Vessel (3 tonne)	—	6.5 + 1.3/tonne
Pump erection (100 hp)	—	14
Motor erection	—	14
Bedplate	—	4
Fit coupling	—	10
Align coupling	—	25
Prefab. Piping (Scheme 40)	—	—
6-inch suction prep.	—	0.81/end
4-inch discharge prep.	—	1.6/butt2.41
Suction welds	—	1.44/flange
4-inch discharge prep.	—	0.62/end
Discharge welds	—	1.27/butt1.89
Discharge slip-on	—	1.14/flange
Pipe erection	10 inch	0.79 × 1.15 = 0.90/m
Pipe erection	8 inch	0.70 × 1.15 = 0.80/m
Pipe erection	6 inch	0.61 × 1.15 = 0.70/m
Pipe erection	4 inch	0.51 × 1.15 = 0.59/m
Site butt welds	10 inch	2.83 × 1.15 = 3.25/butt
	8 inch	2.41 × 1.15 = 2.77/butt
	6 inch	2.0 × 1.15 = 2.30/butt
	4 inch	1.59 × 1.15 = 1.82/butt
Slip-ons	10 inch	3.25 × 0.9 = 2.92/butt
	8 inch	2.77 × 0.9 = 2.49/butt
	6 inch	2.30 × 0.9 = 2.07/butt
	4 inch	1.82 × 0.9 = 1.64/butt
Fit valves	10 inch	2.1 × 1.15 = 1.045/item
	6 inch	0.9 × 1.15 = 1.04/item
	4 inch	0.45 × 1.15 = 0.51/item
Flanged connection	10 inch	0.78 × 1.15 = 0.90/connection
	8 inch	0.43 × 1.15 = 0.50/connection
	6 inch	0.38 × 1.15 = 0.44/connection
	4 inch	0.32 × 1.15 = 0.37/connection
Supports		1.25 × 1.15 = 1.44/support
Hydro test	Set up	6 × 1.15 = 6.9
	Fill and drain	2 × 1.15 = 2.3
	Joint check	0.2 × 1.15 = 0.23/joint
	Blinds	0.5 × 1.15 = 0.58/blind
Hydrotest total = 6.9 + 2.3 + (0.23 × 12) = 9.2 + 2.76 = 11.96 (say 12)		

Table 48.2 Total float.

M SMAC no.	Duration (days)	Backward pass TL$_e$	Forward pass TE$_e$	TE$_e$	Total float	Welding activity
10	14	10	4	0	6	
11	1	11	5	4	6	
12	4	4	4	0	0	
13	4	8	8	4	0	
14	2	11	10	8	1	
15	1	8	5	4	3	
16	1	9	9	8	0	
17	1	10	6	5	4	X
18	1	10	10	9	0	X
19	1	9	5	4	4	
20	1	10	9	8	1	
21	1	11	6	5	5	X
22	1	11	10	9	1	X
23	1	11	11	10	0	
24	1	12	11	10	1	
25	1	12	12	11	0	X
26	1	13	13	12	0	
30	1	5	1	0	4	
31	1	7	2	1	5	
32	1	8	3	2	5	
33	1	9	4	3	5	
34	1	8	1	0	7	X
35	1	8	1	0	7	X
36	1	10	5	4	5	
37	1	11	6	5	5	X
38	1	12	7	6	5	
39	1	10	5	4	5	
40	1	11	6	5	5	X
41	1	12	7	6	5	
50	1	6	2	1	4	
51	1	7	3	2	4	
52	1	8	4	3	4	
53	1	9	5	4	4	
54	1	9	2	1	7	X
55	1	9	2	1	7	X
56	1	10	6	5	4	
57	1	11	7	6	4	X
58	1	12	8	7	4	
59	1	10	6	5	4	
60	1	11	7	6	4	X
61	1	12	8	7	4	
62	1	15	15	13	0	

Table 48.3 EVA analysis.

	EVA no.	EVA budget man-hours	Day 5 A	Day 5 %	Day 5 V	Day 10 A	Day 10 %	Day 10 V	Day 15 A	Day 15 %	Day 15 V
Erect vessel steelwork	10	62	70	100	62	70	100	62	70	100	62
Erect vessel	11	11	12	100	11	12	100	11	12	100	11
Erect bridge sect. A	12	62	60	100	62	60	100	62	60	100	62
Erect bridge sect. B	13	62	40	50	31	65	100	62	65	100	62
Erect stairs	14	30	—	—	—	35	100	30	35	100	30
10-inch suct. head. erect A	15	9	10	100	9	10	100	9	10	100	9
10-inch suct. head. erect B	16	8	—	—	—	8	100	8	8	100	8
10-inch suct. head. welds A	17	15	—	—	—	18	100	15	18	100	15
10-inch suct. head. welds B	18	4	—	—	—	5	100	4	5	100	4
8-inch disch. head. erect A	19	6	6	80	5	6	100	6	6	100	6
8-inch disch. head. erect B	20	10	—	—	—	11	80	8	12	100	8
8-inch disch. head. welds A	21	3	—	—	—	3	100	3	3	100	3
8-inch disch. head. welds B	22	3	—	—	—	—	—	—	3	100	3
Suction header supports	23	6	—	—	—	7	60	4	8	100	6
Discharge header supports	24	6	—	—	—	—	—	—	6	100	6
Final connection	25	1	—	—	—	—	—	—	1	100	1
Hydro test	26	12	—	—	—	—	—	—	10	100	12
Base plate pump 1	30	4	3	100	4	3	100	4	3	100	4
Fit pump 1	31	14	14	100	14	14	100	14	14	100	14
Fit motor	32	14	12	100	14	12	100	14	12	100	14
Fit coupling 1	33	10	12	100	10	12	100	10	12	100	10
Prefab. suction pipe 1	34	6	10	100	6	10	100	6	10	100	6
Prefab. discharge pipe 1	35	5	4	80	4	5	100	5	5	100	5
Erect suction pipe 1	36	7	—	—	—	8	100	7	8	100	7
Weld suction pipe 1	37	7	—	—	—	5	100	7	5	100	7
Support suction pipe 1	38	4	—	—	—	4	80	3	5	100	5

Continued

Table 48.3 EVA analysis.—cont'd

	EVA no.	EVA budget man-hours	Day 5			Day 10			Day 15		
			A	%	V	A	%	V	A	%	V
Erect discharge pipe 1	39	6	5	70	4	7	100	6	7	100	6
Weld discharge pipe 1	40	4	—	—	—	4	100	4	4	100	4
Support discharge pipe 1	41	4	—	—	—	2	50	2	3	100	4
Basic plate pump 2	50	4	3	100	4	3	100	4	3	100	4
Fit pump 2	51	14	14	100	14	14	100	14	14	100	14
Fit motor 2	52	14	12	100	14	12	100	14	12	100	14
Fit coupling 2	53	10	10	100	10	10	100	10	10	100	10
Prefab. suction pipe 2	54	6	10	100	6	10	100	6	10	100	6
Prefab. discharge pipe 2	55	5	6	100	5	6	100	5	6	100	5
Erect suction pipe 2	56	7	5	60	4	8	100	7	8	100	7
Weld suction pipe 2	57	7	—	—	—	5	100	7	5	100	7
Support suction pipe 2	58	4	—	—	—	2	40	2	4	100	4
Erect discharge pipe 2	59	6	6	70	4	8	100	6	8	100	6
Weld discharge pipe 2	60	4	—	—	—	5	100	4	5	100	4
Support discharge pipe 2	61	4	—	—	—	3	70	3	4	100	4
Align couplings 1 and 2	62	50	—	—	—	—	—	—	16	10	20
Totals		**530**	**324**	**56**	**297**	**482**	**84**	**448**	**525**	**94**	**500**

Table 48.4 EVA calculations.

	Day 5	Day 10	Day 15
Budget man-hours	530	530	530
Actual man-hours	324	482	525
Value man-hours	297	448	500
Percent complete	297	448	500
	530	530	530
	=56%	=85%	=94%
Est. final man-hours	324	482	525
	0.56	0.85	0.94
	=597	=567	=559
Efficiency	297	448	500
	324	482	525
	=92%	=93%	=95%

A = actual man-hours
B = budget man-hours
V = value man-hours
V = value man-hours = percent complete × B of activity

$$\Sigma \text{ Percent complete} = \frac{\Sigma B}{\Sigma V}$$

$$\text{Efficiency} = \frac{V}{A}$$

$$\text{Est. final} = \frac{A}{\text{Percent complete}}$$

Activities shifted: 17, 12, 22, 35, 55, 19

12. Place the number of men per week against the activities on the bar chart.
13. Add up vertically per week and draw the labour histogram and S-curve.
14. Carry out a resource-smoothing exercise to ensure that labour demand does not exceed supply for any particular trade. In any case, high peaks or troughs are signs of inefficient working and should be avoided here (Fig. 48.5). (Note: This smoothing operation only takes place with activities which have float).
15. Draw the project control curves using the weekly EVA analysis results to show graphically the relationship between:
    a. budget hours
    b. planned hours
    c. actual hours
    d. value hours
    e. predicted final hours (Fig. 48.6)
16. Draw control curves showing:
    a. percent complete (progress)
    b. efficiency (Fig. 48.6)

The procedures outlined earlier will give a complete control system for time and cost for the project as far as site work is concerned.

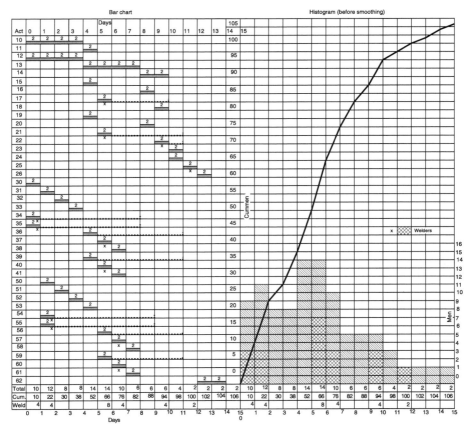

Figure 48.4 Bar chart.

Cash flow

Cash flow charts show the difference between expenditure (cash outflow) and income (cash inflow). As money is the common unit of measurement, all contract components such as man-hours, materials, overheads and consumables have to be stated in terms of money values.

It is convenient to set down the parameters that govern the cash-flow calculations before calculating the actual amounts. For example, consider the following:

1. There are 1748 productive hours in a year (39 hours/week × 52) − 280 days of annual holidays, statutory holidays, sickness and travelling allowance and induction.
2. Each man-hour costs, on average, £5 in actual wages.
3. After adding payments for productivity, holiday credits, statutory holidays, course attendance, radius and travel allowance, the taxable rate becomes £8.40/hour.
4. The addition of other substantive items such as levies, insurance, protective clothing and non-taxable fares, and lodging increases the rate by £2.04 to £10.44/hour.

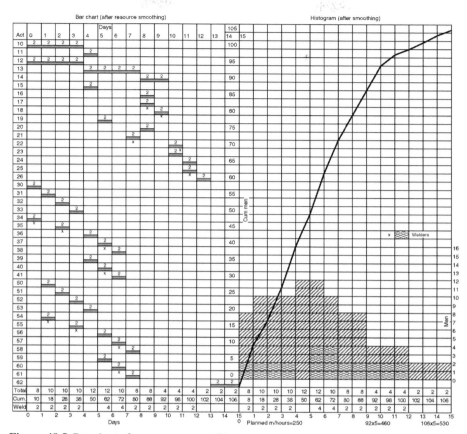

Figure 48.5 Bar chart after resource smoothing.

5. The ratio of other substantive items to taxable costs is $\dfrac{2.04}{8.40} = 0.243$.

6. An on-cost allowance of 20% is made up of:

Consumables	5%
Overheads	10%
Profit	5%
Total	**20%**

7. The total charge-out rate is, therefore, $10.44 \times 1.2 = £12.53$/hour.

8. In this particular example:

    a. The men are paid at the end of each day at a rate of £8.40/hour.

    b. The other substantive items of £2.04/hour are paid weekly.

    c. Income is received weekly at the charge-out rate of £12.53/hour.

9. A week consists of five working days.

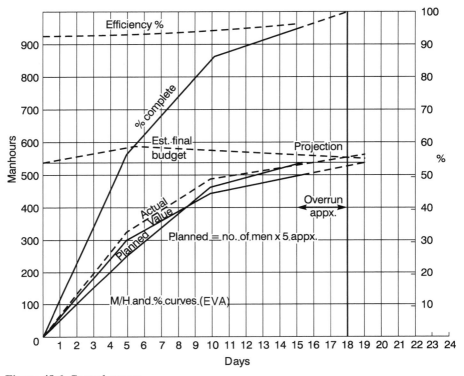

Figure 48.6 Control curves.

To enable the financing costs to be calculated at the estimate stage, cash-flow charts are usually only drawn to show the difference between *planned* outgoings and *planned* income.

However, once the contract is underway, a constant check must be made between *actual* costs (outgoings taken from time cards) and *valued* income derived from valuations of useful work done. The calculations for days 5 and 10 in Table 48.5 show how this is carried out. When these figures are plotted on a chart as in Fig. 48.7, it can be seen that for:

Days 0–5	The cash flow is negative (i.e. outgoings exceed income).
Days 5–8	The cash flow is positive.
Days 8–10	The cash flow is negative.
Days 10–15	The cash flow is positive.
On day 15	The total value is recovered assuming there are no retentions.

The planned costs of the other substantives can be calculated for each period by multiplying the planned cumulative outgoings by the ratio of 0.243.

Table 48.5 Cash values.

Activity	EVA no.	Duration (days)	EVA (budget) man-hours	Planned cost at £8.40 per hour	Planned price at £12.53 per hour	Day 5				Day 10			
						Actual man-hours	Actual cost at £8.40	Value hours	Value (price) at £12.53	Actual man-hours	Actual cost at £8.40	Value hours	Value (price) at £12.53
Erect vessel steelwork	10	4	62	521	777	70	588	62	777	70	588	62	777
Erect vessel	11	1	11	92	138	12	101	11	138	12	101	11	138
Erect bridge sect. A	12	4	62	521	777	60	504	62	777	60	504	62	777
Erect bridge sect. B	13	4	62	521	777	40	336	31	388	65	546	62	777
Erect stairs	14	2	30	252	376	–	–	–	–	35	294	30	376
10-inch suct. head. erect A	15	1	9	76	113	10	84	9	113	10	84	9	113
10-inch suct. head. erect B	16	1	8	67	100	–	–	–	–	8	67	8	100
10-inch suct. head. welds A	17	1	15	126	188	–	–	–	–	18	151	15	188
10-inch suct. head. welds B	18	1	4	34	50	–	–	–	–	5	42	4	50
8-inch disch. head. erect A	19	1	6	50	75	6	50	5	63	6	50	6	75
8-inch disch. head. erect B	20	1	10	84	125	–	–	–	–	11	92	8	100
8-inch disch. head. welds A	21	1	3	25	38	–	–	–	–	3	25	3	38
8-inch disch. head. welds B	22	1	3	25	38	–	–	–	–	–	–	–	–

Continued

Table 48.5 Cash values.—cont'd

Activity	EVA no.	Duration (days)	EVA (budget) man-hours	Planned cost at £8.40 per hour	Planned price at £12.53 per hour	Day 5				Day 10			
						Actual man-hours	Actual cost at £8.40	Value hours	Value (price) at £12.53	Actual man-hours	Actual cost at £8.40	Value hours	Value (price) at £12.53
Suction header supports	23	1	6	50	75	–	–	–	–	7	59	4	50
Discharge header supports	24	1	6	50	75	–	–	–	–	–	–	–	–
Final connection	25	1	1	8	13	–	–	–	–	–	–	–	–
Hydro test	26		12	101	150	–	–	–	–	–	–	–	–
Base plate pump 1	30	1	4	34	50	3	25	4	50	3	25	4	50
Fit pump 1	31	1	14	118	175	14	118	14	175	14	118	14	175
Fit motor 1	32	1	14	118	175	12	101	14	175	12	101	14	175
Fit coupling 1	33	1	10	84	125	12	101	10	125	12	101	10	125
Prefab. suct. pipe 1	34	1	6	50	75	10	84	6	75	10	84	6	75
Prefab. discharge pipe 1	35	1	5	42	63	4	34	4	50	5	42	5	63
Erect suction pipe 1	36	1	7	59	88	–	–	–	–	8	67	7	88
Weld suction pipe 1	37	1	7	59	88	–	–	–	–	5	42	7	88
Support suction pipe 1	38	1	4	34	50	–	–	–	–	4	34	3	38
Erect discharge pipe 1	39	1	6	50	75	5	42	4	50	7	59	6	75
Weld discharge pipe 1	40	1	4	34	50	–	–	–	–	4	34	4	50

		1											
Support discharge pipe 1	41	1	4	34	50	—	—	—	—	2	17	2	25
Base plate pump 2	50	1	4	34	50	3	25	4	50	3	25	4	50
Fit pump 2	51	1	14	118	175	14	118	14	175	14	118	14	175
Fit motor 2	52	1	14	118	175	12	101	14	175	12	101	14	175
Fit coupling 2	53	1	10	84	125	10	84	10	125	10	84	10	125
Prefab. suction pipe 2	54	1	6	50	75	10	84	6	75	10	84	6	75
Prefab. discharge pipe 2	55	1	5	42	63	6	50	5	63	6	50	5	63
Erect suction pipe 2	56	1	7	59	88	5	42	4	50	8	67	7	88
Weld suction pipe 2	57	1	7	59	88	—	—	—	—	5	42	7	88
Support suction pipe 2	58	1	4	34	50	—	—	—	—	2	17	2	25
Erect discharge pipe 2	59	1	6	50	75	6	50	4	50	8	67	6	75
Weld discharge pipe 2	60	1	4	34	50	—	—	—	—	5	42	4	50
Support discharge pipe 2	61	1	4	34	50	—	—	—	—	3	25	3	38
Align couplings 1 and 2	62	2	50	420	627	—	—	—	—	—	—	—	—
			530	4455	6640	324	2722	297	3719	482	4049	448	5613

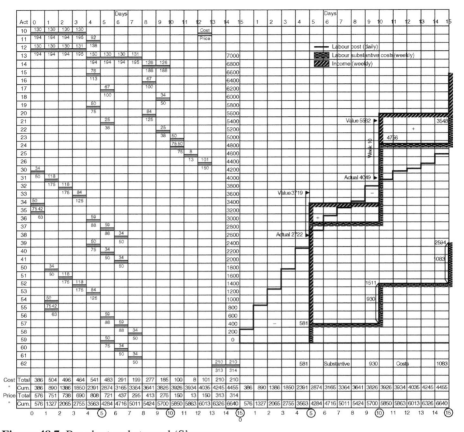

Figure 48.7 Bar chart and stepped 'S' curve.

For day 5	The substantive costs are 2391 × 0.243 = £581.
For day 10	The substantive costs are 3826 × 0.243 = £930.
For day 15	The substantive costs are 4455 × 0.243 = £1083.

These costs are plotted on the chart and, when added to the planned labour costs, give total planned outgoings of:

£2391 + 581 = 2972 for day 5.

£3826 + 930 = 4756 for day 10.

£4455 + 1083 = 5538 for day 15.

To obtain the *actual* total outgoings, it is necessary to multiply the *actual* labour costs by 1.243, for example, for day 5, the *actual* outgoings will be:

2722 × 1.243 = £3383,

and for day 10 they will be:

4049 × 1.243 = £5033.

The total planned and actuals can therefore be compared on a regular basis.

Worked example 3: Motor car

Chapter outline

The example in this chapter shows how all the tools and techniques described so far can be integrated to give a comprehensive project-management system. The project chosen is the design, manufacture and distribution of a prototype motor car, and while the operations and time scales are only indicative and do not purport to represent a real-life situation, the example shows how the techniques follow each other in a logical sequence.

The prototype motor car being produced is illustrated in Fig. 49.1 and the main components of the engine are shown in Fig. 49.2. It can be seen that the letters given to the engine components are the activity identity letters used in planning networks.

An oversight of the main techniques and their most important constituents are discussed as follows.

As with all projects, the first document to be produced is the *business case*, which should also include the chosen option investigated for the *investment appraisal*. In this exercise, the questions to be asked (and answered) are shown in Table 49.1.

It is assumed that the project requires an initial investment of £60 million and that over a 5-year period, 60,000 cars (units) will be produced at a cost of £5000 per unit. The assumptions are that the discount rate is 8%. There are two options for phasing the manufacture:

1. That the factory performs well for the first 2 years but suffers some production problems in the next 3 years (option 1).
2. That the factory has teething problems in the first 3 years but goes into full production in the last 2 (option 2).

The *discounted cash flow* (DCF) calculations can be produced for both options as shown in Tables 49.2 and 49.3.

Project Management, Planning and Control. https://doi.org/10.1016/B978-0-12-824339-8.00049-3

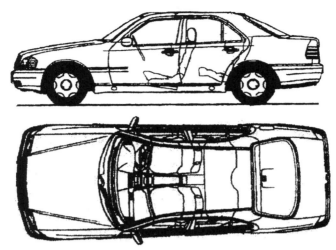

Figure 49.1 Motor car.

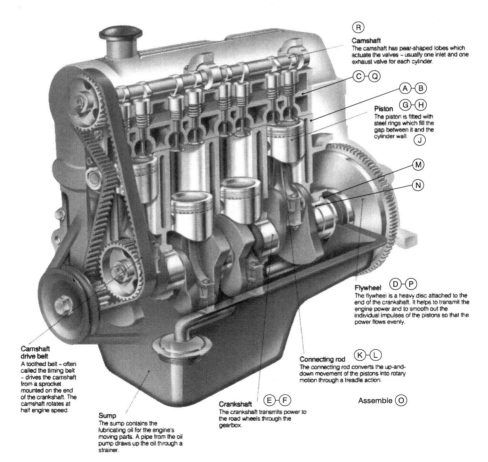

Ⓡ

Camshaft
The camshaft has pear-shaped lobes which actuate the valves – usually one inlet and one exhaust valve for each cylinder.

Ⓒ–Ⓠ

Ⓐ–Ⓑ

Piston Ⓖ–Ⓗ
The piston is fitted with steel rings which fill the gap between it and the cylinder wall.

Ⓙ

Ⓜ

Ⓝ

Flywheel Ⓓ–Ⓟ
The flywheel is a heavy disc attached to the end of the crankshaft. It helps to transmit the engine power and to smooth out the individual impulses of the pistons so that the power flows evenly.

Camshaft drive belt
A toothed belt – often called the timing belt – drives the camshaft from a sprocket mounted on the end of the crankshaft. The camshaft rotates at half engine speed.

Connecting rod Ⓚ–Ⓛ
The connecting rod converts the up-and-down movement of the pistons into rotary motion through a treadle action.

Assemble Ⓞ

Sump
The sump contains the lubricating oil for the engine's moving parts. A pipe from the oil pump draws up the oil through a strainer.

Crankshaft Ⓔ–Ⓕ
The crankshaft transmits power to the road wheels through the gearbox.

Figure 49.2 The parts of an overhead-camshaft engine.

Table 49.1

Business Case
Why do we need a new model?
What model will it replace?
What is the market?
Will it appeal to the young, the middle aged, families, the elderly, women, trendies, yobos?
How many can we sell per year in the UK, the USA, EEC and other countries?
What is the competition for this type of car and what is their price?
Will the car rental companies buy it?
What is the max. and min. Selling price?
What must be the max. manufacturing cost and in what country will it be built?
What name shall we give it?
Do we have a marketing plan?
Who will handle the publicity and advertising?
Do we have to train the sales force and maintenance mechanics?
What should be the insurance category?
What warranties can be given and for how long?
What are the main specifications regarding:
Safety and theft proofing?
Engine size (cc) or a number of sizes?
Fuel consumption?
Emissions (pollution control)?
Catalytic converter?
Max. speed?
Max. acceleration?
Size and weight?
Styling?
Turning circle and ground clearance?
What 'extras' must be fitted as standard?
ABS
Power steering
Airbags
Electric windows and roof

Continued

Table 49.1

Cruise control
Air conditioning
What percentage can be recycled
Investment appraisal (options)
Should it be a saloon, coupé, estate, people carrier, convertible, 4 × 4 or mini?
Will it have existing or newly designed engine?
Will it have existing or new platform (chassis)?
Do we need a new manufacturing plant or can we build it in an existing one?
Should the engine be cast iron or aluminium?
Should the body be steel, aluminium or fibreglass?
Do we use an existing brand name or devise a new one?
Will it be fuelled on petrol, diesel, electricity or hybrid power unit?
DCF of investment returns, NPV, cash flow?

To obtain the *internal rate of return* (IRR), an additional discount rate (in this case 20%) must be applied to both options. The resulting calculations are shown in Tables 49.4 and 49.5, and the graph showing both options is shown in Fig. 49.3. This gives an IRR of 20.2% and 15.4%, respectively.

It is now necessary to carry out a cash-flow calculation for the distribution phase of the cars. To line up with the DCF calculations, two options have to be examined. These are shown in Tables 49.6 and 49.7, and the graphs are shown in Figs. 49.4 and 49.5 for options 1 and 2, respectively. An additional option 2a, in which the income in years 2 and 3 is reduced from £65,000 to £55,000K, is shown in the cash-flow curves of Fig. 49.6.

All projects carry an element of *risk* and it is prudent to carry out a risk analysis at this stage. The types of risks that can be encountered, the possible actual risks and the mitigation strategies are shown in Table 49.8. A risk log (or risk register) for five risks is given in Fig. 49.7.

Once the decision has been made to proceed with the project, a *project life cycle* diagram can be produced. This is shown in Fig. 49.8 together with the constituents of the seven phases envisaged.

The next stage is the *product-breakdown structure* (Fig. 49.9), followed by a combined *cost-breakdown* and *organization-breakdown structures* (Fig. 49.10). By using these two, the *responsibility matrix* can be drawn up (Fig. 49.11).

It is now necessary to produce a programme. The first step is to draw an *activity list* showing the activities and their dependencies and durations. These are shown in the first four columns of Table 49.9. It is now possible to draw the *critical path network*

Table 49.2

DCF of investment returns (net present value)

Initial investment £60,000K 5 year period

Total car production 60,000 units @ £5,000/Unit

Option 1

Year	Production units	Income (£K)	Cost (£K)	Net return (£K)	Discount rate (%)	Discount factor	Present value (£K)
1	15,000	100,000	75,000	25,000	8	0.926	23,150
2	15,000	100,000	75,000	25,000	8	0.857	21,425
3	10,000	65,000	50,000	15,000	8	0.794	11,910
4	10,000	65,000	50,000	15,000	8	0.735	11,025
5	10,000	65,000	50,000	15,000	8	0.681	10,215
Totals				**95,000**			**77,725**

Net present value (NPV) = 77,725 − 60,000 = £17,725K

Profit = £95,000K − £60,000K = £35,000K

Average rate of return (undiscounted) = £95,000/5 = £19,000K per annum

Return on investment = £19,000/£60,000 = 31.66%

Table 49.3

DCF of investment returns (net present value)

Initial investment £60,000K 5 year period

Total car production 60,000 units @ £5,000/Unit

Option 2nd

Year	Production units	Income (£K)	Cost (£K)	Net return (£K)	Discount rate (%)	Discount factor	Present value (£K)
1	10,000	65,000	50,000	15,000	8	0.926	13,890
2	10,000	65,000	50,000	15,000	8	0.857	12,855
3	10,000	65,000	50,000	15,000	8	0.794	11,910
4	15,000	100,000	75,000	25,000	8	0.735	18,375
5	15,000	100,000	75,000	25,000	8	0.681	17,025
Totals				**95,000**			**74,055**

Net present value (NPV) = 74,055−60,000 = £14,055K

Profit = £95,000K − £60,000K = £35,000K

Average rate of return (undiscounted) = £95,000/5 = £19,000K per annum

Return on investment = £19,000/£60,000 = 31.66%

Table 49.4

Internal rate of return (IRR)

Option 1

Year	Net return (£K)	Discount rate (%)	Discount factor	Present value (£K)	Discount rate (%)	Discount factor	Present value (£K)
1	25,000	15	0.870	21,750	20	0.833	20,825
2	25,000	15	0.756	18,900	20	0.694	17,350
3	15,000	15	0.658	9,870	20	0.579	8,685
4	15,000	15	0.572	8,580	20	0.482	7,230
5	15,000	15	0.497	7,455	20	0.402	6,030
Totals	**60,000**			**66,555**			**60,120**
Less investment				−60,000			−60,000
Net present value				£6,555K			£120K

Internal rate of return (from graph) = 20.2%

Table 49.5

Internal rate of return (IRR)

Option 2

Year	Net return (£K)	Discount rate (%)	Discount factor	Present value (£K)	Discount rate (%)	Discount factor	Present value (£K)
1	15,000	15	0.870	13,050	20	0.833	12,495
2	15,000	15	0.756	11,340	20	0.694	10,410
3	15,000	15	0.658	9,870	20	0.579	8,685
4	25,000	15	0.572	14,300	20	0.482	12,050
5	25,000	15	0.497	12,425	20	0.402	10,050
Totals	**60,000**			**60,985**			**53,690**
Less investment				−60,000			−60,000
Net present value				£985K			−£6,310K

Internal rate of return (from graph) = 15.4%

Graph to obtain IRR

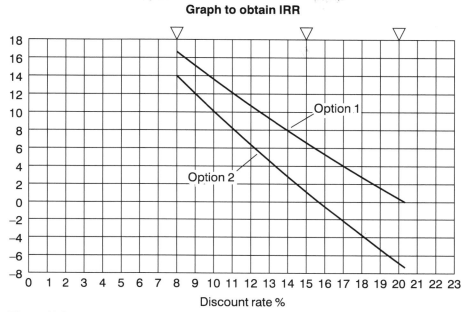

Figure 49.3 IRR curves.

in either activity on node format (Fig. 49.12), activity on arrow format (Fig. 49.13) or as a Lester diagram (Fig. 49.14).

After analysing the network diagram, the *total* and *free floats* of the activities can be listed (Table 49.10).

In addition to the start and finish, there are four milestones (days 8, 16, 24 and 30). These are described and plotted on the *milestone slip chart* (Fig. 49.15).

The network programme can now be converted into a bar chart (Fig. 49.16) on which the resources (in men per day), as given in the fifth column of Table 49.9, can be added. After summating the resources for every day, it has been noticed that there is a peak requirement of 12 men in days 11 and 12. As this might be more than the available resources, the bar chart can be adjusted by utilizing the available floats to *smooth* the resources and eliminate the peak demand. This is shown in Fig. 49.17 by delaying the start of activities D and F.

In Fig. 49.18, the man-days of the unsmoothed bar chart have been multiplied by 8 to convert them into man-hours. This was necessary to carry out *earned value analysis*. The daily man-hour totals can be shown as a *histogram*, and the cumulative totals are shown as an S-curve. In a similar way, Fig. 49.19 shows the respective histogram and S-curve for the smoothed bar chart.

It is now possible to draw up a table of *actual man-hour* usage and *percent-complete* assessment for reporting day nos. 8, 16, 24 and 30. These, together with the *earned values* for these periods are shown in Table 49.11. Also shown are the

Table 49.6

Cash flow

Option 1

Year		1	2	3	4	5	Cumulative
Capital	£K	12,000	12,000	12,000	12,000	12,000	
Costs	£K	75,000	75,000	50,000	50,000	50,000	
Total	£K	87,000	87,000	62,000	62,000	62,000	360,000
Cumulative		87,000	174,000	236,000	298,000	360,000	
Income	£K	100,000	100,000	65,000	65,000	65,000	395,000
Cumulative		100,000	200,000	265,000	330,000	395,000	

Table 49.7

Cash flow Option 2							
Year		*1*	*2*	*3*	*4*	*5*	*Cumulative*
Capital	£K	12,000	12,000	12,000	12,000	12,000	
Costs	£K	50,000	50,000	50,000	75,000	75,000	
Total	£K	62,000	62,000	62,000	87,000	87,000	360,000
Cumulative		62,000	124,000	186,000	273,000	360,000	
Income	£K	65,000	65,000	65,000	100,000	100,000	395,000
Cumulative		65,000	130,000	195,000	295,000	395,000	

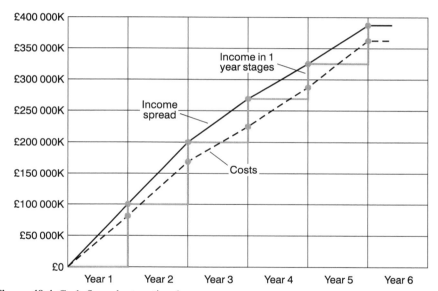

Figure 49.4 Cash-flow chart, option 1.

efficiency [cost performance index (CPI)], schedule performance index (SPI), and predicted final-completion costs and times as calculated at each reporting day.

Using the unsmoothed bar chart histogram and S-curve as a *planned man-hour* base, the actual man-hours and earned value man-hours can be plotted on the graph in Fig. 49.20. This graph also shows the percent *complete* and percent *efficiency* at each of the four reporting days.

Finally, Table 49.12 shows the actions required for the *closeout* procedure.

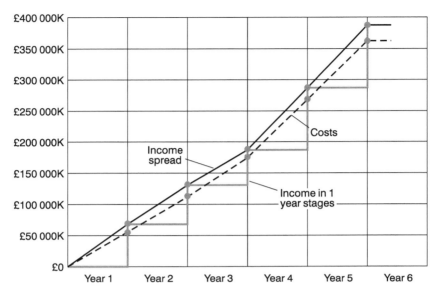

If income falls to £55 000K in years 2 and 3:

Income £K   =   65 000   55 000   55 000   100 000   100 000
Cumulative =   65 000   120 000   175 000   275 000   375 000

Figure 49.5 Cash-flow chart, option 2.

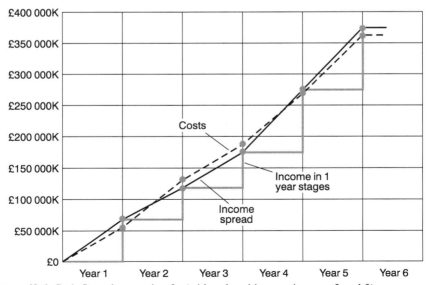

Figure 49.6 Cash-flow chart, option 2a (with reduced income in years 2 and 3).

Table 49.8

Risk analysis	Training problems
Types of risks	Suppliers unreliable
Manufacturing (machinery and facilities) costs	Rustproofing problems
Sales and marketing, exchange rates	Performance problems
Reliability	Industrial disputes
Mechanical components performance	Electrical and electronic problems
Electrical components performance	Competition too great
Maintenance	Not ready for launch date (exhibition)
Legislation (emissions, safety, recycling, labour, tax)	Safety requirements
Quality	Currency fluctuations
Possible risks	*Mitigation strategy*
Would not sell in predicted numbers	Overtime
Quality in design, manufacture, finish	More tests
Maintenance costs	More research
Manufacturing costs	More advertizing/marketing
New factory costs	Insurance
Tooling costs	Re-engineering
New factory not finished on time	Contingency

Summary

Business case

Need for a new model; what type of car? min./max. price; manufacturing cost; units per year; marketing strategy; what market sector is it aimed at? main specifications; what extras should be standard? name of new model; country of manufacture.

Investment appraisal

Options: saloon, coupé, estate, convertible, people carrier, 4 × 4; existing or new engine; existing or new platform; materials of construction for engine and body; type of fuel; new or existing plant; DCF of returns, net present value (NPV) and cash flow.

RISK LOG

Project: Key: H – High; M – Medium; L – Low										Prepared by: *A.L.*		Reference: Date: 12.12.2000	
Type of Risk	Description of Risk	Probability			Impact				Risk Reduction Strategy	Contingency Plans	Risk Owner
		H	M	L	Perf.	Cost	Time				
R1 *Manufact.*	*Factory not finished on time*			10%		1M	3 months delay	*Work more overtime*	*Cancel launch of car*	*PM*
R2 *Quality*	*Window mechanism faulty*		50%		*Not serious*	5K	1 week to rectify	*Test motor*	*Use manual window*	*Chief Designer*
R3 *Safety*	*Air bags may explode*			1%	*Serious*	10K	1 week to rectify	*Run more tests*	*Remove Air bags*	*Chief Designer*
R4 *Legislation*	*Emission levels will be reduced*		50%		*Serious*	3M	1 year to modify CC	*Increase research*	*Buy another proven conv.*	*Chief Engineer*
R5 *Sales*	*Sales forecasts will not be met*		30%		*Very serious*	10M		*Increase advertising*	*Reduce price*	*PM*

Figure 49.7 Copyright © 1996 WPMC Ltd. All rights reserved.

Product and Project Life Cycle

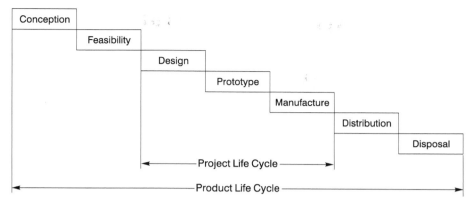

Phases

Conception: Original idea, high level discussions, preliminary market research

Feasibility: Consumer survey, market survey, type and size of car, production run and costs

Design: Vehicle design, tool design, development, component tests

Prototype: Tooling, production line, limited production, arctic and desert testing

Manufacture: Mass production, operator training, spares build up, customizing

Distribution: Deliveries, staff training, sales conferences, marketing, advertising, exhibitions

Disposal: Dismantling production line, selling tools, negotiating licences for spares

The phases could overlap.

The end of each phase could be a decision point to stop or proceed.

Figure 49.8

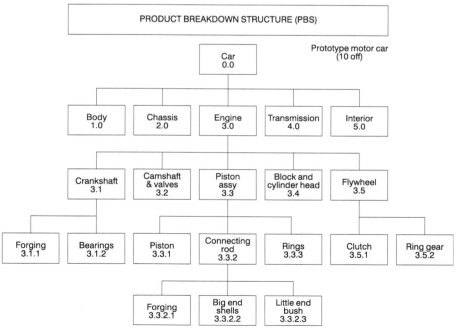

Figure 49.9

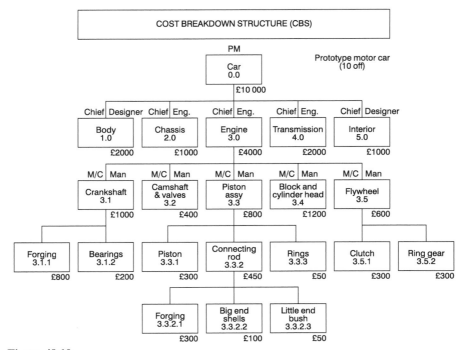

Figure 49.10

Responsibility Matrix

	Sponsor	Project manager	Chief designer	Chief engineer	M/C shop manager	Chassis manager	Styling manager
Body	B	B	A	D	D	D	C
Chassis		B		A	C	C	
Engine	B	B		A	C	D	
Transmission		B		A	C	C	
Interior	B	B	A	D	D	D	C

A Main responsibility

B Must be advised

C Must be consulted

D Requires updates

Figure 49.11

Table 49.9 Activity list of motor car engine manufacture and assembly (10 off), 8 hours/day.

Activ. Letter	Description	Dependency	Duration days	Men per day	Man-hours per day	Total man-hours
A	Cast block and cylinder head	Start	10	3	24	240
B	Machine block	A	6	2	16	96
C	Machine cylinder head	B	4	2	16	64
D	Forge and mc. flywheel	E	4	2	16	64
E	Forge crankshaft	Start	8	3	24	192
F	Machine crankshaft	E	5	2	16	80
G	Cast pistons	A	2	3	24	48
H	Machine pistons	G	4	2	16	64
J	Fit piston rings	H	1	2	16	16
K	Forge connecting rod	E	2	3	24	48
L	Machine conn. rod	K	2	2	16	32
M	Fit big end shells	L	1	1	8	8
N	Fit little end bush	M	1	1	8	8
O	Assemble engine	B, F, J, N	5	4	32	160
P	Fit flywheel	D, O	2	4	32	64
Q	Fit cylinder head	C, P	2	2	16	32
R	Fit camshaft and valves	Q	4	3	24	96
Total						1312

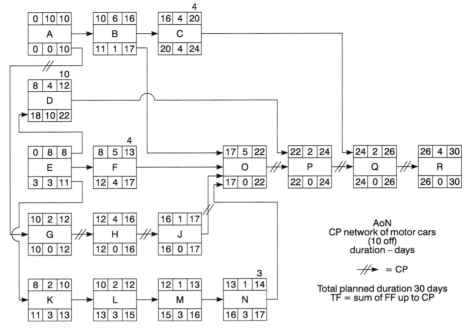

Figure 49.12 AoN network.

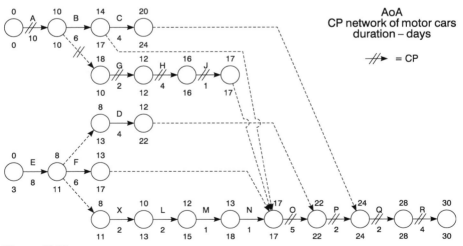

Figure 49.13

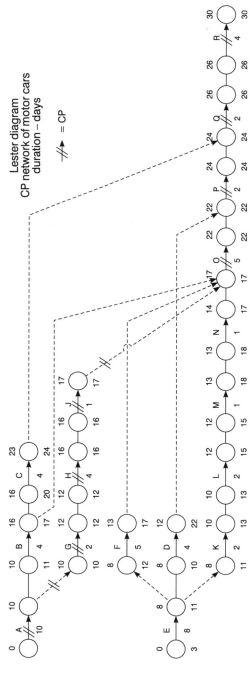

Figure 49.14

Table 49.10 Activity floats from CP network.

Activ. Letter	Description	Duration	Total float	Free float
A	Cast block and cylinder head	10	0	0
B	Machine block	6	1	0
C	Machine cylinder head	4	4	4
D	Forge and mc. flywheel	4	10	10
E	Forge crankshaft	8	3	0
F	Machine crankshaft	5	4	4
G	Cast pistons	2	0	0
H	Machine pistons	4	0	0
J	Fit piston rings	1	0	0
K	Forge connecting rod	2	3	0
L	Machine conn. rod	2	3	0
M	Fit big end shells	1	3	0
N	Fit little end bush	1	3	3
O	Assemble engine	5	0	0
P	Fit flywheel	2	0	0
Q	Fit cylinder head	2	0	0
R	Fit camshaft and valves	4	0	0

Milestones

Milestone 1 Forge crankshaft (E)   Day 8
Milestone 2 Machine pistons (H)   Day 16
Milestone 3 Fit flywheel (P)   Day 24
Milestone 4 Completion   Day 30

Milestone slip chart   **Programme**

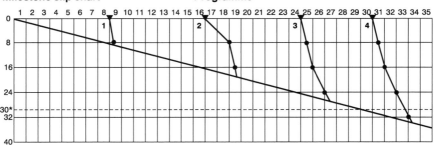

Assume:

* Reporting periods (8, 16, 24 and 30)
Milestone 1 slips ½ day
    "   2   "   2 days, then ½ day
    "   3   "   ½ day, then ½ day, then 1 day
    "   4   "   ½ day, then ½ day, then 1 day, then 1 day

Figure 49.15

Bar chart of prototype motor cars (10 off)

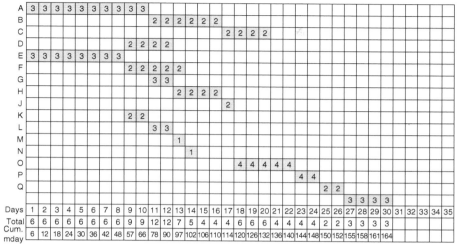

Figure 49.16 Unsmoothed.

Bar chart of prototype motor cars (10 off)

After moving D to start at day 18
and moving F to start at day 12

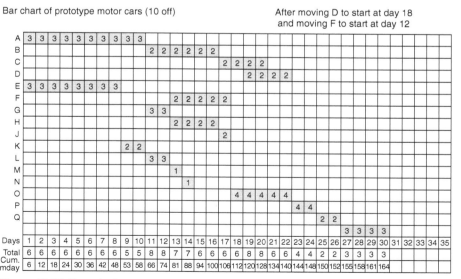

Figure 49.17 Smoothed.

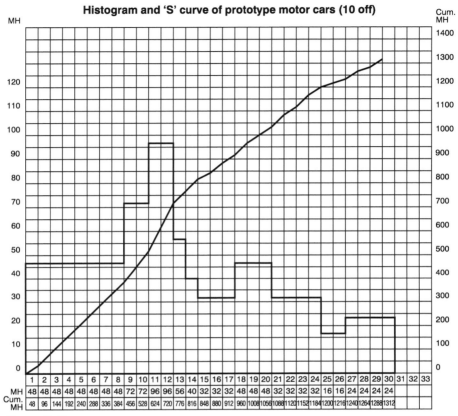

Figure 49.18

Project and product life cycle

Conception	Original idea, submission to top management
Feasibility	Feasibility study, preliminary costs, market survey
Design	Vehicle and tool design, component tests
Prototype	Tooling, production line, environmental tests
Manufacture	Mass production, training
Distribution	Deliveries, staff training, marketing
Disposal	Dismantling of plant, selling tools

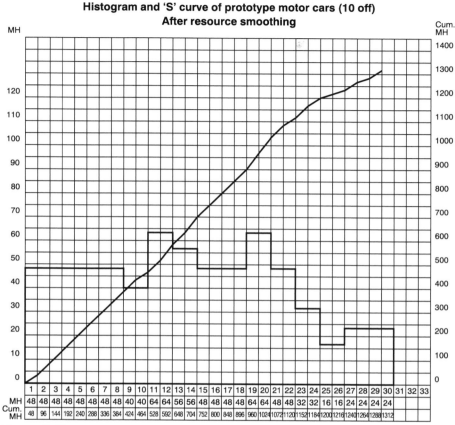

Figure 49.19

Work- and product-breakdown structures

Design, prototype, manufacture, testing, marketing, distribution, training.
Body, chassis, engine, transmission, interior and electronics.
Cost-breakdown structure, organization-breakdown structure, responsibility matrix.

AoN network

Network diagram, forward and backward pass, floats, critical path, examination for overall time reduction, conversion to bar chart with resource loading, histogram, reduction of resource peaks, cumulative S-curve; milestone slip chart.

Table 49.11 Man-hour usage of motor car engine manufacture and assembly (10 off) (unsmoothed).

Period		Day 8			Day 16			Day 24			Day 30		
Act.	Budget M/H	Actual cum.	% Complete	EV	Actual cum.	% Complete	EV	Actual cum.	% Complete	EV	Actual cum.	% Complete	EV
A	240	210	80	192	260	100	240	260	100	240	260	100	240
B	96				30	20	19	110	100	96	110	100	96
C	64							70	100	64	70	100	64
D	64				60	50	32	80	100	64	80	100	64
E	192	170	80	154	200	100	192	200	100	192	200	100	192
F	80				70	80	64	90	100	80	90	100	80
G	48				54	100	48	60	100	48	60	100	48
H	64				60	80	51	68	100	64	68	100	64
J	16							16	100	16	16	100	16
K	48				52	100	48	52	100	48	52	100	48
L	32				40	100	32	40	100	32	40	100	32
M	8				6	80	6	8	100	8	8	100	8
N	8				6	80	6	8	100	8	8	100	8
O	160							158	90	144	166	100	160
P	64										80	100	64
Q	32										24	60	19
R	96										52	40	38
Total	1312	380		346	838		738	1220		1104	1384		1241

% Complete	26.3	56.2	84.1	94.6
Planned man-hours	384	880	1184	1312
Efficiency (CPI) %	91	88	90	90
Est. Final man-hours	1442	1491	1458	1458
SPI (cost)	0.90	0.84	0.93	0.96
SPI (time)	0.90	0.86	0.92	0.89
Est. Completion days	33	36	32	31

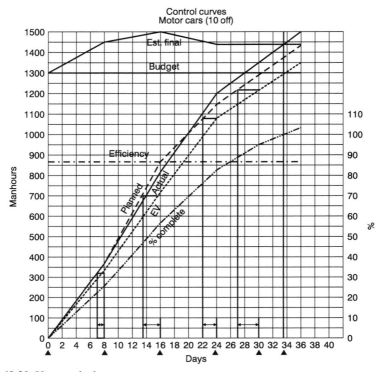

Figure 49.20 Unsmoothed resources.

Table 49.12

Close-out
Close-out meeting store standard tools
Sell special tools and drawings to Ruritania
Clear machinery from factory
Sign lease with supermarket that bought the site
Sell spares to dealers
Sell scrap materials
Write report and highlight problems
Press release and photo opportunity for last car
Give away 600,000th production car to special lottery winner

Risk register

Types of risks: manufacturing, sales, marketing, reliability, components failure, maintenance, suppliers, legislation, quality; qualitative and quantitative analysis; probability and impact matrix; risk owner; mitigation strategy and contingency.

Earned value analysis

EVA of manufacture and assembly of engine, calculate earned value, CPI, SPI, cost at completion, final-project time, draw curves of budget hours, planned hours, actual hours earned value, percent complete and efficiency over four reporting periods.

Close-out

Close-out meeting
Close-out report
Instruction manuals
Test certificates
Spares lists
Dispose of surplus materials

Worked example 4: Battle tank

Chapter outline

Business case for battle tank top secret

Memo: From: General Johnson.
           To: The Department of Defence.
1 September 2006.

Subject: new battle tank

It is imperative that we urgently draw up plans to design, evaluate, test, build and commission a new battle tank (Figs. 50.1—50.13).

The 'What'

A new battle tank which

1. Has a 90 mm cannon;
2. Has a top speed of at least 70 mph;
3. Weighs less than 60 tonnes fully loaded and fuelled;
4. Has spaced and active armour;
5. Has at least two machine gun positions including the external turret machine gun;
6. Has a crew of not more than four men (or women);
7. Has a gas turbine engine and a fuel tank to give a range of 150 miles (240 km);
8. Has the cost not exceeding $5,500,000 each;
9. Has 500 units ready for operations by February 2008.

Project Management, Planning and Control. https://doi.org/10.1016/B978-0-12-824339-8.00050-X

Figure 50.1 Battle tank.

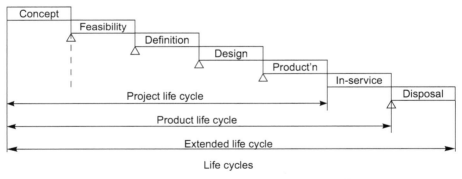

Figure 50.2 Life cycles and phases.

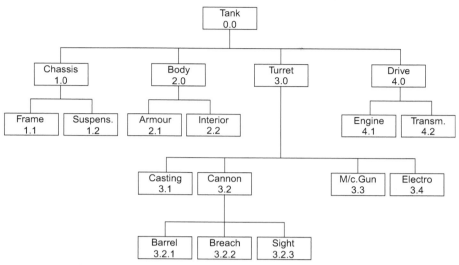

Figure 50.3 Product breakdown structure (PBS).

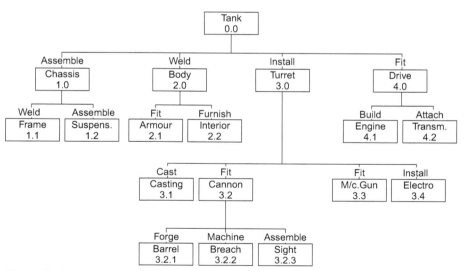

Figure 50.4 Work breakdown structure (WBS).

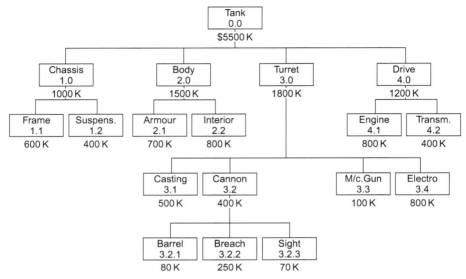

Figure 50.5 Cost breakdown structure (CBS).

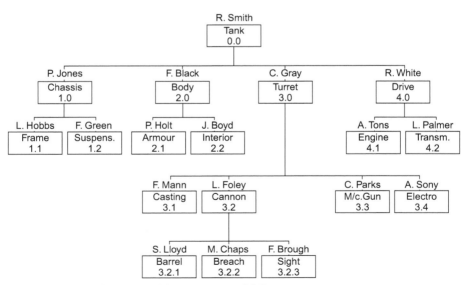

Figure 50.6 Organization breakdown structure (OBS).

	C. Gray	F. Mann	L. Foley	C. Parks	A. Sony	S. Lloyd	M. Chaps	F. Brough	
Turret	R	A	A	A	A	A	A	A	
Casting	C	R	C	C	C	A	A	C	
Cannon	C	C	R	A	C	C	C	C	
M/c Gun	A	C	C	R	C	A	A	–	
Electro.	C	C	C	–	R	–	–	C	
Barrel	A	C	C	–	–	R	C	–	
Breach	A	C	C	–	–	C	R	–	
Sight	A	C	C	–	C	A	–	R	

R = Responsible
C = Must be consulted
A = Must be advised
– = Not affected

Figure 50.7 Responsibility matrix.

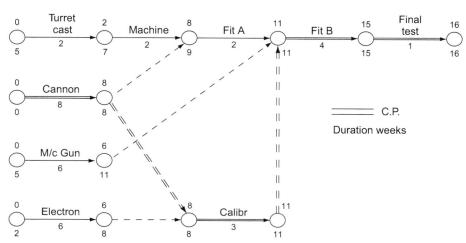

Figure 50.8 Activity on arrow network (AoA).

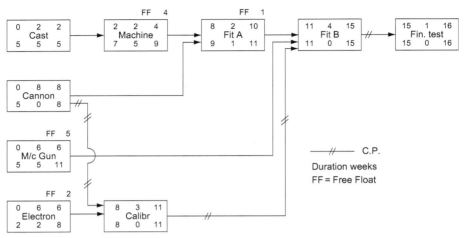

Figure 50.9 Activity on node network (AoN).

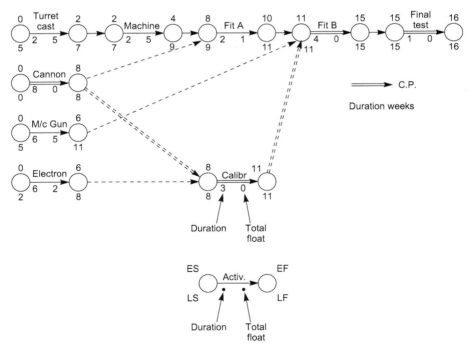

Figure 50.10 'Lester' diagram.

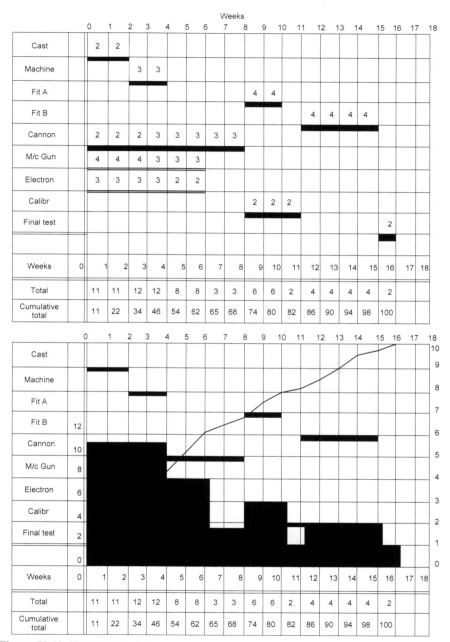

Figure 50.11 Histogram and 'S' curve.

	M/H	Week 4				Week 8				Week 12				Week 16			
Activity	Budg	Plan	Act	%	EV	Plan	Act	%	EV	Plan	Act	%	EV	Plan	Act	%	EV
Casting	160		180	100	160		180	100	160		180	100	160		180	100	160
Machine	240		180	80	192		200	100	240		200	100	240		200	100	240
Fit A	320		—	—	—		—	—	—		300	80	256		340	100	320
Fit B	640		—	—	—		—	—	—		80	10	64		600	50	320
Cannon	840		600	50	420		780	90	756		850	100	840		850	100	840
M/C Gun	840		500	60	504		700	90	756		820	100	840		820	100	840
Electronic	640		300	40	256		620	80	512		700	100	640		700	100	640
Calibrate	240		—	—	—		—	—	—		140	60	144		250	100	240
Test	80		—	—	—		—	—	—		—	—	—		110	20	32
Total	4000	1840	1760		1532	2720	2480		2424	3440	3270		3184	4000	4050		3632
% Complete		$\frac{1532}{4000}$	=	38.3%		$\frac{2424}{4000}$	=	60.6%		$\frac{3684}{4000}$	=	79.6%		$\frac{3632}{4000}$	=	90.8%	
CPI (Efficiency)		$\frac{1532}{1760}$	.870	=	87%	$\frac{2424}{2480}$	.977	=	98%	$\frac{3184}{3270}$	.973	=	97%	$\frac{3632}{4050}$	.896	=	90%
SPI (Cost)		$\frac{1532}{1840}$	=	.832		$\frac{2424}{2720}$	=	.891		$\frac{3184}{3440}$	=	.925%		$\frac{3632}{4000}$	=	.908	
SPI (Time)		$\frac{3.2}{4}$	=	.80		$\frac{6.7}{8}$	=	.837		$\frac{10.3}{12}$	=	.858%		$\frac{13.7}{16}$	=	.856	
Final cost		$\frac{4000}{.870}$	=	4598		$\frac{4000}{.977}$	=	4094		$\frac{4000}{.973}$	=	4110		$\frac{4000}{.896}$	=	4462	
Final time	(Cost)	$\frac{16}{.832}$	=	19.2		$\frac{16}{.891}$	=	17.9		$\frac{16}{.925}$	=	17.3		$\frac{16}{.908}$	=	17.6	

Budget man hours = duration × no. of men × 40 hrs/week   Duration in weeks

Figure 50.12 Earned value table.

The 'Why'

1. The existing battle tanks will be phased out (and worn out) in 2008.
2. Ruritania is developing a tank which is superior to our existing tanks in every way.
3. The existing tank at 80 tonnes is too heavy for 50% of our road bridges.
4. The diesel engine is too heavy and unreliable in cold weather.
5. The armour plate on our tanks can be penetrated by the latest anti-tank weapon.
6. A new tank has great export potential and could become the standard tank for NATO.

Major risks

1. The cost may escalate due to poor project management.
2. The delivery period may be later than required due to incompetence of the contractors.
3. The fuel consumption of the gas turbine may not give the required range.

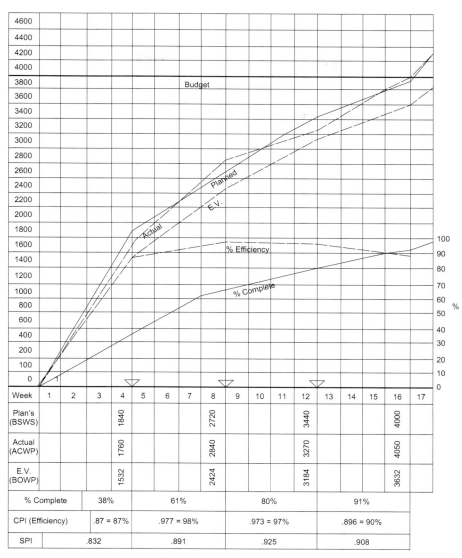

Week	1	2	3	4	5	6	7	8	9	10	11	12	13	14	15	16	17
Plan's (BSWS)			1840					2720				3440				4000	
Actual (ACWP)			1760					2840				3270				4050	
E.V. (BOWP)			1532					2424				3184				3632	

% Complete	38%	61%	80%	91%
CPI (Efficiency)	.87 = 87%	.977 = 98%	.973 = 97%	.896 = 90%
SPI	.832	.891	.925	.908

▽ Report dates

Figure 50.13 Control curves.

4. Ruritania will have an even better tank by 2008.
5. No matter how good our tank is, NATO will probably buy the new German Leopard Tank.
6. Heavy tanks may eventually be replaced by lighter airborne armoured vehicles.

Primavera P2

Chapter outline

Evolution of project management software

Early project management software

Early project management software packages were mostly scheduling engines. Users of the packages were specialists, and the software did automate some low-level tasks, such as calculating early and late dates, but overall the users had to understand how these worked in order to manipulate the software and make sense of the results.

As more refinements were introduced, they were typically geared towards refining the detailed understanding of the work, resource and cost plans. As the models were getting more precise, the user-base also had to become more skilled to use these extra functions.

Also, simple IT limitations, such as price and availability of computer memory, usually meant that each project had its own files. This made it difficult to spread best practices through the organizations, led to duplication of effort to redefine data structures for each project and also made it difficult and time consuming to aggregate reporting to all levels of the organization.

Project Management, Planning and Control. https://doi.org/10.1016/B978-0-12-824339-8.00051-1

Although those tools achieved a good modelling of projects, in accordance with principles described in this book, they had limitations for a wider use, in particular:

The production of reports to all levels of the organization was labour intensive, used different tools and data sets for different reporting levels, and the full process was time consuming, meaning that decisions frequently had to be based on information that was already out of date.

Different specialists worked with different data sets, leading to disconnects between various plans. For instance, many companies had planning engineers and cost engineers working in different packages. This sometimes led to having multiple versions of the truth.

The need for a specialized user workforce restricted them mainly to large projects. Organizations working on smaller projects were frequently scared by the perceived complexity of the packages and therefore avoided them.

The enterprise-level database

To address the first point mentioned earlier, software companies started to look at ways to get all levels of information into a central database. As networks developed, it became feasible to have people sitting in different locations either publish all their information to a central data repository or even work straight into a central database.

This helped reporting at various levels of the organization, as well as preventing wasting effort by recreating data structures for each project. Although this improves the timeliness of high-level reports while reducing their labour intensiveness, the improvement is not significant for individual projects.

Systems integration

Working with centralized systems allowed companies to link them with automated interfaces. This makes it possible for each participant in the management of the project to work in the tool that best matches their needs while providing views and reports that combine information from those different systems. This helps with the second point mentioned earlier, by making sure that the information has a single point of entry and that all reports are using the same original data set.

Getting systems that were designed and configured separately to share information usually requires making some compromises. Each system will lose some flexibility, and decisions made when initially setting up the systems may have to be reversed to ensure data compatibility. Additionally, upgrading any of the systems involved can only be done after making sure the integration still works or is upgraded. As separate systems have separate upgrade cycles, this maintenance can be costly.

The scalable integrated system

Some project management software vendors now provide systems that in their design, although developed individually to provide good functionality in their respective domains, include connexion points with the other packages contained in the solution.

This makes integration between the systems much easier to build and maintain. Each package can be implemented and run individually, but when used with other packages, the link is natural enough to feel like the solution was developed as a single system.

These solutions are scalable both in terms of size of the content − from a single project to a department, or even all projects in a company − and in terms of functions to be covered. This also makes their implementation more flexible as benefits come quickly from the first functions implemented and other modules can be added later on.

Oracle Primavera P6

Oracle Primavera P6 (referred to as P6 in the rest of this chapter) is an integrated system. As of version 8, Oracle has embedded in P6 a number of other systems. Each of those can be implemented separately and will work as a standalone solution, but when used together they will behave as a single integrated system.

Project planning

The core of P6 is its scheduling engine. This is based on the planning tools developed by Primavera Systems since 1983. The scheduling engine uses the critical path method to calculate dates and total float. While incorporating advanced scheduling functions, Primavera made sure each of these functions remained simple enough so that the main skills required to properly analyse the project information relate to project management rather than the software.

Work breakdown structure and other analysis views

The default reporting in P6 is based on the work breakdown structure (WBS). Summarization at all levels of the WBS is automatic, and managers can easily access this information to the level that is relevant to them. However, cost, resource usage and day-to-day organization of the work on the project sometimes require viewing the project from different angles. For this, P6 lets users define as many coding structures as they like, at project, resource or activity level. These codes can be hierarchical or flat.

With simple ways to summarize, group and filter activities, this allows building views of the project plan relevant to any actor of the project. This goes from a client view relating to the contract to all the way down to a resource-specific view by system, location or any other relevant way to organize the information.

These views can be saved for an easy access in the future and can also be shared with any other user of the system. The same principles apply at all levels of the system so that a user fluent with detailed activity-level views can also build portfolio or resource assignment level ones.

Resource usage

P6 offers a resource pool shared between the projects to make it possible to analyse resource requirements at any level, from a single WBS element to a complete portfolio of projects, or even for the company as a whole.

P6 allows to model resources as labour, equipment, facilities and material. Once described in the dictionary, resources can be assigned to activities in the project. Once the activities have been scheduled, this provides profiles of resource requirements.

Resources can be assigned unit prices, should the company decide to model costs on the schedule based on the usage of the resources modelled. Costs can also be modelled as expense items, which are direct cost assignments that do not require resources to be created in the dictionary. Expense items can relate to vendors, cost category, cost account or any other analysis angle deemed desirable for analysis or reporting.

Should there be a need to identify resources more precisely than they are known at the time the baseline is taken (e.g. analysis of named individual time is required, but only the skills needed are known at the time the baseline is taken), P6 also allows resourcing of activities by role. This allows the user to book a budget on the schedule without knowing exactly which resource is going to do the job.

Baseline and other reference plans

P6 can save any number of versions of a schedule to be used as references and compare to the current schedule when required. This can include the baseline, any re-baseline following a change order, any periodic copy of the schedule or what-if analysis versions of the schedule that someone may want to compare with the current schedule.

Each reference plan contains a full copy of all the information contained in the schedule. It is, therefore, possible to compare schedule information — such as dates, duration and float — resource information — such as requirements or actual usage — and cost information — such as budget, earned value or actual cost.

Baselines can also be updated by copying a selected subset of the current project information into an existing baseline. This makes it much easier to adjust the content of the baseline to the current scope without affecting the part of the project that was not affected by change management. As any number of reference plans can be maintained, this enables earned value analysis against the original baseline, or against a current baseline reflecting the current scope of the project.

Progress tracking

P6 has very flexible rules for tracking progress to allow each organization to track only information deemed to be relevant. Progress-related quantities such as remaining duration, actual and remaining units, and cost can be linked or entered separately. This allows each company to decide on what information should be tracked, while making reporting available for all based on the level of detail that can be gathered.

Depending on the requirements, progress can be displayed based on the current schedule or mapped on the baseline in the form of a progress line.

As different organizations find different information to be relevant, it is possible to choose from many percentages to report progress, such as (but not limited to) physical percentage complete, labour unit percentage complete, material cost percentage complete or cost percentage of budget.

P6 also calculates the variance between the baseline and the current schedule in terms of duration, dates and units of resources by type or costs.

Earned value analysis

As P6 tracks dates, resource usage and costs at detailed level, and as reference plans, including the baseline, are maintained with the same level of detail, earned value analysis can be performed at any level required. As for any information in the system, earned value information can be summarized according to any angle considered useful for analysis, be it through the use of the WBS or any coding.

P6 calculates and aggregates earned value information automatically. This can be displayed as easily as any current plan information. If the decision is made to track the relevant information, earned value fields available in P6 include for both labour units and total cost: actual, planned value, earned value, estimate to complete, estimate at completion, cost performance index, schedule performance index, as well as cost, schedule and variances at completion.

Based on the level of confidence in progress tracking, it is possible to define different rules for the calculation of earned value. It is also possible to consider the cost performance index and schedule performance index to date in the calculation of the estimate to complete.

Risk management

As standard, P6 includes a risk register. As with the rest of P6, this risk register can be configured to contain the relevant information for a company. Probability, impact type and impact ranges can be defined to reflect the important factors for a specific company. The system allows the definition of as many impact categories as needed, to help reflect quantifiable impacts, such as schedule and costs, as well as other impacts, such as image, health and safety or environment.

Once qualified with levels or probability and impacts, P6 will rate the risk based on a configurable risk-rating matrix. The overall rating of a risk can combine the impact ratings in different ways, by selecting the highest one, the average of the impacts or the average of the impact ratings.

Running a Monte Carlo analysis on a schedule requires Primavera risk analysis. It is possible to store 3-point estimates in the P6 schedule though, so the uncertainty can be maintained within the main schedule.

Multi-project system

Even though schedules are split in projects in the database, P6 handles multiple projects as if they were just subsets of activities, part of the same total group. This means that reporting makes no difference between single and multi-project content, but also that scheduling can be done across all the projects.

Even while opening only one of a group of interdependent projects, the user has a choice between taking into account interdependencies or not during the scheduling calculations. Similarly, one can analyse resource utilization based on only the schedule he or she is working with, or include requirements from other projects. If needed, projects can be prioritized to only consider projects of a high enough priority in the resource analysis.

Role-based access

There are two main ways to access P6: through the web or by using the optional Windows client. The later requires a high bandwidth between the client and the database server, or the use of virtualization technology such as Citrix or Terminal Services. It is a powerful tool, but the high number of functions available makes it feel complex for users who do not have much time to learn it. This makes it a specialist tool, perfect for planners or central project office people, but less appealing to people who only have limited interactions with planning.

The web access of P6 lets users connect to the P6 database through a web browser-based application. It is both simpler and more complete. It is simpler in that, when looking at a specific function, the interface is not quite as busy as the Windows client. Yet, it is more complete in that it provides views and functions that are geared towards the different roles participants of the project may play in the project.

From the resource assigned to a few tasks on a project to a company director interested only in traffic light type reporting on cost and schedule for each project, P6 Web can display relevant information to each person involved based on their involvement. Resources can see the tasks they are assigned to, including detailed information about those tasks and documents that may be attached to the tasks to help complete them. Planners can see the schedule, with similar functionality to what the optional Windows client provides. Project managers can see high-level reporting and analysis on the schedule, costs and resources, with drill-down capability to find where the problem is. Resource managers can view how busy their team is across all the projects in the company, as well as details of what each individual is working on. Executives and directors can have portfolio or programme-level traffic light type reporting, with high-level schedule, cost, resource or earned value figure summarized at any level that makes sense to them.

For each role identified, the administrator can provide a dashboard that will contain easy access to each report and function needed for that role. The content of those reports and function portlets will be based on the individual, the activities he is assigned to, the projects he is in charge of, or the portfolios that are relevant to that user. As users could have several roles, they can subscribe to several dashboards.

Reporting

Most of the reporting out of Oracle Primavera P6 is based on viewing or printing on-screen layouts. In agreement with the role-based access described earlier, P6 offers many ways to present and aggregate the project information based on the person accessing this information. These ways of viewing project information are usually enough for most people. However, should people prefer to get reports delivered in other formats, it is possible to use Oracle BI Publisher, which comes bundled with P6 licences.

With BI Publisher, users can schedule reports to be run at specific times, and either made available on a website or sent by e-mail. These reports can be created in a number of formats, including MS Office tools and Adobe PDF.

Using P6 Through a Project Life Cycle.

The following pages describe the planning, execution and control of a project using P6.

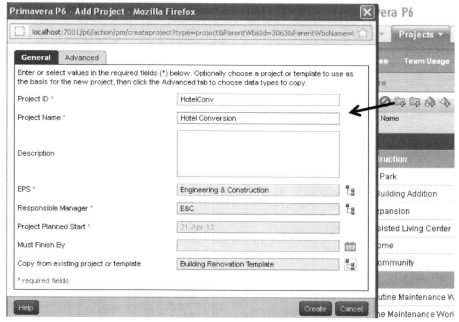

From the EPS tab of the Projects section, select Add a Project. If available, select a template, or copy an existing project to ensure high level preferences are common to all projects.

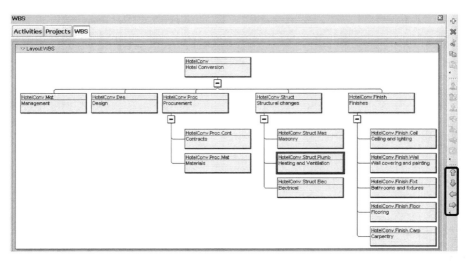

Build the Work Breakdown Structure in the WBS window. If necessary, adjust the structure by using the navigation arrows.

In the Activities tab of the Projects section, select a view or customize your own to simplify adding activities. Grouping by WBS and making sure the relevant columns are displayed help save time.

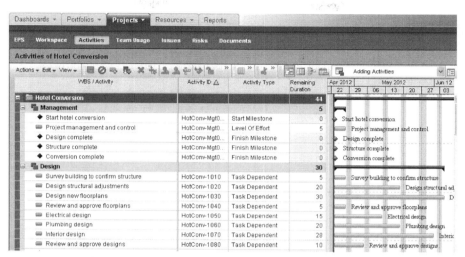

Add the activities necessary to complete each element of the WBS, as well as any milestone that can be useful for tracking the project progress. As the activities have not been scheduled yet, the bars only represent the duration of those activities.

Logic can be built by linking graphically the bars,

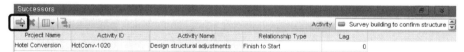

Adding successors in the Successors details tab at the bottom of the view,

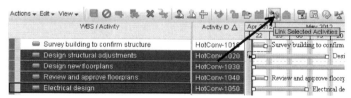

Selecting multiple activities, and then clicking the Link Selected Activities button at the top of the view,

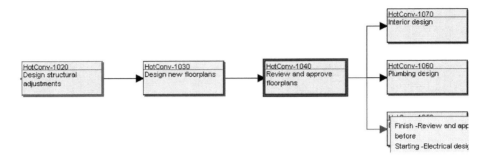

Or drawing the relationships in the activity Network view.

Double click the – sign at the top left corner of a group to collapse the content of that group.

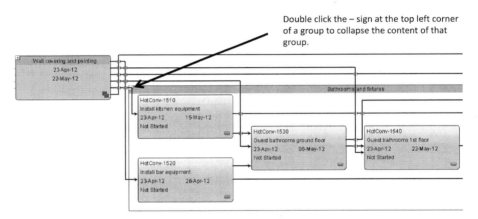

You can review the logic by displaying a PERT view of the project.

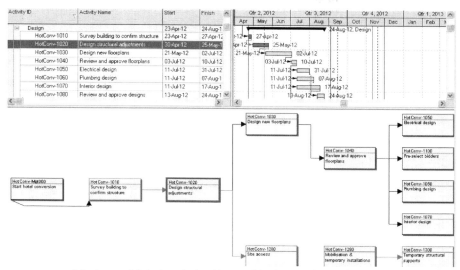

It is also possible to trace logic with a combined Gantt Chart and PERT view.

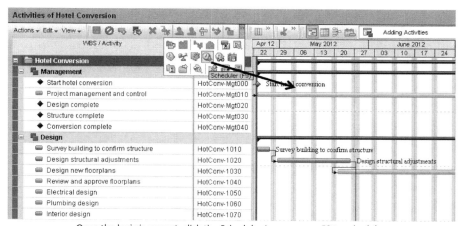

Once the logic is correct, click the Scheduler icon or press F9 to schedule the project.

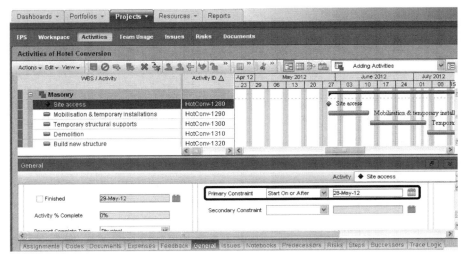

If required, assign constraints on activities in the General tab, to reflect external constraints on the project.

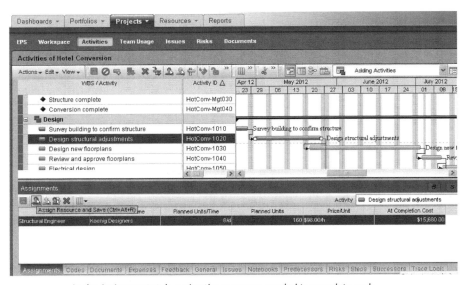

In the Assignments tab, assign the resources needed to complete each activity.

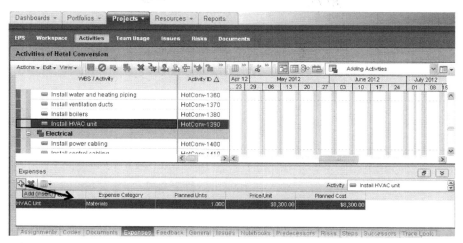

For resources that do not need to be tracked in detail or for subcontracts, it is possible to just assign a cost in the Expenses tab.

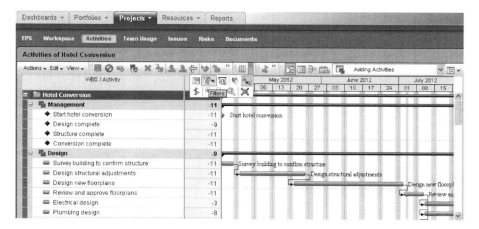

From the filters menu, run the critical activities filter and make required adjustments to the schedule for the project to finish on time.

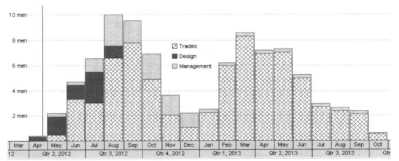

Analyse resource usage. If necessary, adjust the schedule.

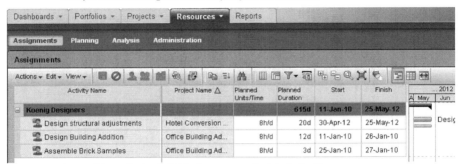

It is possible to display individual resource schedule in the Resources
view. If resources are working on multiple projects, it is possible to show
usage on any or all projects.

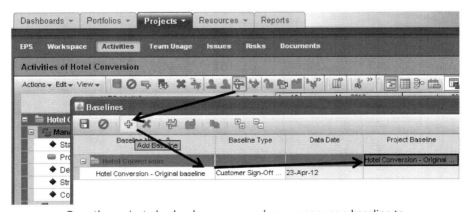

Once the project plan has been approved, save a copy as a baseline to
keep as a reference and compare with the current schedule.

Track activity progress by entering actual dates for completed activities, actual start and percent complete for activities in progress,

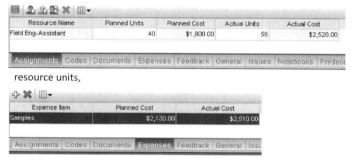

resource units,

and expenses in the corresponding tabs.

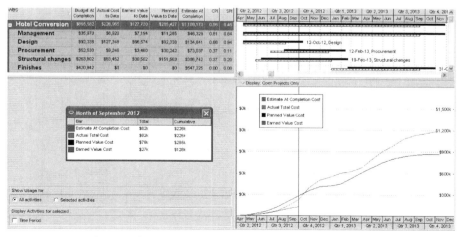

Cost and earned value data can be analysed from many angles, including time-distributed tabular presentation,

totals at any relevant level, bar chart of schedule progress vs the baseline, and as S-curves.

WBS	Budget At Completion	Actual Cost to Date	Earned Value to Date	Planned Value to Date	Estimate At Completion	CPI	SPI		Apr-12	May-12	Jun-12	Jul-12	Aug-12	Sep-12	Oct-12
Hotel Conversion	$ 865,582	$ 228,965	$ 127,730	$ 285,427	$ 1,188,173	0.56	0.45	Baseline	$ 3,301	$ 31,466	$ 66,980	$ 126,019	$ 207,786	$ 285,427	$ 359,439
								Earned Value	$ 1,928	$ 16,150	$ 35,074	$ 66,397	$ 101,221	$ 127,730	
								Actual Cost	$ 2,021	$ 17,806	$ 36,586	$ 69,719	$ 144,062	$ 228,965	
								Estimate At Completion	$ 2,021	$ 17,806	$ 36,586	$ 69,719	$ 144,062	$ 226,455	$ 400,951
Management	$ 35,970	$ 8,920	$ 7,194	$ 11,285	$ 46,329	0.81	0.64	Baseline	$ 605	2,821	$ 4,937	$ 7,053	$ 9,370	$ 11,285	$ 13,602
								Earned Value	$ 385	1,799	$ 3,147	$ 4,496	$ 5,974	7,194	
								Actual Cost	$ 478	2,230	$ 3,902	$ 5,575	$ 7,407	8,920	
								Estimate At Completion	$ 478	2,230	$ 3,902	$ 5,575	$ 7,407	8,920	$ 11,817
Design	$ 92,338	$ 127,348	$ 86,574	$ 92,338	$ 134,841	0.68	0.94	Baseline	$ 2,697	$ 25,831	$ 41,847	$ 76,012	$ 92,338	$ 92,338	$ 92,338
								Earned Value	$ 1,543	$ 14,351	$ 28,374	$ 39,051	$ 61,286	$ 86,574	
								Actual Cost	$ 1,543	$ 15,576	$ 30,804	$ 42,321	$ 82,555	$ 127,348	
								Estimate At Completion	$ 1,543	$ 15,576	$ 30,804	$ 42,321	$ 82,555	$ 127,348	$ 134,841
Procurement	$ 52,530	$ 9,246	$ 3,460	$ 30,242	$ 73,037	0.37	0.11	Baseline				$ 3,460	$ 17,688	$ 30,242	$ 40,651
								Earned Value					$ 3,460	$ 3,460	
								Actual Cost					$ 6,736	$ 9,246	
								Estimate At Completion					$ 6,736	$ 6,736	$ 38,506
Contracts	$ 28,034	$ 6,736	$ 3,460	$ 19,113	$ 38,682	0.51	0.18	Baseline				$ 3,460	$ 13,848	$ 19,113	$ 20,335
								Earned Value					$ 3,460	$ 3,460	
								Actual Cost					$ 6,736	$ 6,736	
								Estimate At Completion					$ 6,736	$ 6,736	$ 26,393

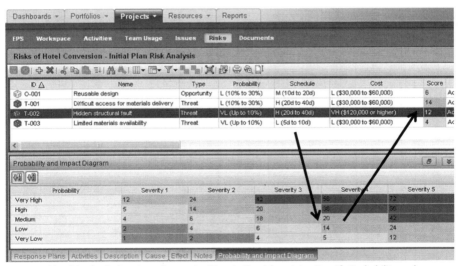

List risks, both threats and opportunities, in the risk register. Specify Probability and impact, which will be used to calculate the score.

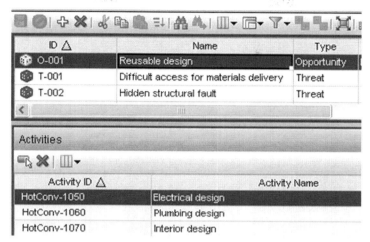

Link each risk with the activities it impacts in the schedule.

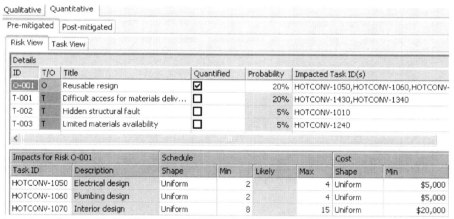

Impacts can be quantified in PRA.

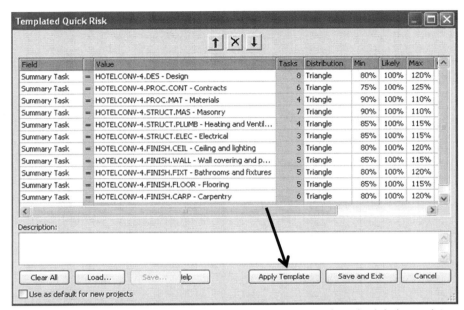

Define uncertainty in Templated Quick Risk and assign it to the schedule by applying the template.

Run the Monte Carlo simulation to analyse its results, such as date distribution analysis, or probabilistic cash flows.

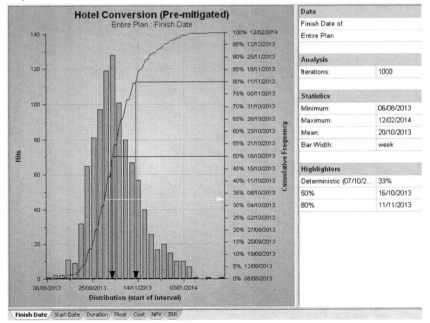

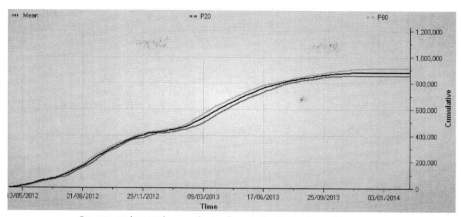

Reports can be sent by e-mail using formats familiar to the recipients

Building Information Modelling (BIM)

Chapter outline

Introduction

Technology to access and control construction information is constantly changing based upon cost or time-saving requirements, and the relevant project information is always required to be accessible and editable on various devices and platforms. For example, even preparing this chapter has involved storing information on The Cloud, interfaced from a smartphone, tablets, various laptops and PCs, all on different platforms, and this is just for controlling text and a few graphical files. The construction industry is moving to digital integrated design team project delivery on platforms such as Trimble Connect (www.connect.trimble.com) or Bentley ProjectWise, with

Project Management, Planning and Control. https://doi.org/10.1016/B978-0-12-824339-8.00052-3

all of the required information (specification, models, point clouds, drawings, reports, etc.) being available at any stage of the contract and beyond. Each project design team can be formed from many different players depending on the actual work requirements and content. For example, a design team could include the client; architect; engineer; mechanical, electrical and piping (MEP); contractors; etc. One of the processes that support these developments is building information modelling (BIM). BIM is not a single 3D application, but a process that streamlines the product model content and delivery.

It is worth concentrating on the 'Information' part of B'I'M before addressing the actual BIM process itself. Information always needs to be collected and contained in a physical object or system. From a historical point of view, this could have been on many drawings or nowadays a 'project library' could be made available for the reader or viewer. This really is the same with BIM; however, the whole project information is not just available as an online encyclopaedia but broken down to object-level nuggets of information for ease of reference.

A product model can be built or defined with a 3D or 4D application, where the latter is the 3D information which also encapsulates the relevant time information. This can also include manufacturing information, start and completion dates, maintenance information and even the required special demolition procedures, as the model can support the full life cycle of the building or project. However, the BIM model is not limited to just 4D information, as costing can also be included, which is sometimes referred to as 5D information, or anything else that the user wants to track (nD information) and control.

Sometimes, BIM is thought to be just another service that provides users with instant online access to an ever-increasing stream of constantly evolving, instantly updating digital data. However, if the processes are in place then real project control is possible. Many times it is said that: 'Nowadays to change a pump physically on a building site is relatively simple. However, changing all of the 3D models, drawings, sketches, specifications, etc. is the hard and time-consuming part'. Adopting the BIM process will revolutionize this, as the information has only to be changed once with authoring the application, after which the rest of the design team members can simply reinsert the new 'reference model', with all the latest information, ensuring all the models are up to date.

History of BIM

2D drawing systems have been used since the early 1990s only as 'electronic drawing boards', copying and pasting details or 'blocks' to reproduce drawings quicker than the older manual processes. However, as far as interoperability was concerned there were no real advantages. Various applications allowed different drawings or blocks to be imported into the working drawing, using some form of 'XREF' options, which really was the start of the reference model concept.

General computer-aided design (CAD) tools were used for a while, leading to bespoke solutions which were developed to suit each industry or design team member's requirement. For example, tools used by architects would require different functionalities from the toolkit required by engineers. Also, a structural engineering application would have different drawing requirements from a MEP solution. It is interesting to note that in many companies the technicians will produce the drawings, with engineers just resolving typical project sections or sketches.

In the building and construction sector, 'information modelling', is normally defined as the computer representation of a building or structure, including all the relevant information required for the manufacture and construction of the modelled elements. The elements or objects are required to be intelligent and should therefore know what they are; how they should behave in different circumstances and know their own properties and valid relationships. A simple example of this would be the reinforcement in a pad foundation. If the foundation size is modified, then the embedded reinforcement should dynamically update and reposition itself accordingly. This is very different from a computer-generated model, which is constructed for purely visualization purposes, where every object is a non-related element.

The structural steelwork industry is normally recognized as the lead sector in 3D modelling solutions, and their developments can be traced back over the last 30 years with applications such as the forerunner of Tekla Structures. This type of applications allowed the 3D steel frame to be modelled and the connection applied by the user-defined macros, which allowed the automatic production of general arrangement drawings, fabrication details and then, after a few more years, the development of links to computer numerically controlled machines, for cutting and drilling of steel profiles and fittings.

Once the 3D modelling technology was extended to include parametric modelling (true solid objects), clash detection systems were then developed. Hard or soft (allowing for access) clash detection allows application to identify any possible material overlapping conditions or objects existing in the same space.

Nowadays, globally unique ID or globally unique identifier (GUID) is available in most applications, which are unique strings usually stored as a 128-bit integer associated with the objects. The GUID is used to track the objects between applications for change management. Some systems allow the support for internal and external GUIDs. The term GUID usually, but not always, refers to Microsoft's universally unique identifier standard.

In the initial stages of BIM, only drawings and reports were made available to others, as the actual BIM model was always confined to the office where the more powerful computers were situated. However, as computers advanced, the models are now accessible on laptops and, over the last few years even on tablets for site use, either located on the device or available from the cloud.

What is BIM

So what is BIM? It could simply be defined as a rapidly evolving collaboration tool that facilitates integrated design and construction management. The importance of 'I' in BIM should never be underestimated, as this becomes a project or support for the company's enterprise framework and not just a means for 'building models'. This information means that more work can be done earlier in the project to support green issue concepts, as less waste saves both materials and energy.

BIM enables multidimensional models including space constraints, time, costs, materials, design and manufacturing information, finishes, etc. to be created and even allows the support for information-based real-time collaboration. This information can be used to drive other recent technologies including city-sized models, augmented reality equipment used on site, point clouds of existing buildings and equipment, radio-frequency identification tags to track components from manufacture to site and even the use of 3D printers.

It may be useful to consider the players who would want to have access to the BIM models. Not limiting the list they could include − the internal and external clients, local authorities, architects, engineers (structural, civil and MEP), main contractors, steelwork and concrete subcontractors, formwork contractors and all site personnel. Until recent years BIM has just been available as a solution for architects, engineers and steelwork contractors, etc., leaving everyone else just to work with 2D drawings that may be industry-specific, but not being totally readable without knowledge of that environment.

Various references have been made to architects' BIM model or the structural BIM. However, they really are the same, as the boundaries between their models and their content are reducing all the time. Architects' BIM models will include structural member; however, the models that they produce do not normally require including the material grades, reactions and finishes, whereas the model produced by the steelwork contractor will include at least the manufacturing details and all the information necessary to order, fabricate, deliver and erect the members. The MEP contractor could also define the site fixings on their version of the model, as the contractor will want to know when the assemblies will be delivered to site, where it will be fixed or poured, and how much the item costs. The client's view of the same member would be for control and for possible site maintenance. For this reason, various models are created in the 'best-of-breed' authoring applications and shared with other design team members as reference models, which are normally in the form of industry foundation class (IFC) files for all structures except the plant and offshore markets, where CIMsteel Integration Standards (cis/2) and dgn format files are the dominant interoperability transfer model formats.

It is so much easier to work with a BIM model and to explore the building in 3D with rich information, than looking at hundreds or thousands of drawings and having to understand the industry drawing conventions. Now, a user can simply click on to an object and obtain all the information that they require either through the native object, if on the authoring application, or through the reference model or even from a viewer or collaboration tool directly from the cloud.

UK government recommendations

The United Kingdom and other global governments have expressed support for BIM over recent years. However, in March 2011, the UK government published a BIM strategy paper, full details of which can be downloaded from www.bimtaskgroup. org. In fact, the UK government has made it clear that BIM level 2 needs to be adopted on all of their projects by 2016, to save the model objects from being rebuilt many times within the same project by different design team members.

The main aims and objectives of the working group are to:

- Identify how measured benefits could be brought to the construction industry through the increased use of BIM methodologies;
- Identify what the UK government, as a client, needs to do to encourage the widespread adoption of BIM;
- Review the international adoption of BIM including the solutions by the US Federal government; and
- Look into government BIM policy to assist the UK consultant and contractor base to maintain and develop their standing in the international markets.

The general recommendations were to:

- Leave complexity and competition in the supply chain;
- Be very specific with supply-chain partners;
- Measure and make active use of outputs;
- Provide appropriate support infrastructure;
- Take progressive steps; and
- Have a clear target for the 'trailing edge' of the industry.

The report also defines the project BIM maturity levels, from level 0 to level 3 and as a quick summary:

- Level 0 is where just CAD tools have been adopted.
- Level 1 is where 2D and 3D information is used to defined standards.
- Level 2 is where BIM applications are used with fully integrated model collaboration.
- Level 3 is where BIM models are used for project/building lifecycle management.

The UK Government's main targets in the report were:

- 20% reduction in costs.
- Level 2 BIM by 2016.
- COBie information should be available for decision-making at critical points in the design and construction process.

COBie information is a formal schema that helps design teams to organize architectural BIM object information into spreadsheets that are normally shared with other project players for facility management.

Also, with the initiative from the BIMForum (www.bimforum.org), there is currently much interest in the level of development (LOD) of the BIM model objects, so the client can obtain a clear picture of their expected BIM deliverable quality. For full information see www.bimforum.org/lod.

In very general terms, these levels are the following:

- LOD 100 — A graphically represented object mainly used for cost.
- LOD 200 — A generic object with approximate size, shape and location.
- LOD 300 — Design-specified object with full specific, size, shape and location of brackets and corbels, etc.
- LOD 350 — Actual object model with full specification, size, shape, location.
- LOD 400 — Similar to 350, plus all relevant project information and finishes including rebars and accessories for concrete members and all fittings for steel members.
- LOD 500 — Similar to 400 including a site-verified reports, test certificates and CE marking information.

How BIM is applied in practice

If a BIM model is created or being amended on the authoring application, it is normally referred to as a physical or native model, which can be enhanced using normal authoring tools. If the model is required to be fixed (not editable) by one design team member, then an IFC file or other reference models are normally adopted, where objects can be commented upon but not changed by other members of the design team.

Tekla Structures

Tekla Structures is a multi-material BIM software tool that streamlines the construction design and delivery process from the planning stage through design and manufacturing, providing a collaborative solution for the cast-in-place (in situ)/precast concrete, steelwork, timber, engineering and construction segments.

The structural BIM is the part of the BIM process where the majority of multi-material structural information is created and refined. These are normally created by the structural engineer as the architects generally just work with space, mass, texture and shapes, and not with building objects in the same way as defined in the structural BIM. However, the connection between the architects models and structural BIM is a very obvious way to help in the future development of intelligent integration, and these should be always available in the form of reference models in the same way that the XREF function is used in a 2D drawing. These reference models could also be used as 2D information for collaboration with non-BIM applications.

The model starts to evolve during the engineering stage, where conceptual decisions of the structural forms are made. It is sometimes thought that the design portion of analysis and design (A&D) is just the pure physical sizing of the structural elements. It is in fact more than that, as it should also include the engineering and the 'value engineering' of the project, including all materials, their relationships and their reference to the architectural and service objects, together with possible links to other design application using .NET technology to form an application programming interface, or API as it is more commonly known.

In its basic form, .NET is a flexible programming platform for connecting information, people, systems and devices together using a programming environment and tools based upon the Microsoft Visual Studio .NET developments. There are over 30 programming languages that are .NET-enabled which allow true object information to be seamlessly transferred between systems. So, for example, element information and geometry can be passed from modelling applications to any other .NET-enabled system. These could be A&D systems, management information systems (MIS), cost control or just systems used for internal bespoke company development down to the humble excel spreadsheet.

One of the principal advantages of the structural BIM is that the project is built for the first time in the memory of the computer, before any physical materials are involved. This allows the scheme to be refined to a greater extent, allowing full clash checking facilities, automatic drawing production, bar-bending schedules and report preparation. The modern application also allows the drawings and reports to be automatically updated should the model be amended, thus change management can be controlled and different design solution scenarios can be considered at any time together with having links back to the A&D systems if required.

For further information on the Tekla Structures or the applications authors Trimble Solutions Corporation, see www.tekla.com. For web tutorials on Tekla Structures or for general BIM lessons information, see www.campus.tekla.com.

Linking systems through open .NET interfaces

Sometimes, using industry-standard files is not appropriate when tight linking of applications is required. For example, a user may want to share design information amongst modelling systems and say an A&D application; MIS; enterprise solutions; connection and fire engineering applications; project management systems or planning systems. In such a case, the objects, a simplified form of the object or just the object attributes, can be passed between systems using the Open .NET interface as defined earlier. This same interface could also be used for model transfer or for repetitive processes such as drawing and report creation.

Tekla BIMsight and other collaboration tools

Tekla BIMsight is a free BIM collaboration tool which is available for anyone to download and install. It runs on a normal PC environment and is an easy-to-access application that presents the complete project including all necessary building information from different construction disciplines. It is also much more than a viewer as the user can communicate using the various IFCs and other models not just from Tekla Structures.

With Tekla BIMsight the user can perform the following:

- Combine multiple models and file formats from a variety of BIM applications into one project;
- Share building information for coordination between different trades and deliverables;
- Identify and communicate problem areas: check clashes; manage changes; approve comments and assign work in 3D by storing a history of different view locations and descriptions in the model;
- Measure distances directly in the model to verify design requirements and construction tolerances;
- Control the visualization and transparency of different types of parts in the model to make it easier to understand complex and congested areas of the project; and
- Query properties such as profile, material grade, length and weight from parts.

Tekla BIMsight can be downloaded from www.teklabimsight.com which also includes video tutorials and a user forum.

In addition to Tekla BIMsight, Trimble Solutions Corporation also produces Field3D (www.teklabimsight.com/tekla-field3d), which is an enhanced collaboration tool that runs on both Apple and Android devices and also has a cloud service option.

Savings with BIM

Recent reports from a number of global organizations have confirmed that around 50% of structural engineers are adopting BIM on a regular basis.

It is always hard to establish the return on investment and project savings with regard to any software systems. However, as the BIM project is created within the memory of the computer before involving any materials or site personnel's time the change is inexpensive. Various reports and white papers regarding the cost of change and constant remodelling was one of the reasons for the UK Governments BIM initiative.

Sample BIM projects

Alta Bates Summit Medical Centre — by DPR Construction Inc.

A $289 million, 13-storey, patient-care pavilion and future home of over 200 licenced beds is a new addition to the Alta Bates Summit Medical Campus (ABSMC) and the facility is due to open in January 2014. DPR's highly collaborative integrated project delivery team was faced with many challenges during this project. The use of Tekla Structures and Tekla BIMsight has grown significantly since the start of construction in 2010. Modelling scopes within Tekla Structures includes structural steel, cast-in-place concrete, reinforcing bar, miscellaneous steel and light gauge drywall framing.

Seismic design requirements set forth by the Office of Statewide Healthcare Planning and Development in addition to a hybrid steel and concrete shear wall structure required in-depth coordination between rebar and steel-fabrication models.

These coordination efforts between multiple trade partners could not have been achieved without the use of this highly collaborative and detailed BIM platform. Since subcontractors Herrick Steel and Harris Salinas Rebar were both detailing with Tekla, it was very useful to provide the rebar detailer with the steel model during detailing to identify constructability issues with anchor bolts, stiffener plates and other connection details that were not shown in the engineering design model.

More recently, the ABSMC Project had implemented the concept of dynamic detailing within Tekla Structures on both the reinforcing steel and drywall framing. By referencing in the structural steel Tekla model along with IFC models of the MEP&FP systems, Harris Salinas Rebar and DPR Self-Perform Drywall were able to identify conflicts during the process of modelling, as opposed to the traditional method of modelling in a silo and having to rework the modelling after clash detection. This workflow results in a more efficient, streamlined workflow with fewer chances for modelling errors.

The DPR Self-Perform Drywall detailer, Robert Cook, has developed an efficient workflow to detail the rebar using multiple custom components. Robert Cook is also creating drywall framing assembly spool sheets that can be printed to $11'' \times 17''$ or viewed on an iPad to help the site team to increase efficiency and quality while reducing rework. As all the ductwork is fabricated and assembled directly from the model, DPR drywall is now able to install framing around the duct and pipe openings prior to MEP installation, and be confident that the framing is in the right location to align with the prefabricated ductwork and piping.

In addition to Tekla Structures, Tekla products have been used to convey differences between site conditions and the design model related to the exterior skin and expansion joints. The ABSMC project requires seismic expansion joints where surrounded on three sides by the existing hospital campus and around a 5-storey pedestrian bridge. The comment tool, dimensioning tool and photo attachments were used to convey coordination and constructability issues to the designer, fabricator and installer.

Actual drywall framing around the MEP equipment.

Model of the drywall framing around the MEP equipment.

Tekla BIMsight showing model to site comparisons.

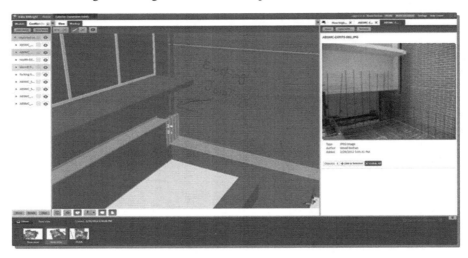

Drywall assembly 'spool sheet' from Tekla Structures.

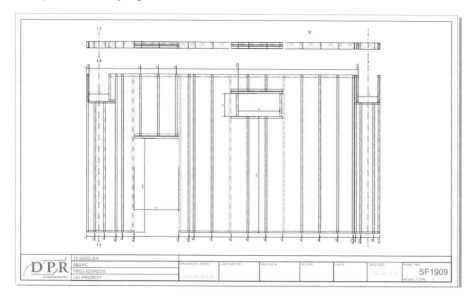

PCP tower adjacent to the existing hospital.

Architect's rendering of the Alta Bates Summit Medical Centre.

The National Museum of Qatar (Arup)

The National Museum of Qatar in Doha is the flagship project for an important series of cultural and educational projects which were commissioned by the Qatari government. The project, which drew inspiration from the desert rose, now started on site and was in the planning phase since 2008. The desert rose is a crystalline formation found below ground in saline regions of the desert. When imagined as a building, the result is a 4-storey, $300 \times 200 \text{ m}^2$ sculpture of intersecting discs which are up to 80 m in diameter.

The evolved structural solution consists of radially and orthogonally framed steel trusses, supporting fibre-reinforced concrete cladding panels to create the required aesthetic and performance characteristics of the building envelope.

Challenges

The key challenges for the design were the highly complex geometry of the disc interactions. No two discs were the same and no two discs intersected each other in precisely the same way. The galleries and other key spaces in the building were created by the interstices between the discs; any alteration to the architecture involved moving discs and thereby moving the structure within the discs. This led to an evolution of systems and processes which were required to handle, manipulate and develop geometric ideas from the architects, so that engineering solutions could be established before communicating these in their most useful form to the wider project community.

For this reason, the following structural modelling (for analysis, design and manufacturing and construction) is needed to address the following requirements:

- Position the elements in the correct place in the 3D space within the cladding envelope.
- Generate and model elements as efficiently and automatically as possible in order to keep up with iterations of the architectural arrangement.

- Facilitate cross-discipline coordination, both with the Arup MEP design and 3D modelling teams in London, plus the architectural team based in Geneva and Paris together with the client in Qatar.

National Museum of Qatar.

Showing the complex building geometry.

Leeds Arena, UK, Fisher Engineering Limited

Leeds Arena is the United Kingdom's first purpose-built, fan-shaped arena. Using this form of geometry allows every spectator to have a perfect view of the centre stage. The main facades are rounded and have a domed effect which terminates with a flat roof. Formed with two columns, one sloping outwards and the second spliced to the top and cranked inwards, these curving elevations are clad with a honeycomb design of glazed panels which contain lights of changing colours.

Leeds Arena project.

The steel framed structure of the roof is supported by a series of 13 trusses spanning up to 70 m across the auditorium with five central trusses being supported over the stage area by a 170 tonne trussed girder and plated columns, which form the 54 m long × 10.5 m deep proscenium arch. The proscenium arch truss was delivered to site in 32 separate sections; a total of nine trailer loads. Assembling the truss took 3 weeks, using two large mobile cranes.

Section through the Leeds Arena project.

The bowl terraced seating is formed from precast concrete units supported on a radial steelwork structure, braced and tied into two main concrete stair cores which provide the required stability. Acoustic resistance is a major design factor for a venue of this size, which is situated within a city centre and required the structure to be shrouded in a skin of precast concrete wall panels with a concrete roof topping on a metal deck.

Typical general arrangement drawing.

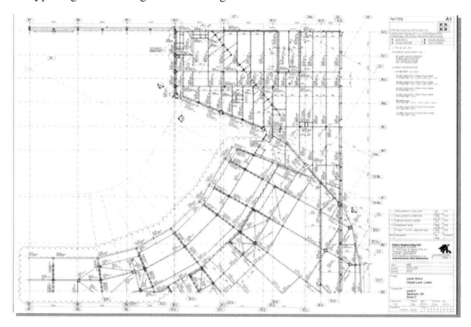

Steelwork arrangement detail.

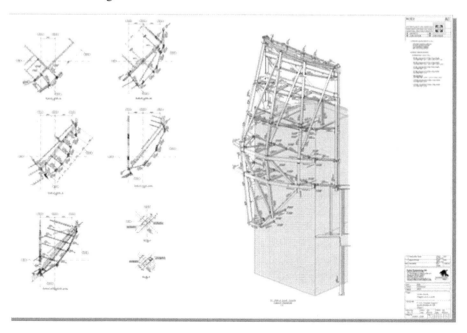

A BIM strategy was essential for efficiency

With so many subcontractors providing major structural element, all of which required prefabricated connections to interface with the complex geometry, it was clear from the start of the project that a BIM strategy was essential to obtain the necessary project efficiencies through the evolving design and collaboration process. The BIM model proved invaluable to all parties involved as it was passed between the design team and contractors for clash detection and for the resolution of incomplete design issues. This greatly assisted the project programming, sequencing and general constructability.

Model of the bowl terraced seating.

Arena during construction.

Technical detail

For readers who are interested in more technical details, the following section may be of interest. The various industry abbreviations and acronyms are explained below.

Interoperability and principle industry transfer standards

3D interoperability between various building and construction applications is generally achieved through industry-standard formats, such as dwg; DXF; SDNF; cis/2 and IFC, with the older systems being listed first. Other bespoke links have been adopted in the past based upon XML (extended mark-up language which is basically an extension of HTML used for creating websites) or special file formats. Excel sheets have also been used in the form of reports or to enhance various applications. It is generally accepted to adopt the full BIM process, as only IFC files are advanced enough to support all of the objects of building models.

DXF, DWG, DWF and DGN formats

DXF (drawing exchange format) was developed by Autodesk for enabling data interoperability between AutoCAD and other programs. As the file format does not contain any form of part ID, it is not possible to track changes between different physical objects contained within different versions of a file.

DWG — a file format used for 2D and 3D CAD data and is the standard file format for Autodesk products.

DWF (design web format) — a secure file format developed by Autodesk for efficient distribution and communication of rich design data, normally created with DWG drawings. However, it is rarely seen within the BIM environment.

DGN — the standard reference file transfer between plant design programs. Originally developed by MicroStation, which is now part of Bentley Systems Inc., it is similar to DWG in that it is only a graphical data format, but does contain part ID's unique for the given model.

IGES and STEP

Initial graphics exchange specification (IGES) defines a neutral data format that allows the digital exchange of information amongst CAD systems. It was defined by the US National Bureau of Standards and is largely been replaced by the Standard for the Exchange of Product Model Data (STEP) over recent years.

The International Standardization Organization (ISO) is concerned with creating the standards for the computer interpretable representation and exchange of product manufacturing information, so STEP files are available across many manufacturing industries. In the construction market, it is normal to only see files relating to ISO 103003 AP230 and these are generally treated as reference files.

SDNF format

Steel detailing neutral file (SDNF) was originally defined for electronic data exchange between structural engineers, A&D and design systems to steelwork modelling systems. Version 3.0 is the latest format supported by the software industry and this format has been used for many years for transferring even complex plant structures between system such as Tekla Structures and plant design systems such as Intergraph's PDS or Aveva's PDMS applications.

As a quick overview, the SDNF files are split into packets and records. The main packets are defined as follows and generally not all items are supported by all applications:

Packet 00 − Title packet.
Packet 10 − Linear members.
Packet 20 − Plate elements.
Packet 22 − Hole elements.
Packet 30 − Member loads.
Packet 40 − Connection details.
Packet 50 − Grids.
Packet 60 − Curved members.

CIS/2 format

The CIS (CIMsteel Integration Standards) is one of the results of the European Eureka CIMsteel project. The current version 'cis/2' is an extended and enhanced second-generation release of the format, which was developed to facilitate a more integrated method of working through the sharing and management of information within, and amongst, the companies involved in the planning, design, analysis and construction of steel-framed buildings and structures. There are a number of different format versions, analysis, physical and manufacturing formats for steel structures, and the physical format has been widely used in the structural steelwork sector in the past.

The only downside of this format is that true multi-material objects cannot be defined as the standard really just concentrates on steel objects. However, an extended cis/2 format has been adopted by Intergraph and Tekla to link the Smart 3D plant design application to Tekla Structures within the last few years.

IFC format

The latest and most complete transfer standard used within the BIM environment is the IFC as defined by the buildingSMART organization (www.buildingsmart.com) which was formally called the International Alliance of Interoperability. The organization defines itself as 'buildingSMART is all about the sharing of information between project team members and across the software applications that they are commonly used for the design, construction, procurement, maintenance and operations'. Data interoperability is the key enabler to achieving the goal of a buildingSMART process. Building SMART has developed a common data schema that makes it possible to hold and exchange relevant data between different software applications. The data schema comprises interdisciplinary building information as used throughout its life cycle. The current version of the standard is 2×3, while the next version IFC4 has been defined and many applications are currently being undated to adopt this latest standard.

True building objects as defined by architectural, engineering, MEP and other systems can be shared with IFC. This allows the users to use the systems that they know, or that are best for creating the objects which are normally referred to as 'best-of-breed' systems. Adopting the IFC standard allows true object information to be shared between the major modelling applications. Different IFC formats and flavours are available, so the users need to know which one is adopted, and this can normally be determined just by looking at the header part of the IFC file.

Further reading

Briscoe, D. (2015). *Beyond BIM*. Routledge.
Kumar, B. (2016). *A practical guide adopting BIM in construction projects*. Whittles.
Paterson, G. (2015). *Getting to grips with BIM*. Routledge.
Sauchez, A. X. (2016). *Delivering value with BIM*. Routledge.
Saxon, R. (2016). *BIM for construction clients*. NBS.

Virtual Design and Construction (VDC)

Chapter outline

Introduction

Virtual Design and Construction (VDC) is a process of collecting, organizing and sharing data from the design phase through the construction phase and into the life cycle of assets. VDC was coined from construction entities due to the misnomers or pre-conceived ideas of building information management (BIM). There may be debates about BIM and VDC, but in general terms they perform similar functions. Construction utilizes VDC within their project execution.

VDC + project management

VDC plays a very large part in project management, in remembering that VDC is a process of information flowing through the project, and it starts in the design phase. A project is first desired by an owner with a need for something to be constructed. A concept will be created and then sent out for Bids. A Design firm will be awarded the project to design it per owner specifications. As the design happens, it is generally done within a 3D modelling environment. As the 3D model is designed, the model objects are created with information that can participate in different aspects of the project.

The following are some of the acronyms and abbreviations used in VDC:

AIM — Asset Information Management/Modelling
BIM — Building Information Management

Project Management, Planning and Control. https://doi.org/10.1016/B978-0-12-824339-8.00053-5

BrIM — Bridge Information Management/Modelling
CDE — Common Data Environment
CIM — Civil Information Management/Modelling
COBie — Construction Operation Building information exchange
DE — Digital Engineering
IDC — Intelligent Design and Construction
PIM — Project Information Management/Modelling
VDC — Virtual Design and Construction

VDC software

In reality, there is no VDC software as VDC is a process of information flowing through projects. But there is software created to enhance VDC capabilities, and some of those are InEight, Autodesk, Bentley, Hexagon and more. Some BIM-capable software have 3D design capabilities and then stay within their own environment with limited interoperability with other VDC-capable software. InEight is not a 3D design authoring software but imports 3D models into its ecosystem and also enhances the VDC capabilities for other project management needs (estimating, control budget, planning, schedule, field progress, completions and more).

Some software is referred to as Point Solutions, a solution resolving one or two needs. A few examples of point solutions would be a model used for clash detection in one software, a model used for estimating in another software, and a model used with schedule in a different software.

InEight Model introduction

InEight Model is a platform for cross-discipline collaboration, focused on models and information aggregation to support: clash detection and issue tracking, documents linked to model objects, model-based take-off, advanced work packaging, work face planning, completions and visual reporting. Provides real-time updates and maintains a complete history of all changes for many users utilizing the same version project model at the same time in different time zones all over the world.

InEight Model supports VDC through project coordination, field management, visual scheduling, sensor monitoring and facilities management. CAD models and .ifc type files are published from authoring tools into InEight Model.

VDC + coordination

Coordination within InEight Model provides the ability to aggregate all discipline models into one project model. Utilize *Master Presets* to create a one-click recall of saved views for project focus areas and topics, such as coordination areas per schedule

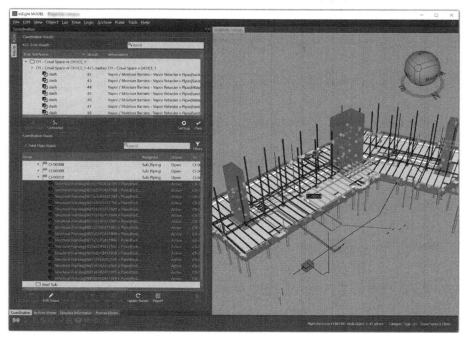

Figure 53.1 Coordination for clash detection and design issue tracking.

to help the team resolve any coordination issues before finding them in the field. Issues found by the field crew result in delays and schedule setbacks, which cost time and money to the project. *Coordination Master Presets* will help the team focus on project areas per schedule.

Coordination Rules are custom flexible rules to create distinct hard clashes and soft clashes to address regulations and codes. *Coordination Rules* can be saved into custom groups to execute many rules at once possibly broken up by discipline and/or intent. The *Coordination Rules* will identify *Coordination Issues*, which can be assigned to a responsible person or group to resolve. The *Coordination Issue Panel* tracks the responsible person, the status of the issue, when it was created, notes and more. It can filter and search on any of the metadata to easily address outstanding issues through completion.

The *Coordination* process identifies issues before construction crews get into the field, which creates project efficiencies and provides the project team a means to resolve issues before they become a problem (Fig. 53.1).

VDC + model-based take-off

3D design models are created to deliver construction drawings and are not innately structured for construction and owner use. All 3D models are not created equal, some provide a lot of data to support the VDC process and some do not provide useful data at all.

InEight *Model* brings together disparate models and provides the ability to normalize and clean data: clean-up inconstancies and incorrect data, make data quantifiable, address model nomenclature, parse data and concatenate data. Add custom data: construction work areas, labour codes, schedule IDs. Add quantities not modelled, utilizing existing model objects for representation: formwork, waterproofing, doorknobs and hinges, sheetrock, paint, etc.

Once the models have been normalized and cleaned-up then they are ready to be used for quantities, whether that is just a quantity take-off or an integration with InEight Estimate. The integration from *Model* to *Estimate* provides the ability to push Cost Items from *Estimate* to *Model*, then *Model* can align specific model objects to the cost items and sync the quantities back into *Estimate*.

A model integrated with an estimate can save the project time versus traditional take-off processes. Additional savings is noticed with *Model* revisions. In the traditional process, possibly hundreds of drawings are mind through to identify the changes and effects of design changes. *Model* can analyse model changes, run the model normalization and clean-up rules, then sync updated model quantities to *Estimate* to track the quantity changes for the project (Fig. 53.2).

VDC + work packaging

VDC work packaging is basically a breakdown of the schedule and how construction crews are going to execute the work in smaller chunks than what the hire level schedule shows. Models can participate in project management by creating visual work packs. The project would be broken up into construction work areas (CWA), then into construction work packs (CWP) and then into installation work packs (IWP).

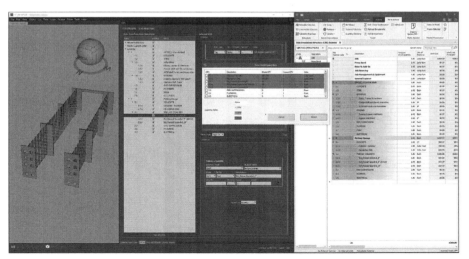

Figure 53.2 InEight Model syncing quantities into InEight Estimate.

Some industries use this work packaging process formally on their projects for project management and follow the advanced work packaging process. Informally, all projects are broken down to chunk size to execute the project effectively as planned.

Experienced trades are used to looking at 2D drawings and seeing 3D as to the work they are going to execute. As models are becoming more prevalent for projects, the 3D visual work pack process provides unrealized gains and efficiencies.

The Superintendent will receive the project schedule from the schedulers, then the Superintendent and/or the Forman will create a breakdown of the overall schedule into smaller chunks referred to as work packs. Work packs will include safety, quality, environmental, activity components, materials, equipment, tools, budget, lessons learned, labour, work sequence, goals and more.

In general, the budget, the schedule and the work packs are disparate softwares, and the information, if possible, by the software capabilities, is shared through import/export functions. Information in disparate softwares that are not integrated introduces quality issues, possible project delays, duplicate work and pertinent information not available for team members when they need it.

A VDC-capable software that is driven by interoperability will mitigate the quality issues, the duplicate work, project delays and accentuate the right information into the team members' hands at the right time. To provide an example of what this may look like, let us look at InEight. InEight provides interoperability between *Model*, *Plan*, *Control* and *Schedule* for integrated work packaging. All are synced and can participate in the work package structure continually throughout the project.

Projects can start the work pack process in *Model* or *Plan* or *Schedule*, depending on what project information is available first.

Model can create and participate in the CWA/CWP/IWP structure for project management. Group model objects into *Material Components* (items that will be delivered and installed) and link the model *Material Components* to IWPs.

Plan can create and participate in the CWA/CWP/IWP structure for project management. Group *Material Components* and *Activity Components* into IWPs and link safety, quality, environmental, activities, materials, equipment, tools, budget, lessons learned, labour, work sequence, goals, etc. into work packs. *Plan* is also integrated with *Control* (the Budget).

Schedule can create and participate in the CWA/CWP/IWP structure for project management. Break down the overall schedule into CWPs and then IWPs, linking associated dates shared with *Plan* and *Model*.

This interoperability saves the project team members from duplicate data entry into different systems, eliminating quality issues and increasing efficiency, providing one truth of information at the fingertips of all team members continuously throughout the project. Right information at the right time (Fig. 53.3).

VDC + visual reporting

As 3D models are more prevalent in design and construction, 3D snapshots are used throughout a construction site providing a visual of the end goal. Snapshots and

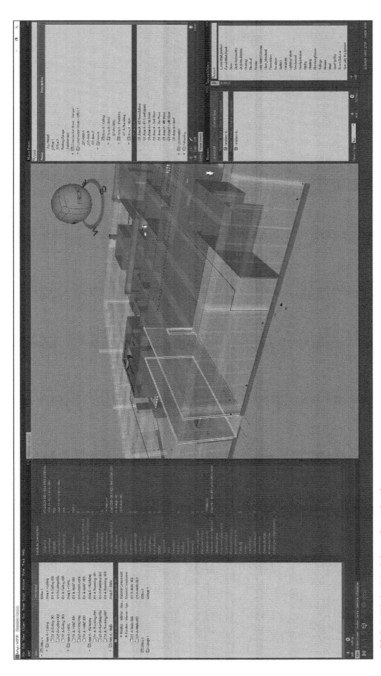

Figure 53.3 InEight Model work packaging.

sections of models printed out and posted throughout a construction site to provide a reference point for crews executing work in that area is the norm for some companies.

In today's world, some projects are utilizing the model to indicate construction progress. Construction progress has always played a part in project management, and a VDC-capable software accentuates those capabilities providing an up-to-date clear and comprehensive visualization.

Some use cases for visualization: work not started, work included/not included in a work pack, work pack status, material purchased/on-site, work installed, work issues, superintendents/foremen/sub-contractors responsibilities, model data validation, equipment needed for special lifts, risk areas, etc. All project information could potentially be visually communicated.

A VDC-capable software that is driven by interoperability can emphasize and communicate project information. If we look at InEight's integrated systems, the status of the executed work packs within *Plan* can be visualized within *Model* to provide and communicate to all stakeholders daily of construction progress. The risk areas identified within *Schedule* can be visualized within *Model* to give the project team a clear picture of what needs attention to mitigate the risks. The issues identified within checklist and the completeness of systems within *Completions* can be visualized within *Model*. Custom project data external to integrated systems can be visualized within the model to communicate project information to all *project* stakeholders (Fig. 53.4).

VDC + facility management

Facility management has been a necessary practice for eons and for years loads of paper and digital format of data has been handed over to owners at the end of a project. As owners and projects are trying to go paperless, digital PDFs and data are turned over from contractor to owner.

Facility managers are still access drawings and asset data from disparate systems to manage their facilities. Today, owners generally require Construction Operation Building information exchange (COBie) to be part of the turnover. COBie is generally imported into the owners' Computerized Maintenance Management System so that they can utilize the data to participate in the maintenance of their assets and facility.

A VDC-capable software that is driven by interoperability can make the turnover from construction to owner cleaner and less overwhelming, plus the information is already organized and ready to be used for the facility mangers. If we look at InEight's integrated systems as an example, data can be captured during the *Completions* activities of the project; like the serial numbers of the assets, the last/next inspection dates and information, warranty information, installed date, commissioned date, etc. QR codes for assets and facility areas can be generated and attached appropriately. The facility maintenance personnel could scan the QR code to access the latest and up-to-date information and related documents or select an asset within *Model* and access the desired information (Fig. 53.5).

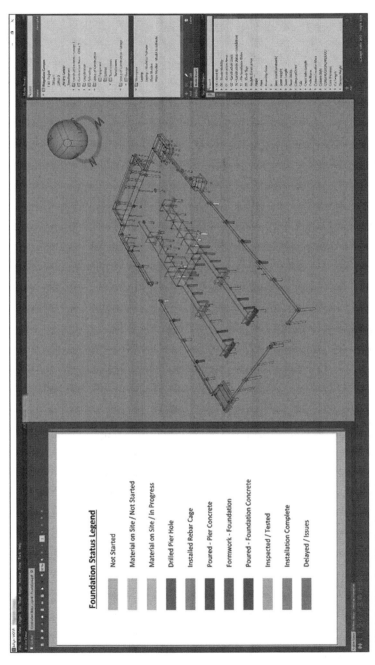

Figure 53.4 InEight Model visual reporting.

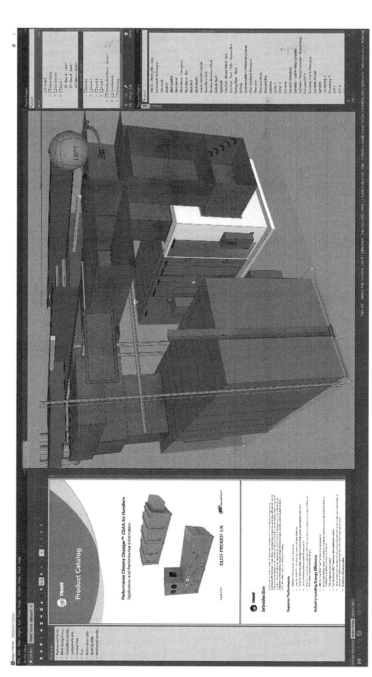

Figure 53.5 InEight Model participating in facilities management.

Sustainability

54

Chapter outline

There are many different definitions of sustainability, but the neatest one is the definition suggested by the World Commission on Environment and Development:

'A process of change in which the exploitation of resources, the direction of investments, the orientation of technological development and institutional change are all in harmony and enhance both current and future potential to meet human needs and aspirations'.

This means that biological systems have to be studied and developed to ensure that the environment, in the form that we desire it, can endure indefinitely. The preservation of existing ecosystems is not only the responsibility of governments and international agencies, but also of corporations and indeed any organization whose activities impact on the environment or the pool of natural resources.

Programme and project managers must therefore ensure that the projects under their control create minimum (ideally zero) pollution of the atmosphere, water resources including river and seawater, and minimum contamination of minerals, soils and vegetation including forests and grasslands. This involves environmental-impact assessments being carried out at the planning stage and developing a mechanism which ensures that the methods and procedures agreed at these meetings are actually carried out.

For corporations, this may (as with safety) require a change in the organization's culture which must flow down from the boardroom to the design and construction staff as well as the operators of the facility. There has to be general awareness that the health and availability of resources of future generations cannot be endangered. In practice, all these safeguards cost money, and unless we allocate sufficient funds *now* to protect the environment and ensure the conservation of limited resources, future generations will pay the price.

It has been widely accepted that there are three pillars of sustainability which both interconnect and overlap. Indeed, the metaphor can be expanded as a structure with the

Project Management, Planning and Control. https://doi.org/10.1016/B978-0-12-824339-8.00054-7

three pillars being connected at their base by a common foundation and at their head by a portico. These pillars are as follows:

(1) Economy
(2) Society
(3) Environment

Economy

This area covers the methods and techniques such as waste reduction, elimination of industrial pollution (liquid or gaseous), improvement of efficiency in the manufacture and distribution of products and services while at the same time ensuring that the businesses and indeed the wider economy generate sufficient profit to enable them to continue to operate and perform their essential functions for the benefit of all the stakeholders. This may require a company or other profit-orientated organization to change its fundamental concepts from generating long-term rather than short-term profits. The effects on climate change by industrial pollution have been known for many years as shown by the increase in damage caused by toxic discharges into rivers and waterways as well as the world's oceans. Many companies are now aware of their responsibility to society and have instituted strategies to deal with these issues. These corporate responsibility strategies (CRS) should be represented and sponsored at the Board level to ensure implementation, and it is up to the project manager to incorporate these strategies in the project management plan.

There are optimistic signs that industrial and financial corporations are taking steps to reduce their polluting emissions and their carbon footprint. Indeed, one very influential investment company has asked the CEOs of their larger clients to set out how their companies intended to ensure a net-zero economy by the year 2050. They even threatened that unless these companies took active steps in this respect they would reduce their investments in these corporations.

Recently, it was reported in the technical press that 30 of the largest civil engineering consultants and contractors had published their own targets to achieve carbon-neutral status by periods that varied between 2030 and 2050. Steps to achieve this covered

operational emission reductions, designing structures using building materials with less embodied carbon, reducing business travel, investing in electric vehicles and ensuring that their supply chain meets similar criteria.

Society

To survive on this planet we need clean air and clean water to stay healthy and clothes to keep warm when it is cold. For years, the means to provide these requirements has been at the expense of polluting both the air and the water, and steps have to be taken to reduce or eliminate what are collectively known as 'greenhouse gases'.

The increase in consumerism and the proliferation of more and more electronic devices give rise to a greater and greater demand on the earth's resources. It is up to people and society in general to curb this plunder by demanding a more sustainable form of extraction and utilization. A balance has to be struck between demand and supply, but too often the demand is being created deliberately by companies (usually by vigorous advertising) to sell their products. One obvious example is the clothing industry, which has fostered the fashion concept in order to keep the factories producing clothes which are deemed to be out of fashion half a year later. This undoubtedly enriches the manufacturers, distributors and retailers, but it results in the premature waste of the products and the inevitable diminution of the earth's resources such as oil or gas, from which many of the fabrics are made. Even the more sustainable fibres such as wool or cotton should not be squandered on garments whose lifespan is no more than a few weeks. It can be seen therefore that the scope for society to improve sustainability in this industry alone is enormous.

Another area where people can bring pressure to bear is packaging. This is in some ways even more in need of reform, because the pollution to our waterways and oceans from discarded plastic products is also killing or poisoning marine life which we need to feed the world's ever-expanding population. It could also of course be argued that this potential over-population should be controlled, and efforts have been made in some countries to tackle this problem. However, it has been found that legislation to limit the sizes of families is not a long-term solution. The most effective way to do this is by education and changes to outdated customs and practices.

As previously mentioned, electronic devices consume rare metals which will have to be recycled to a much greater extent than currently to ensure the future production of the very appliances in which they have been incorporated. Whether some of these devices are really necessary is an open question. Does a healthy human really require a servo to throw a simple light switch? How far can we go with the 'Internet of Things' without losing the use of our muscles which we then have to re-build using another electronically controlled device in a gym. The scope for a re-assessment is huge. Many products are manufactured in such a way that it is either impossible or uneconomical to repair them. If it were possible to easily replace individual components instead of scrapping the whole device, huge savings could be made in terms of basic materials, even it means spending a little more time on the repair.

The recycling of domestic refuse is now well established in most western economies.

Metals, paper, glass, plastics, garden waste and food waste are now separated by the householder and collected by the local authority for recycling. The residue can then be sent to landfill or ideally burnt in a refuse incineration plant which produces steam for district heating and/or electricity generation. In one case, the separation of waste by the residents of one German local authority was so effective that a new state-of-the-art combined incineration and power generating plant was shut down because the remaining waste material delivered to the plant from the catchment area had insufficient calorific value.

Environment

The earth we live on is an ecosystem which has limited resources; that is, a limited biocapacity, which has been defined by the World Wide Fund as follows:

'the capacity of the ecosystem to produce useful biological materials and to absorb waste materials generated by humans, using current management schemes and extraction technologies'. In other words, we have to put back what we take out.

The aim of all companies should be to be net carbon-free. Now, at last, a start has been made by the electricity generating industry which has become almost sustainable with the phasing out (in the western world) of coal, oil or gas-fired power stations and replacing them with renewable energy forms such as solar, wind and nuclear power. Other high polluting industries such as steel and cement producers have published forecast dates by which they expect to be net carbon neutral.

However, it is not only the power supply which has to be de-carbonized, One of the greatest polluters of our towns and cities is the motor car, and there has been a call for many urban areas to provide clear deadlines for the banning of diesel or petrol-powered vehicles from their town or city centres. Replacing these modes of transport with electric vehicles will not only require a massive investment in new green power plants but also creating and supporting the necessary infrastructure for distributing this power.

Reducing carbon emissions in the transport and construction industries can also be achieved by substituting petrol or diesel fuels with hydrogen. This is especially suitable for site machinery, heavy lorries, buses and short-range railway locomotives.

All materials, including building materials such as steel, aluminium, concrete and timber, have a carbon footprint which includes not only the method of extraction or harvesting but also their manufacture, transportation and installation. It has been estimated that the global use of concrete alone is responsible for 8% of all carbon dioxide emission, while 1.83 tonnes of carbon dioxide are vented into the atmosphere for every tonne of steel produced.

Some organizations have already compiled a table of building materials giving the carbon content and a carbon rating. One such table was published in the July 2020 issue of 'The Structural Engineer', which gives the embodied carbon factor (ECF) in $kgCO_2e/kg$ for the most common building materials. This, however, only includes the embedded CO_2 in the raw material and that expended during the manufacturing and transportation stages. To this must be added the ECF of the construction, operation (usage) and end-of-life phases to obtain the total ECF. The methods of calculating embodied carbon are discussed more fully in the guide issued by the IStructE. There is a trend to replace steel or concrete with timber, and indeed some prestigious wooden structures using laminated timber components have been built in a number of countries, but this must be balanced against the long-term resistance to decay, the risk of fire and the high premiums charged by insurance companies for such structures. Another way to reduce the need for steel is to reduce the spans of beams so that the load can be carried by a component (or a number of components) made of laminated timber.

Modern methods of construction and standardization will undoubtedly replace some of the more traditional methods. It has been clearly demonstrated that off-site manufacturing is more energy efficient and hence less polluting.

Local production and transportation is another method by which clients, developers, designers and contractors can reduce the carbon footprint of the materials used in the project. Sooner or later, all construction projects will come to the end of their useful life and will have to be demolished and replaced. It is important, therefore, if one wants to consider the whole life carbon footprint to include the demolition and material recycling stages. This has been set out in 2015 by the UN member states as part of their Sustainable Development Goals.

From the foregoing discussion, it can be seen that there is an obligation for project and programme managers to ensure that all the facets of sustainability are considered (and eventually incorporated throughout the life cycle) at the start of the project. To the three basic project management criteria of cost, time and quality/performance, must now be added the fourth criterion of sustainability. Future awards for construction and building projects will include the life cycle sustainability factor as well as the conventional time, cost and quality/performance criteria.

In effect, the well-known project management 'Iron Triangle' will be replaced by the new project management 'Wooden Rhomboid'.

Carbon Credits

A Carbon Credit is a permit that allows the company that holds it to emit a certain amount of carbon dioxide or other greenhouse gases. One credit permits the emission of a mass equal to one ton of carbon dioxide (CO_2).

At the 1997 United Nations' Intergovernmental Panel on Climate Change in Kyoto, an agreement known as the 'KyotoProtocol' was reached to set carbon emissions for all participating countries in an endeavour to reduce emissions globally. At this meeting, a mechanism called 'Carbon Credits' or Certified Emission Reduction (CER) was devised which allows a company to buy sufficient Carbon Credits to enable it to continue emitting greenhouse gases up to a set limit. All the 170 countries which have signed up for the Kyoto Protocol are allotted CERs (also known as Assigned Amounts) according to their pollution levels. A National Registry in each country sets quotas on the emissions on polluting companies or other organizations which are approved and monitored by the United Nations Framework Convention on Climate Change. In the United Kingdom, the Environmental Agency acts as the national administrator of the Registry and also distributes allowances to the market by auction.

If the company emits less CO_2 than the set limit they can sell the surplus Carbon Credits to another company who need them to cover their own emissions. If they emit more gases than their set limit they are fined. The receipts of these fines are then invested in projects that reduce pollution, such as planting forests which absorb CO_2. Over time the set limits are reduced until the company achieves net zero, which means that they are removing as many emissions as they produce.

As a result, a trading system developed called 'Cap and Trade' which, while aiming to reduce the overall gas emissions also gives any company an incentive to reduce their emissions (or be fined if they exceed their limit) and thereby also obtain additional income by selling their surplus credits. The Carbon Credits (known as Carbons) can be traded on the open market through banks, specialist traders or exchanges that list these products.

There are of course people including 'Friends of the Earth' who do not believe that trading Carbon Credits actually reduces the emissions because, they argue, the polluting emissions are merely transferred to another polluter who finds it more profitable to pay for the certificates than investing in less polluting processes. They also claim that the projects which were funded from the sale of the certificates are often short-term operations instead of long-term schemes such as re-forestation.

Project assurance

Chapter outline

It has been estimated that over 40% of major projects are either delivered late or exceed their budget or both. An effective system of project assurance must therefore be implemented to give early warnings of deviations from the project plan to top management and the sponsor.

Project, programme or portfolio assurance can be defined as follows:

'An independent assessment at regular intervals into the structure and performance of a project, programme or portfolio which critically examines the planned and applied processes, systems and controls in relation to financial, technical and operational risks and performance'.

Purpose of project assurance

The main function of these assessments is to provide confidence to the sponsor and top management that the project or programme will meet the agreed strategic objectives and benefits in terms of cost, delivery, performance and sustainability. For this reason, the assurance process should be carried out by an independent team which can be part of the organization tasked with delivering the project (or projects) or an external specialist company familiar in the type of work being undertaken.

The team, department or organization carrying out the planned assessments is accountable to top management and the project sponsor.

The scope of project assurance should cover project governance as well as a risk assessment of the financial, technical and operational aspects. Of particular importance is the need to check that realistic contingency allowance has been allocated to the major areas of operation. A blanket overall allowance should not be accepted.

Project Management, Planning and Control. https://doi.org/10.1016/B978-0-12-824339-8.00055-9

Operation of project assurance

One of the first tasks a project assurance assessor must undertake is to carefully study and become familiar with the project management plan which is the project 'Bible' (see Chapter 14) and the baseline against which all the topics are assessed. At the same time, any discrepancies or omissions discovered must be brought to the attention of the project manager.

However, no realistic assurance exercise can be carried out unless the fundamental systems and procedures for project execution are in place.

These include systems and processes for cost and time control, quality control and risk analysis. The assurance assessments must be carried out at regular intervals and at the arrival at pre-determined control gates.

Contractors, subcontractors and suppliers must sign up to implementing the control systems of the client or the managing contractor. This must be a contractual obligation, and all must be made aware that non-compliance will be construed as breach of contract. Where specialist training is envisaged, this must be undertaken well in advance of the time the specialists are needed and the assessor must check that these training periods, as well as adequate time frames for the parallel off-site testing of systems (especially new software), have been incorporated in the critical path (CP) network.

Part of the assessment must be the financial stability of contractors, their access to adequate funding and labour, material and plant resources.

Also included is a check that contractors and subcontractors who are required to submit drawings or other data for use by others have been given clear instructions that late or non-delivery of these documents is subject to liquidated damages claims as if they were hardware or computer software.

It must be emphasized that the project assessment team should not be involved with the day-to-day operation but ensure that the agreed control systems which have been set up are correctly implemented.

Procedures to be assessed

For time control, there is probably no better system than a critical path network which has been drawn up and agreed by the project team including the major stakeholders and lead subcontractors. This programme must be monitored and updated weekly or monthly and reported to the project manager. One of the functions of the project assurance team is to ensure that this process is being carried out.

Cost and budget control, whether by the application of earned value analysis (EVA) or regular re-measurement, can be implemented by careful monitoring of the EVA control curves or comparison with original Bills of Quantities or costed milestones. Inspection of curves generated by an earned value analysis will give an immediate state of health of the project (see Chapter 33). A drooping of the earned value curve indicates a potential time overrun and a steepening of the actual cost curve (often associated with a descending EV curve) shows that the final cost risks being exceeded. Either of these phenomena should trigger an inquest. There must always be a reason for the deviations from the original time or cost plan.

Such an inquest should be linked with the monitoring of the programme so that the assurance team will be able to assess the performance of these two criteria at the same time and report any divergence from the plan to top management and sponsor at the earliest possible time. Only then will it be possible to carry out a risk analysis (should this be required) and discuss and formulate the necessary mitigation strategy to bring the project back under control. There is no more appropriate proverb in project assurance than 'A stitch in time saves nine'.

The quality control processes must be checked as well as the compliance with quality and performance procedures including the inspection of all issued acceptance and rejection certificates. An excess of rejections clearly signifies that a supplier/subcontractor may have to be replaced or operators require additional training.

Change control procedures including the administration of contract variations must be in place from day one as well as the procedure for recording any delays caused by external organizations, industrial disputes or natural forces (rain, snow, wind, tremors).

Where disputes have arisen between some of the stakeholders, they must be brought to the attention of top management as soon as possible so that any agreed alternative dispute resolution procedures can be implemented.

Conformity with legislation and regulations

Compliance with safety regulations and national/local government legal requirements have to be checked. This is especially the case with overseas projects.

Safety must always be paramount as accidents will invariably delay the project and the increase in cost could even destroy the reputation and credibility of the operator or contractor. Site diaries and the keeping of contemporary records of accidents and important occurrences should be examined at regular intervals and good record keeping must be made obligatory.

Checks must be made that local public consultations and hearings have been set up in good time and that steps have been taken to satisfy the requirements of local authorities and public utilities. All these restraints should appear on the CP network.

Arrangements for complying with environmental requirements and the implementation of pollution control procedures must be checked as are the steps being taken to reduce waste.

Security, whether on-site or in the offices is of utmost importance and any suspected weaknesses or breaches of any of the established systems must be reported to the project manager and head of security.

Project stakeholders and personnel

A successful project is only as good as the staff and personnel employed. The welfare and cooperation of all stakeholders must be monitored and any signs of a lack of enthusiasm or lack of commitment to meeting the project objectives must be reported as these can be a sign of inadequate management in one or more areas.

A good picture of the state of the project can often be obtained by site visits or visits to contractors', subcontractors' or suppliers' premises. These can be scheduled or unscheduled. When the latter option is chosen, not more than one day's notice should be given so that while it is enough time to prepare the documents asked to be inspected it is probably not long enough to 'cook the books'. Every effort must be made to instil into all stakeholders and staff a sense of honesty and openness.

On large projects, it may be necessary to employ a number of independent assurance providers and it then falls on the chief assessor to co-ordinate the work and reports of these organizations.

On large projects, the following support staff may be necessary to assist the project assurance assessor in covering the large range of activities: auditor, quantity surveyor, quality inspector and programmer. Some of these could be temporarily recruited from the project office or external agencies.

Appendix 1: Agile project management

Graham Collins
Faculty of Engineering Sciences, University College London, London, United Kingdom

Having an adaptable process of development that can respond to rapidly changing economic conditions is one way organizations can compete effectively. Agile project management enhances the ability of teams and organizations to react to these changes. Traditional approaches to project management often entail following a set plan, and if there is any divergence from this plan the project manager is often expected to correct this or seek approval for any revised plan. Agile approaches, however, recognize that goals will inevitably change and that achieving value for the client should be the most important consideration.

It is often the case that during a development process the requirements will have substantially changed. The longer the time interval from requirements gathering to delivery, the more likely the client will indicate that what has been developed does not meet their needs. Also, it is unlikely that many of the requirements will still be considered important after a few months or even at the start of development. During workshops with clients, it often occurs that every participant volunteers at least one requirement, but if after outlining this list there is a voting opportunity for the priority of these requirements, it is likely many of them are not voted for at all. It is not only that the client team needs to see a prototype, often to understand what they truly want, but it is also likely that any software developed may change the business processes to such an extent as to make the original requirements even more obsolete.

One way to reduce the risk of development being detached from what is actually required by the client is to provide a more frequent feedback to the client including what is being developed. Agile project management accelerates the feedback cycle and actively involves the customer in the prioritization of the requirements and design of the product. Delivering a product after 12 or 18 months runs the risk that the business needs have changed or that the client team realizes on viewing the software it is not what they actually want. It is better to have a set regular feedback cycle continually prioritizing the most valuable functionality and delivering some thread of working software for the client to comment on. Producing a tangible software or product at regular intervals and having a continual communication cycle, involving the client, is at the heart of agile methods.

Traditional planning is typically based on the concept of delivering a project within a set budget within a set schedule. The agile philosophy encompasses delivering high-value products or software as rapidly as possible. Ensuring that this delivery benefits from regular feedback enhances this value. It also ensures that within a fixed time the greatest value in terms of the functionality as prioritized by the client is delivered. The shift for both the organization and the project manager is one from delivery of a project to schedule and budget to one of delivery of the highest value within the time and other constraints set.

It is through a cycle of iterations and release, with continual working software of product developments, that trust is built, and the client can see that every release is providing increased functionality and business value.

The paradox

Managers typically wish to know how much and how long a project will take and yet they still want to have the flexibility to respond to the business environment and embrace innovation. How can we achieve flexibility and respond to change, and at the same time follow a plan? The answer is partly in granularity, considering the capabilities at a high level of abstraction for planning and allowing development teams to define specific tasks according to the needs of the project. Agile is inherently measurable because of the clear regular cycles and internal and external measures. At a high level of abstraction, these are the regular releases defined by the needs of the project, domain and agile method, often every 90−120 days. Within these, there are the iterations of 1 month or shorter cycles. Furthermore, within a day there is the daily cycle identifying what has been delivered, what are the problems and what will be done next. Within this, especially in software development, there are cycles that are achieved within even shorter periods by development teams using test-driven development and automated testing techniques.

Definitions of success are part of the problem

If we measure success in terms of achieving the original specifications, then measuring agile projects that are designed to incorporate changes in goals to achieve maximum business value to the client are bound to be problematic. What is needed is to base measures on what the client considers of value and for this to be updated continually. In agile methods, and particularly with the Scrum method, this is achieved via the product owner who is the client representative on the team. Typically, the product owner is part of the development team who represents and works closely with the client or client team to determine the most valuable capabilities and constituent features. At the start of each iteration, they will all be involved in prioritizing features from capabilities they have identified into a finer detail of functionality expected from the system. Dependent on the method, the requirements may be in the form of user stories to gain some idea of how users will use the system. It is through this process that the product owner with the development team prioritizes the most valuable stories for the coming iteration. This cycle continues so that at any given point the project has delivered the highest value functionality as defined by the client.

One advantage of reducing cycle time is that the team soon learns what is working and what is not, and can correct the development as necessary. Another advantage is that valuable working software or products are brought into use earlier, starting to contribute economic benefits, so that the project reaches the breakeven point and provides a return on investment sooner.

What is agile?

Jim Highsmith (2010) outlines that being agile is not a silver bullet to solving your development or project management problems. He characterizes agile in two statements: 'Agility is the ability to both create and respond to change in order to profit in a turbulent business environment. Agility is the ability to balance flexibility and stability' (Highsmith, 2002).

The concept of iterative and incremental development is not new, and developers in the 1980s and 1990s were designing light methods aimed at involving the development teams and customers in closer collaboration. An alliance of these developers met in Snowbird, Utah, in February 2001 to see if there was anything in common between the various methodologies being used at the time. They agreed on an agile manifesto and values (below) supporting the manifesto. The latest updates to this can be found at www.AgileAlliance.org:

- Our highest priority is to satisfy the customer through early and continuous delivery of valuable software.
- Welcome changing requirements, even late in development. Agile processes harness change for the customer competitive advantage.
- Deliver working software frequently, from a couple of weeks to a couple of months, with a preference to the shorter time scale.
- Business people and developers must work together daily throughout the project.
- Build projects around motivated individuals. Give them the environment and support they need, and trust them to get the job done.
- The most efficient and effective method of conveying information to and within a development team is face-to-face conversation.
- Working software is the primary measure of progress.
- Agile processes promote sustainable development.
- The sponsors, developer and users should be able to maintain a constant pace indefinitely.
- Continuous attention to technical excellence and good design enhances agility.
- Simplicity — the art of maximizing the amount of work done — is essential.
- The best architectures, requirements and designs emerge from self-organizing teams.
- At regular intervals the team reflects on how to become more effective, then tunes and adjusts its behaviour accordingly.

One of the most important aspects of agile processes is the continual reflection and adaption. Many of these values have also been adopted by agile project management approaches, especially the concepts embedded within one of the methods, Scrum, including set iteration lengths termed sprints, daily stand-up meetings and reviews, as part of the process. At the end of each sprint, a tangible product is delivered and at the end of a series of sprints, a release, the current working version of the product is released to the customer for detailed review.

Lean

Lean processes are typified by reduced inventories and cycle times. There are many concepts that agile and lean have in common particularly in processes to remove waste and rework. The background to the lean movement can be seen in the Japanese manufacturing sector and also in the Six Sigma quality improvement initiatives. As with many agile methods and approaches, some of the ideas have their roots in previous practices.

Grant Rule (2011), during his guest lecture at UCL, 2010, outlined the similar concepts used in the production line techniques of building warships at the height of the Venetian empire, several hundred years prior to the concept of automotive production lines. During this period of expansion, the wooden warships were built to protect the trading interests of the Venetians. Due to the limited space in the Arsenale the inventory and waste was kept to a minimum and the pressure of conflicts meant that they had to reduce cycle times and release ships to protect their routes and territories as frequently as possible. This would be the equivalent of producing working software frequently to customers. There is evidence of continual improvement in the process although whether this allowed teams' time for reflection and self-organisation is highly unlikely.

The importance of frequent releases of software is central in agile project management. This enables feedback from the client to build trust and an effective product that provides the highest value within the given time. As requirements often fit into a profile similar to Pareto analysis with only 20% of the functions used 80% of the time, delivering the highest value and functionality first will often be sufficient as many of the additional requirements may be obsolete, seldom used or need to be changed.

To take full advantage of agile methods, lean practices need to be adopted across the organization (Shalloway, Beaver, & Trott, 2010). It is certainly the case that the major challenges in an organization are often the necessary cultural changes, which include the empowerment of teams, creating an environment of trust to allow the teams to determine their own approach to development.

Agile is not a panacea. There is no set recipe to follow, but there are some patterns that perhaps all projects should follow. One pattern is the idea of time-boxing every aspect of the project cycle. This is not just a defined time but as short as possible time to speed up the feedback and associated learning cycle. This includes the discussions with clients as well as the review meetings.

Two levels of planning

Agile techniques encourage planning at two levels of abstraction. The customer or client, represented by the product owner, usually has an initial idea for the capabilities of the product being developed. This allows for a high-level capability plan to be developed in which the capabilities can be valued. Dean Leffingwell (2007) suggests a value feature approach in which the features are assigned priority and value, giving immediately some indication of the value of the project as well as an approach to track progress in terms of value at a high level.

At an iteration level, the capabilities need to be decomposed further into features and stories and be prioritized. It is this that gives the second level of tracking and can be achieved in terms of stories and size estimated in terms of story points. The team then decides how these tasks are broken down and commit to a daily level of delivery of this work.

Terminology

As with any professional practice, there is often terminology that may be difficult for the outsider to interpret. In law, Latin phases are often used, although increasingly in other areas such as medicine it appears that the language is converging to more readily understandable terms. A glossary of agile terms can be found in the Guide to Agile Practices http://guide.agilealliance.org/. Perhaps it is that sometimes those not involved in agile project management and development are somewhat perplexed by terms such as scrum and planning poker. Is this some kind of game we are playing? Well, part of the answer is yes, as the highly interactive and collaborative planning process has been termed a 'game' by some leading agile proponents (Cockburn, 2002). A game involving the customer at its core and delivering the highest business value within a given period.

How does a generic agile development and project run?

A workshop may be needed to determine the overall process and project management approach. Different agile techniques favour different approaches, and use different terms. All outline the need for prioritization of requirements. In some methods, such as feature-driven development (FDD), these are known as features and are similar to other approaches that incorporate an understanding of how the user makes use of the system, such as by user stories.

User stories tend to have more clarity if developed in conjunction with the client, and time is spent on considering how they will be tested. The user stories are stored in a product backlog and the product owner or representative from the client organization determines the value of each of these and prioritizes. The development team then determines which are done within the next iteration. This is achieved by an initial planning meeting at the start of the iteration often lasting only 2 hours in which the stories are estimated by the development team as to how long these will take. The team also takes into account the priority within the product backlog as to which stories are to be developed for the next iteration, and reviews this when the iteration is completed.

Stand-up meeting

At the start of each day, there is a stand-up meeting. The idea is to keep the meeting short to a maximum of 15 minutes and to encourage comments to be kept succinct. Here, team members outline briefly what they did yesterday, what were the impediments to getting certain tasks done and what they are going to achieve today. Tasks and who is doing them are summarized on the whiteboard, and because these commitments are made to the group there tends to be motivation to achieve what individuals have outlined as their tasks. After the short meeting, any technical problem discussed will often immediately be resolved by others in the group depending on whether it is an architecture, programming or resource issue. There is regular feedback. Typically, there is a retrospective or review at the end of each iteration to outline what went well and what did not go as well, but there are also daily reviews and the stand-up meeting helps highlight problems early. The key is a built-in system of reviews at every level.

With my research groups after their projects, I asked them, if they were allowed to repeat their projects how long would they take to achieve the same results? Most had been working nearly 6 months and agreed that to repeat their work and achieve the same results would take less than half the time. Much of this is due to the learning curve, working out how to solve a problem, deciding the valid metrics, developing effective testing procedures, designing the tests and introducing automated testing procedures wherever possible. This is the importance of speeding up the cycle time to get results quickly and learn from them. To develop code and deliver this to the client after six months without a review with the client is a recipe for disaster. They will invariably say that this is not quite what they had envisaged. The feedbacks are necessary both for team learning and also to ensure that the teams are delivering the right product for the customer and ensuring that what they deliver is of the highest value and quality (Collins, 2013).

It is important for the research groups to consider the value of the project management approaches and consider them as assisting rather than being an overhead. Techniques in testing clarify the goals and if testing is automated should accelerate the process. A more facilitative approach to project management is required both in research and development. Within the Scrum method, this is supported by the role of the ScrumMaster whose function is to ensure that the processes determined by the group are followed.

Estimation

In traditional approaches, managers estimate and try to establish predictable plans, and any deviation from these is seen as a problem with the development team. This approach, often combined with a waterfall sequence of requirements for gathering, design, development and testing, is often a failure. The long list of requirements that may have taken months to gather is often out of date before the design is started. This is why it is so important to build something for the client to see at regular intervals to check that what is being built meets the requirements.

Mike Cohn (2004) humorously pointed out in his book on user stories, using an analogy based on shopping, that seemingly trivial tasks often take several hours to complete. This is a way of introducing several important points about over-optimism in estimation, and not allowing enough time for a technical development or thorough testing.

Of course, there are times when project managers may deliberately tell senior management what they want to hear, and agreeing to an unrealistically short schedule that the development team has no hope of achieving. With agile project management, these aspects are substantially reduced. First, the 'death march' scenario of no time and unrealistic planning from the project manager is avoided because the teams estimate their own work. The priorities determined by the client in fact decide what should be drawn from the product backlog and incorporated into the next iteration. This backlog can then be used as a clear measure to protect the team from unreasonable additional demands. If further tasks are added to the product backlog, it is the priority of each task that determines the urgency. As only a certain number of stories in terms of relative difficulty and duration in story points can be achieved in any iteration, there is an immediate indication of whether it may be feasible or even if it is a priority to add the new stories into the current iteration. The work is then delivered at a sustainable pace in a predictable way, and the project schedule is fine-tuned as more process and project metrics become available.

One aspect of agile is using the 'wisdom of crowds', tapping into the fact that groups can estimate with better predicted outcomes and faster than individual team members. Earlier approaches have used Delphi techniques based on averages and giving higher ranking to central values. In recent years, a popular technique that has been adopted by the agile community is 'planning poker'. This allows the rapid estimation of user stories and will allow, based on these estimates, the team to plan a realistic set of user stories and constituent tasks for the current iteration. It is based on the concept that those doing the work are best placed to do the estimation. Also, as many of the benefits of estimation are quite quickly achieved, putting in more effort has decreasingly diminishing returns on accuracy.

Each of the development teams has a set of cards marked with a series of Fibonacci numbers. These are made up of the previous two numbers added, 1, 2, 3, 5, 8, 13, with the larger card not necessarily fitting in the sequence, say 100. There are other variants of this series that begin with 0 and 0.5, others based on multiples of 10 and others even with a question mark for the decisions that are difficult to initially make. The idea is that it is relatively easy to determine how long a task will take relative to another task, but this becomes more difficult as the size increases. As it becomes more difficult to differentiate between consecutive numbers as the task becomes bigger, the widening gap makes it easier to estimate (e.g. it is easier to determine whether a task will take closer to 8 or 13 units than say 11 and 13 time units).

For each user story, each member of the team estimates in story points the relative effort it is going take to develop and places their selected card with the value face down. As soon as everyone has done this, they turn their cards over to reveal their estimate at the same time. There may be consensus on the number range say 3, 3, 3 and 5 in which case the card representing the generally agreed value is taken. However, if there is variation, a low and a high number say, those who selected these values outline their perspective, and may be in light of this there are assumptions and technical issues that the others have not considered, and there is another round of voting. If there is consensus the team can move on to the next story. Rounds without discussion, as outlined by Amr Elssamadisy (2009), allow reflection and consensus, but may have the consequence of individuals being too influenced by the perceived leaders in the group.

If the team is working with user stories, then they need to estimate the relative size of these stories at the beginning of the iteration as a team. Mike Cohn (2004) points out this can be done in story points, which can be relative to the team's own working practices and experience. However, the true measure of working becomes apparent from the team's own measures of velocity, how many of their story points they are

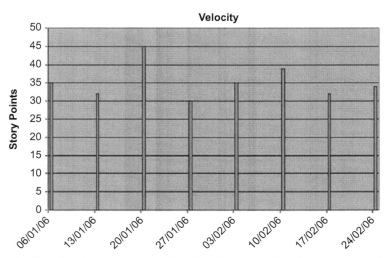

Figure A.1 Chart showing velocity or work achieved in story points during each iteration. This can help the team to determine the amount of story points they can realistically achieve in future iterations, and when the backlog is likely to be complete.

completing within a week or iteration. This measure can then be used as yesterday's weather, to provide improved estimates of the rate of work, and better baseline as the measures for completion of user stories and their constituent tasks to accomplish these (Cohn, 2006) (Fig. A.1).

The team uses its own measures such as velocity and 'burndown' charts that show the remaining stories to be developed. As capabilities and features become increasingly clear and stable from management, and the team acquires a sustainable development rate, the test data will support the figures to show when the project will be complete. The test data including percentage passing acceptance tests, the amount of code changes (churn) and the defect rates are going to be of particular importance in software development and will be used in conjunction to establishing realistic estimates for the overall completion of the project.

Technical debt

During the iteration planning process, it is essential to consider the quality of the product and ensure attributes such as scalability and security. These may not always be outlined by the customer or the product owner, but to avoid technical debt and the code being difficult to maintain and extend these issues need to be addressed. It is important in the planning process that architectural issues are dealt with. One way to help achieve this is to consider not just the user functionality or user stories but also tasks that have to be achieved in order to keep the architecture aspects required. This can be achieved through a dependency mapping as outlined by Brown, Nord, and Ipek (2010).

Defining the architecture of the system

The architecture of a system describes its overall structure, components, interfaces and behaviour. Definitions vary but often include the perspectives or views of the structure (Bass, Clements, & Kazman, 2003). One area that is emphasized by one instance of the generic unified process and rational unified process is the concept of requirements. In this case, it is often user functionality via use cases and subsets of scenarios that show what functionality a user or actor requires of the system.

Some agile methods such as extreme programming favour the use of understanding of the system in an exploratory way, via development of code, improving the design without affecting the behaviour, that is, refactoring. So, for instance, developing a security check in a banking system would necessitate understanding the structure and coding this feature would verify and quickly establish an architecture, if a model is not already available.

Unless the architectural issues are addressed, there will be inconsistencies in performance as the system is scaled. These defects will require ever more refactoring to avoid design problems. Therefore, some consideration of the design is necessary to avoid later problems. This process of consideration of the planning of the architecture is termed architectural runway, allowing for a smooth transition and rework and to avoid technical debt. Planning ahead, the architecture can be allocated on a planned process as advocated in agile architecture provisioning (Brown et al., 2010). Here, architectural elements that are necessary for quality attributes (non-functional) such as security are allocated to the iteration backlog and provisioned within this period to carefully outline both functional and non-functional requirements, which are so important in determining scaling and performance factors. The rationale for architecture decisions should be recorded and can be incorporated within iteration planning as shown in Fig. A.2. An alternative but similar approach would be to include design

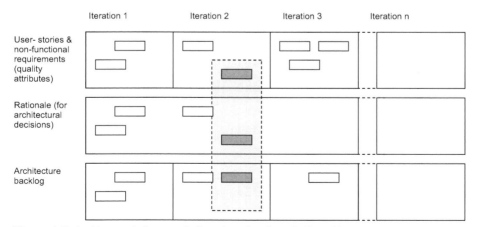

Figure A.2 Architectural elements in iteration planning (Collins, 2011a).
From Brown, N., Nord, R., & Ipek, O. (November/December 2010). Enabling agility through architecture. CrossTalk. https://apps.dtic.mil/sti/pdfs/ADA552111.pdf

spikes at the last responsible moment to ensure the flexibility of the architecture. When there is a choice of architecture, this can then allow for a different approach to the estimation of value. Instead of using a static cost–benefit analysis, which is normally based on static estimates and architecture, an investigation of alternative options, and their relative future changes can be investigated and interpreted via real-options analysis (Bahsoon & Emmerich, 2004). This approach, which was originally developed in the financial markets, is increasingly used to determine the dynamically changing nature of viable options in areas such as provisioning of resources as in cloud provisioning within the IT sector (Collins, 2011b). The nature of engineering is changing and increasingly 'composing systems from open source, commercial, and proprietary components' (Bosch, 2011), and in agile environments where the focus is on early exploration the 'selection and trade-off decisions' should be captured including the rationale that will help to understand why the product is better and why it is being built (CMMI, 2010). Pareto optimality and how this can be applied to balancing requirements as well as trade-off decisions for goals within project and programme management is a current area of research within the Faculty of Engineering Sciences at UCL.

Sharing knowledge and reflection

Research teams are well known for their synergy. Jeff Sutherland (2005) outlines this enthusiasm in agile projects and Dean Leffingwell (2007) outlines these concepts including how teams can create new points of view and resolve problems through dialogue. Within workshops, particularly in scientific research and innovation, it is useful to allow dialogue, not having to defend your idea, and further time to explore possibilities. This is refreshing, as often the stress is on discussions to resolve issues within a specified time period, that is, time-boxing. Leffingwell also points out that in the spirit of scrum, amongst its many attributes including commitment and autonomy, 'leaders provide creative chaos'. This is precisely the concept that Sir Paul Nurse, during the BBC David Dimbleby lecture in 2012, was conveying when he was outlining how to create a collaborative environment for researchers to excel at the future Crick Institute.

For each project, the degree of understanding of goals, emergent design needs to be considered and appropriate patterns need to be in place. Agile development and research both need process and governance frameworks.

Earned value

With traditional project management progress is based on tracking the intermediate tasks, such as production of the requirements document and the design artefacts, which can be achieved without a demonstrable or working product. Allowing measures to be unrelated to working products can give a false sense of progress.

With agile, the focus is on whether the software works, whether it is what the client intended or if it is of value. This is done through continual feedback via releases to demonstrate and allow feedback to improve the product and value.

To track progress towards goals in technology and IT projects where there is emergent design, this needs to be achieved at a higher level of abstraction. In research projects, a clear idea of initial goals or exploration of concepts still needs careful planning; otherwise, it will be unlikely that funding will be granted.

There may be an exploratory phase equivalent to a feasibility study in which one or two workshops are planned to outline strategy or develop architecture to develop the technology. This initial phase will have clear goals and should have a clear rationale for those invited. As this is bounded by time, and consultants and facilitator costs, the budget can be easily ascertained. This is likely if parties agree to an exploratory design phase, then a development phase.

During the initial phase, in order to define the scope or boundaries of the research or development, it will cover the goals that will in many instances then be broken down further to capabilities according to the type of innovation or development project.

Capabilities, a term often used in military projects, determine what is required without determining how this will be achieved. If this is a software development project, these may be subdivided into features and stories. Sometimes, the term epic in agile development is often used for a larger aggregation of features.

User stories represent what the client or product owner requires. These requirements are often written as a short outline on a card and then decomposed further to tasks. The product owner discusses with the team the prioritization and order of development with regard to development issues including software architecture.

The use of earned value management (EVM) is well developed in certain sectors of project management including the construction industry and is being increasingly mandated in defence projects in the United Kingdom. This can be applied to agile and software development. The essence of agile development is this shared collaborative communication between client and development team, ensuring value for the client and a motivated development team. Although management may be used to using earned value measures, this needs to be developed for the agile process so that the development team does not see this as an overhead, and both groups can work collaboratively together. One criticism, however, of EVM is that this does not give any indication of the quality. One benefit of adapting this approach in conjunction with agile project management is that processes that improve quality, such as refactoring, are often incorporated into agile development practices.

If EVM is required by senior management or through the governance process of the project, one approach is to create a reporting structure on two levels; one for the capability tracking and one based on stories as shown in Fig. A.3, which can be generated with minimal overhead as part of the planning process.

In agile development, there is continual prioritization at the iteration level. Earned value can be applied to these user stories, but there is a subtle difference in application. The reconciliation seems difficult, as the user stories are continually under prioritization according to the client as to what they consider the most valuable user story.

Business value of story	Story points	Points earned	Planned developer hours	Actual logged hours	Earned value
3	10	10	100	120	100
2	8	8	60	60	60
2	4	4	60	80	60
1	2	0	20	0	0
	24	22	240	260	220

Figure A.3 Earned value derived from stories completed at the end of an iteration. Story points are an estimate of the relative time that the development team thinks the work will take to complete.

These stories are stored on a backlog for selection during each iteration, and the team estimates how large they are in terms of story points.

The stories according to prioritization are selected for the iteration by the client on their perceived value. If the table is prefaced with the functional value or business value, the order of priority can be much clearer, although the highest priority should be at the top of the backlog, and the development process then pulls the next set of stories and set of tasks.

As can be seen from the figure, the earned value can be derived from the amount of completed work. For the first story, with the highest priority on the list and business value, the story was completed and therefore earned all the points.

Although the story took slightly longer than expected (i.e. 120 hours), it was complete and therefore earned the full earned value of 100. When the story was not complete, it was allocated zero. From these figures, schedule performance index and cost performance index can be ascertained to give an indication of progress (Collins, 2006).

Some have argued that velocity is more important. This is not the same as earned value. Velocity is the rate at which a team works and is a useful internal measure as are burndown charts, which show the remaining work and can act as a motivator for the team.

However, whilst it can be seen that earned value is valid within an iteration, there are different interpretations of how this could work at higher levels of abstraction and the value to project management at a project level. Craig Larman (2004) suggests the use of estimates in terms of budgets. This is most easily achieved in terms of hours of work. He also outlines the concept of re-calculating the planned work for each iteration, and that the baseline is updated as more information arises. For the tasks to earn value, it is prudent to only consider these when fully completed. However, assessing this needs careful consideration as the development of each story is usually considered finished when all tests including integration and acceptance tests are completed.

It can be seen that using a simple spreadsheet as a by-product of the team's estimates, an earned value system can be used as an indicator of progress in terms of business value.

Emerging trends for agile project management

Agile project management provides the opportunity to rapidly deliver software, services and products to satisfy the demands of ever-changing customer needs.

The need to rapidly develop software and digital processes using agile methods has become increasingly important in order to compete in digital markets, indeed for many organizations it is becoming vital for their survival.

One of the trends is the provision of a continuous delivery of software and services. This approach uses concepts from agile methods and lean manufacturing, incorporating frequent releases and automation wherever possible. Regular feedback from customers ensures the quality of deployments and what is delivered actually meets customer needs.

Continuous delivery can be far more effective with the integration of development teams with delivery teams. Considerable further gains can be made if software, service and project delivery can be integrated with other business processes. The challenge for managers is to ensure that all relevant processes are optimized, providing an opportunity to reduce the cost and complexity of processes across the organization.

Agile projects are becoming increasingly data-driven to ensure that they are meeting the goals of the project and aligned to what the customer requires. The progress of projects is now often tracked in real time, based on working software or solutions and feedback from customers. It is the governance and compliance aspects of project management that are increasing in importance in agile approaches, ensuring that process improvement, business goals and regulations are being met.

The following sections examine some of the major trends and challenges: understanding customer needs, use of data and visualizations, achieving collaboration across teams and providing teams with autonomy so they can select whichever lean and agile methods they consider add value to achieve the business strategy.

Delivering enterprise agility

Although there has been considerable success in software delivery using agile methods, the challenge now, as Rigby, Sutherland, and Takeuchi (2016) outline in their Harvard Business Review article, *Embracing Agility*, is that many organizations need to capitalize on agility across the organization. In an increasingly competitive market, it is necessary to gain the advantages of adaptability provided by agile processes. Evidence for the economic benefit has been shown by surveys extending back to 2007 by the Center for Information Systems Research at MIT Sloan, indicating that agile firms can increase profits by 37% (Weill, 2007). In addition, Peter Weill and Stephanie Woerner (2015) have shown there is a significant increase in revenue and profit margins for organizations that embrace digitalization and understand their customers better.

Research published by McKinsey showed that many organizations are using agile project management to deliver goods and services with greater efficiency (Comella-Dorda, Lohiya, & Speksnijder, 2016). This also brings further challenges from 'always on' customers who expect continual availability and reliability. To take advantage of new market opportunities and adapt dynamically requires not just adaptability within delivery pipelines but across the enterprise. As Ross, Weill and Robertson explain in their book,

Enterprise Architecture as Strategy: Creating a Foundation for Business Execution, to keep up with the changing business environment, it is also necessary to have flexibility within the digital processes and enterprise architecture (Ross, Weill, & Robertson, 2006). It is with agile project management that organizations can take advantage of these new digital business opportunities and accelerate time to market.

Understanding business value and customer value

Forming strategic partnerships and looking for ideas outside the organization is one way to accelerate innovation. This can be aided by more agile approaches allowing teams to decide how to structure these conversations and change how teams share information at the start of projects, via workshops or alternatively hackathons.[1] These approaches provide an insight into the processes and value streams of different groups. For a customer, this may give further insight in their value stream, from their request to fulfilment of their order.

Agile processes are increasingly integrated with different perspectives so that the teams better understand the customer's journey. Customer needs are increasingly incorporated at an earlier stage, not just after the engineering solution, as we now often see, integrated with agile processes is the mapping of the customer experience (CX).[2] Processes need careful design and consideration, whether they are providing value for the business or customer, or both. For example, it is of limited value if, an app helping a customer locate a store selling a product they want (via geo-location on their smart phone), if this system is not integrated into the supply chain and the store takes weeks to order the product. The whole supply chain needs to be considered. Mark Schwartz highlights in his book, *The Art of Business Value*, that managers need to ensure that the meaning of value is considered from the perspective of the business, partners and what the customer values (Schwartz, 2016).

Integration of processes

Many organizations are using agile project management to meet customer needs and improve the integration of their processes. Siemens is an example of a global organization who are not just focused on agile software development but use agile project management to focus on business value, reducing lead times and release times, and getting feedback from their customers. These processes allow flexibility, so that if something does not meet needs of the business, or the customer, it can be adapted.

---

[1] Hackathons are where teams including domain experts, coders (software programmers) and project managers intensively work together and rapidly develop a prototype, often over only 1 or 2 days.

[2] Humble et al. (2015) made a distinction between customers, who pay for software and are often involved in the development, and users who may use the software in their day-to-day work and may also contribute to development through social networks. Different perspectives need to be considered whether this is via (CX) or user experience (UX) or other design approaches.

Siemens also integrate their application lifecycle management to their product lifecycle management (PLM), for example, with the manufacture of embedded systems. The sharing of information across these processes makes the production of these embedded systems seamless and lowers production costs. It also allows requirements to be traced more easily. Integrating PLM software and agile software development practices enhances information systems and is seen as a necessity to help deliver the corporate strategy. This not only allows the consolidation of different systems but also enables global engineering groups to collaborate and launch products as one team, reducing time to market.

Siemens also achieve rapid feedback within their prototyping, for example, within their autonomous robot manufacturing systems, part of Siemens agile manufacturing systems (SiAMS). At Siemens Corporate Technology Centre Princeton campus they are building mechanical spider-like robots to work in hard to access areas, in manufacturing or surface preparation. They are optimizing them to work together using algorithms that mimic the knowledge sharing and collaboration of human agile teams (Siemens, 2016). In fact, these robots are being designed for other attributes you may expect from small agile empowered teams, self-improvement and the ability to share tasks autonomously.

To fully capitalize on agile processes, enterprises do not need just high-performing individual agile projects but take advantage of agile methods across the organization so that the entire supply chain is examined and optimized.

Continuous delivery

Problems typically arise when teams work in isolation. Integrating the teams responsible for development and operations (DevOps) allows for a continuous delivery pipeline. Releases of small functionality can be made available in smaller increments to get feedback from customers and reduce risk. If there is a problem with a release, the fact that only a small amount of software is affected means that this can be quickly rectified. This is further enhanced by what is known as blue-green deployment, having two identical production streams 'blue' and 'green'. Blue will be live and have the production traffic, whilst green will be in the final stage of testing. When green is fully tested and ready, it is switched so that it is live, eliminating downtime. Additionally, if a problem did occur then there could be an immediate roll back to the blue version further reducing risk.

Cultural shift

DevOps can be considered as a cultural shift, encapsulating the philosophy of cross-functional teams, with both development and operations collaborating effectively to improve the flow and delivery of products. To be fully effective this needs to involve other related business units as well. This can be easier to achieve in organizations that fully embrace agile methods than have a bimodal solution, where traditional and agile modes coexist. However, having predictable outcomes and moving to a more exploratory state, if managed effectively, can facilitate the transition from traditional processes to digital processes whilst reducing risk of a 'big-bang', an all at once, transition.

A natural progression in DevOps is to ensure that all relevant aspects of the business and development are involved in the process, including the quality assurance and the project management office (PMO). The PMO supports the delivery of projects in an organization, monitors review of the delivery of projects and provides decision support.

The conditions for project success, as indicated by the Association of Project Management survey (APM, 2015) includes, 'effective governance, the project needs to have clear reporting lines and effective communication between all parties'. Governance frameworks ensure goals and standards are being followed and allow organizations to make investment decisions. When integrating agile practices, the role of the PMO needs to be considered to ensure a suitable fit to the strategy. However, it can be a challenge for the PMO to have the necessary and up-to-date skills to provide advice for continuous delivery teams. Lloyds Bank has recognized that having a technical capability within the PMO is vital, and Sanjeev Sharma, CTO, DevOps Technical Sales and Adoption, IBM, explained at the DevOps Summit London that the banks embed an expert from the DevOps team within the PMO to facilitate this process (Sharma, 2016).

For agile focused organizations, where the priority is to favour digital transformation, there is often a need for a specialized PMO. This structure can be thought of as analogous to the structure of programmes, with PMOs having different specialities supporting relevant groups of projects. This can enhance governance and compliance in global organizations as with IBM, which adopted this kind of structure with specialized PMOs for each of its product lines. However, as Guy Barlow points out in his Oracle blog, the real-time sharing of information becomes even more critical if the role of the PMO is decentralized (Barlow, 2014).

DevOps when integrated with other business units responsible for understanding the customer needs, quality assurance and the PMO can considerably accelerate the successful deployment of software and services. Collaboration across all teams related to delivery and deployment, ensures a reliable and resilient delivery pipeline of software being continually adapted to customer needs.

Data-driven

These changes throw up new challenges so that software and enterprises are increasingly becoming data-driven.

There is a shift in agile project management towards projects being driven by real-time data. It is through this continual experimentation using empirical methods and agile and lean approaches, based on the scientific method, that projects can be managed in fast-moving digital environments. Although the planning may be reduced within exploratory and fast-moving digital projects, feedback as to whether the project is meeting the goals of customers and the business is often enhanced. Organizations are realizing that the project management aspects of agile processes are ideally suited for the necessary governance, compliance and tracking that the strategy and customer needs have been met.

NASA's regular feedback from the development of their customer-centric software from their nightly builds also helps them keep track. The priority for NASA's teams is to place emphasis on tracking the code and their goals through daily feedback from their customers rather than upfront planning (Trimble & Webster, 2013). Their measure of progress is increasingly working code. This analytics is becoming more important in continuous delivery where problems can be identified in real time.

It is the ability to rapidly get feedback from customers that is changing project planning to an emphasis on data-driven processes establishing whether we are meeting customer goals. This may be in the form of A/B testing, where a portion of the web traffic is being driven to one group of customers to see if there is an improvement in sales, and if so then the whole traffic is switched to the new version. Humble, Molesky and O'Reilly outline in their book, *Lean Enterprise*, that this method using data from this method is used by technology organizations, such as Amazon and Microsoft, to see if features in their projects will even add value before it is built (Humble, Molesky, & O'Reilly, 2015).

As organizations move to continuous delivery, this throws up further challenges. Managers need to provide an environment of psychological safety and trust, so that their teams are willing to experiment and continually improve, via processes, such as improvement katas,[3] as Humble et al. (2015) outline. Skelton and O'Dell (2016) point out that it is the cultural issues, supporting teams via manageable workloads and ensuring software architecture aligned to team structure that is important. Joe McKendrick, quoting the CTO of Amazon, Werner Vogel, outlines that Amazon makes it 'easy for developers to "push-button" deploy their application' and release updates for customers, in effect, achieving deployments at a rate equivalent to one every second (McKendrick, 2015). If organizations are to replicate the success of Google and Amazon, and be able to adapt their processes within seconds and meet the ever-changing needs of customers, agile and lean methods incorporating rapid feedback and real-time data are required.

Shared vision

To avoid silos of ideas and to ensure projects take into account all necessary viewpoints, workshops are typically held. Facilitated workshops at the start of projects are integral to many agile approaches, such as the dynamic systems development method (DSDM, 2008). Rally Software, now part of CA Technologies, also integrates its workshops with release planning. Their approach which is termed, 'agile big room

---

[3] Improvement katas are one approach to process improvement. One feature is the setting goals with clear targets, for each iteration.

planning' is designed to ensure business units across the organization are brought together and collaborate from the beginning so that everyone is on the same page. As part of this process, teams co-locate to participate in the planning process. This ensures communication with the business, that the project is aligned with the business goals and that the entire organization is engaged and has an input to delivery of the product or service being developed. Seagate, a data storage organization, adopted this method to identify bottlenecks in its processes and improve predictability, achieving a regular cadence for release planning.

Research by Dingsøyr, Fægri, Dybå, Haugset, and Lindsjørn (2016) has shown that it is important in agile software development that teams have a shared mental model. Having this shared vision is equally important as we scale projects and integrate other units across the organization. This can be enhanced by visualization tools, which can be adopted for DevOps, to bring the tasks for both streams of operation and delivery into better coordination.

Kanban is a lean concept that is becoming increasingly adopted to improve the visibility of progress across teams, reduce waste and ensure that work in progress is kept to a minimum. Tasks are written on coloured sticky notes, which are placed on a whiteboard, called a Kanban board, to track the progress across stages of development. Jaguar Land Rover uses Kanban software to reduce lead time and improve flow. The software also allows continuous feedback, allowing different engineering teams to add comments and improve the decisions. The visibility of progress is further enhanced, by showing the status of each activity, including assessment. When activities are assessed either, 'assessed OK', or 'not OK' is posted. The latter is linked within the software to different categories, why the assessment has not met the required standard and real-time discussions as to further work that needs to be completed. Key decisions are also recorded including architecture hardpoints, such as wheelbase or position of the seat with respect to other features. The discussions and diagrams illustrating any problem arising can also be shown at the same time so that a seamless resolution of issues can be achieved.

Systematic, an international software company has successfully deployed and scaled agile methods for its software development. Agile processes are applied across the entire organization, including their management systems, improving visibility and reducing the document workload. Their teams have been early adopters of visualization and automation and within their continuous delivery pipelines for clients. This has resulted in Systematic being one of the first organizations to be accredited the Capability Maturity Model Integration (CMMI) level 5 showing the high level of repeatability within its processes. This efficiency allows a focus on business-critical areas and customer satisfaction.

Ensuring everyone contributes

Agile teams are increasingly using open source messaging apps, such as Slack, for sharing their communications. This has partly given rise to enterprise agile development, increasingly providing a more interactive environment and often allowing developers to discuss and share ideas in a social way. Communication tools incorporating a social dimension can ensure that a wider range of ideas is explored. Executives are starting to recognize that harnessing the creativity of external stakeholders as well as that of their internal teams can improve knowledge sharing and be a catalyst for innovation. Harrysson and co-researchers showed that social technologies can invigorate collaboration and help develop strategic insights (Harrysson, Schoder, & Tavakoli, 2016).

One of the challenges of scaling agile project management is often the boundary between teams, particularly with external stakeholders (Strode, 2012). Here, the communication strategy needs to be carefully managed. It is often a project manager needed in this role to coordinate these teams, although there are various approaches to scaling and designating roles based on Scrum, including the scaled agile framework outlined by Richard Knaster and Dean Leffingwell (Knaster & Leffingwell, 2016). It is however recognized in methods that a representative with detailed knowledge of the business (or product owner) is central to the prioritization of goals, and should, as Mike Cohn emphasizes, ideally come from the business domain to fully appreciate and communicate the business priorities (Cohn, 2009).

Aligning teams to the architecture of the system can enhance communications between development teams. Aligning teams to traditional functions can cause a considerable amount of additional work with hand-offs, and waiting for teams to deliver work to other teams in the process. It is important to design teams according to the communication structures using Conway's Law: which outlines that organizations are 'constrained to produce designs which are a copies of communication structures of these organisations'. Sam Newman outlines the need for flexible architectures and that these provide opportunities for modular architectures to reflect the structures of teams to improve the efficiency of communications (Newman, 2014).

Nvidia uses agile methods throughout its development processes. For example, in the development of its graphics processors, it arranges real-time interaction with customers to enhance their designs. It not only uses real-time interaction and feedback from customers during the development of the products but also uses its products in the design of the building to facilitate interaction and collaboration. Deborah Shoquist, Nvidia's vice-president for operations, was quoted in the International *New York Times* outlining the use of their 'computational power and technology to model their new building in Silicon Valley' (Markoff, 2016). Nvidia is using its highly interactive rendering software in conjunction with its graphics processors to design its offices. This allows teams of architects, designers and engineers to work in collaboration with state-of-the-art virtual reality. These design teams can make changes to problems that only become apparent when they interact with the 3D visual model, avoiding costly alterations later in the project.

Dingsøyr et al. (2016) have shown the validity of agile principles and clear goals. They have also established that it is important that decisions are discussed. Even discussions where there is disagreement for the processes or technical approach required have been shown to improve teamwork. In addition, research published in the IEEE journal *Software* by Van Heesch, Eloranta, Avgeriou, Koskimies, and Harrison (2014) indicates that it is also important to record key decisions. This, for example, has been shown to be effective in the design of their decision-centric architecture reviews, where not only the decision chosen but also alternatives are documented, the pros and cons for the chosen solution, as well as envisaged future issues which may occur. Those disagreeing with a decision need to have their viewpoints heard if the teams are going to value each other's opinions and accelerate their learning and performance.

Agile does not mean an end to planning

Unfortunately, there is still the misconception amongst some senior executives that agile does not need a process and that agile is tantamount to anarchy. Rigby et al. highlighted this impression within their Harvard Business Review article (Rigby et al., 2016). However, to truly take advantage of agile project management and processes, the reverse is true: clear lines of communication, roles and responsibilities are necessary. There is also a need for a consensus-driven approach that is equitable, inclusive and transparent.

A documented architecture certainly provides clarity but the strategy and business processes still need to be considered in terms of the operating environment and changes in regulation. The throwaway aside by the CEO of the middleware software company Software AG 'if you are agile you don't need a strategy' reported by David Cassidy PCPro, August 2016, suggests that if you are an agile organization and adopt their enterprise software suite a strategy is not required (Cassidy, 2016). It is certainly unlikely that modular software aimed at workflow can fully understand your customer and business needs, security and data storage issues. If anything agile should enable the technical teams to focus in-depth on architecture and development, and therefore free up more time for management to focus on the strategy of their organization. Agile development will also allow teams to rapidly hone-in on the desired needs of customers and test whether the strategy is correct and evolve this to their needs. With the adoption of continuous delivery, you cannot have a monolithic architecture but need a modular architecture, such as microservices: small domain-focused services. A modular architecture certainly helps but still needs careful consideration, if this is going to be resilient and scalable.

Delivering value not bureaucracy

Agile teams are typically provided autonomy so that the team decides which agile practices and tools they should adopt and make them more productive as a team. Spotify reduces bureaucracy by encouraging its teams to prioritize agile principles over specific practices as well as encourage the teams, called 'squads', to select processes they consider to add value. So, for example, a squad may omit burndown charts if they consider this does not add to their effectiveness. Spotify also ensures that members treat each other as equals: every aspect is designed to reflect this. Even the label for the role of ScrumMaster, which can be construed as someone in-charge, rather than mastery of a process, is termed an 'agile coach', as explained by Henrik Kniberg, to emphasize that this role should be a servant leader (Kniberg, 2014). Typically agile teams are empowered and determine what is important to deliver the functionality. Brian Bergstein illustrates, with an example from MIT, that managers also need to recognize that it is the high level of collaboration, and sometimes a willingness to break the rules, which trigger innovations (Bergstein, 2016). Throughout the agile processes, managers and teams need to ensure that bureaucracy is kept to a minimum.

An observation from history provides a reminder of the importance of autonomy. In writing about the industrial production within the *Arsenale*, the shipyard in Venice, during the 17th century, Joanne Ferraro outlines that the government of the time 'was successful in co-opting workers' support by permitting them considerable self-governance' (Ferraro, 2012). This is one aspect that can be applied today in project management in adopting agile processes, providing trust for teams to make their own decisions, but balancing this with direction and clarity of goals. Providing the appropriate balance of autonomy and governance is key if we are to support disparate groups to contribute and deliver as one team.

Summary

In agile development, it is often about trends emerging rather than making guesses about the future. Walker Royce wrote about the indicators for converging on the solution and the indicators for value and progress (Royce, 2011). The way to gain credibility is through working with the client on a joint understanding of a problem, how to measure progress and when to converge on the solution. This creates the real value, not only to the business, but to the self-worth of the team.

Agile project management is about achieving value collaboratively for the project and client team, and the organizations concerned. This is not just about the bottom line but achieving something at work, feeling valued and sharing knowledge with colleagues to achieve that next breakthrough. This is the true value of agile for individuals, the team and the organization.

At UCL at the front of projects are different types of workshops, examples of agile patterns that bring the right mix of researchers and project managers and support staff to solve problems. These vary from the Town Hall meetings where the challenge and opportunity is outlined to more detailed workshops allowing discussions and dialogue. The agile project management and development approaches favour the time-boxed approach to discussions, and estimation and often adding more time does not necessarily give a better outcome. It is these workshops which are often the driver for new innovations and approaches to development. The use of a trained facilitator is often vital with large programmes. It is important that staff with different technical approaches can outline their view. It may be that the idea is rejected in favour of an alternative idea discussed at the meeting. The key is fair process and that the team are deciding the direction. This is the essence of agile allowing the client to work with the development team collaboratively to decide capabilities they wish to develop and prioritizing the functionality and user stories, or in the case of research the investigations (Collins, 2013).

It is self-evident that if one member of a team suggests a technical solution, another may disagree and point out an alternative, and why in certain circumstances it is a better resolution.

Developing real-time modelling agile approaches with colleagues had a significant impact on communication and indirectly in one project resolving political issues by the clear focus on the technical problem rather than individuals. During a consulting assignment my colleagues in a small consulting group were asked to outline our object and business modelling approach. We had been asked for a solution and found on arrival at the client site already strained working relationships with a development project that had been on-going for over a year. It was agreed to use our agile modelling approach to clarify the goals of the project and be clear on the direction. It was clear things were not going well, the project manager complained that he didn't know what kind of project he was working on, as it wasn't properly defined by the programme director. He didn't know whether it was a business transformation project or a business improvement project. The client really wanted the project management problems to be resolved and not upset the director of the consulting firm under contract. It was a mess and an expensive mess with technical teams having worked for a considerable time. Without a real-time modelling solution and experience in forming a unified team this impasse would not have been resolved.

Getting buy-in wasn't just a problem of clarifying the goals but getting support by under- standing each of the stakeholder's goals, communicating the direction over a short iteration with a clear product and defined time and engaging all stakeholders already working on the project.

The leader must be seen by others not to be gaining personally. In the case of the programme that had to be put back on track, it was imperative to listen to other parties and support each of their objectives so that the programme could move forward. Self-interest other than wishing for a successful outcome was not in the cards. Likewise, the leaders in agile project management must lead by trusting their team and allow their team to deliver the project in a self-determined way. Self-organizing teams and allowing them to report on progress are areas that the leader must embrace in agile project management. Much of what has been written on agile has been on what the teams do and how they track their progress. The burndown chart, keeping progress visible and keeping tasks visible on the wall are for the team and the team leader. Keeping key tasks and communications visible, this is the 'white box' of progress reporting. It is the leader in agile project management who needs to understand this iterative process and be a resource provider, to remove all impediments to the team, who must trust his or her team to deliver in the technical approach they consider best. It is this beyond anything that defines the change to a leadership culture in agile.

The challenge in agile project management is not prescriptive plans and practices to follow, but to populate the project-planning process with appropriate patterns that are effective and add real value. For the time being, the challenge must be to balance the planning so that you can achieve the flexibility to deliver increasingly complex projects and rapidly add new developments to enterprises and research establishments.

Bibliography

APM.(2015). Available online https://www.apm.org.uk/sites/default/files/Conditions%20for%20 Project%20Success_web_FINAL_0.pdf.

Bahsoon, R., & Emmerich, W. (2004). Evaluating architectural stability with real options theory. In *Proceedings of the 20th IEEE international conference on software maintenance (ICSM'04)*.

Barlow, G. (September 29, 2014). Has the role of the PMO peaked? [Blog]. In *The future of decentralised management, enterprise portfolio management (EPPM)*. Oracle. Available online https://blogs.oracle.com/eppm/entry/has_the_pmo_peaked_a.

Bass, L., Clements, P., & Kazman, R. (2003). Software architecture in practice. In *SEI series in software engineering*. Addison-Wesley.

Bergstein, B. (2016). *EmTech: A legendary MIT building's lesson's on innovation*. Available online https://www.technologyreview.com/s/531011/emtech-a-legendary-mit-buildings-lessons-on-innovation/.

Bosch, J. (May 16−20, 2011). *In Keynote abstract Saturn Conference San Francisco*. SEI.

Brown, N., Nord, R., & Ipek, O. (November/December 2010). *Enabling agility through architecture*. CrossTalk.

Cassidy, D. (August 2016). If you are agile you don't need a strategy. *PCPro*, (262), 122.

CMMI® for Development. (November 2010). *Version 1.3 CMMI-DEV, V1.3*. SEI.

Cockburn, A. (2002). *Agile software development*. Addison-Wesley.

Cohn, M. (2004). *User stories applied*. Addison-Wesley.

Cohn, M. (2006). *Agile estimating and planning*. Addison-Wesley.

Cohn, M. (2009). *Succeeding with agile: Software development using scrum*. Addison-Wesley.

Collins, G. (June 12−15, 2006). Experience in developing metrics for agile projects compatible with CMMI best practice. In *SEI SEPG Conference Amsterdam*.

Collins, G. (2011a). *Post-graduate course GZ07, academic year 2010−11*. UCL: Department of Computer Science, Faculty of Engineering Sciences.

Collins, G. (2011b). Developing agile software architecture using real-option analysis and value engineering. In *SEI SEPG Conference Dublin June 2011*.

Collins, G. (2013). *Experience as lead consultant on commercial consulting project 1999−2000 included in GZ07 post-graduate teaching for GZ07 course*. UCL: Department of Computer Science, Faculty of Engineering Sciences.

Comella-Dorda, S., Lohiya, S., & Speksnijder, G. (May 2016). *An operating model for company-wide agile development*. McKinsey and Company. Available online http://www.mckinsey.com/business-functions/business-technology/our-insights/an-operating-model-for-company-wide-agile-development.

Dingsøyr, T., Fægri, T. E., Dybå, T., Haugset, B., & Lindsjørn, Y. (2016). Team performance in software development: research results versus agile principles. *IEEE Software, 33*(4), 106−110.

DSDM. (2008). *Facilitation approach handbook*. Available online: https://www.dsdm.org/content/facilitated-workshops.

Elssamadisy, A. (2009). *Agile adoption patterns: A roadmap to organisational success*. Addison-Wesley.

Ferraro, J. M. (2012). *Venice, history of the floating city* (pp. 184−185). Cambridge University Press.

Harrysson, M., Schoder, D., & Tavakoli, A. (2016). The evolution of social technologies. *The McKinsey quarterly, 53*(3), 8−12.

Highsmith, J. (2002). *Agile software development ecosystems.* Addison-Wesley.

Highsmith, J. (2010). *Agile project management* (2nd ed.). Addison-Wesley.

Humble, J., Molesky, J., & O'Reilly, B. (2015). *Lean enterprise: How high performance organisations innovate at scale.* O'Reilly Media.

Knaster, R., & Leffingwell, D. (2016). *SAFe® 4.0 distilled: Applying the scaled agile framework® for lean software and systems engineering.* Addison-Wesley.

Kniberg, H. (March 24, 2014). *Spotify engineering culture [video], spotify.* Available online: http://abs.spotify.com/2014/03/27/spotify-engineering-culture-part-1/.

Larman, C. (2004). *Agile & iterative development: A manager's guide.* Addison-Wesley.

Leffingwell, D. (2007). *Scaling software agility: Best practices for large enterprises.* Addison-Wesley (Chapter 21 with Ken Schwaber).

Markoff, J. (July 19, 2016). *Using virtual reality to create a new corporate headquarters.* International New York Times, Media Technology Business, 15.

McKendrick, J. (March 24, 2015). *How amazon handles a new software deployment every second.* ZDNet. Available online: http://www.zdnet.com/article/how-amazon-handles-a-new-software-deployment-every-second/.

Newman, S. (June 30, 2014). *Demystifying Conway's Law* [blog]. Thought Works. Available online https://www.thoughtworks.com/insights/blog/demystifying-conways-law.

Rigby, D., Sutherland, J., & Takeuchi, H. (May 2016). Embracing agile: how to master the process that's transforming management. *Harvard Business Review*, 40−50.

Ross, J. W., Weill, P., & Robertson, D. C. (2006). *Enterprise architecture as strategy: Creating a foundation for business execution.* Harvard Business Review Press.

Royce, W. (2011). Measuring agility and architectural integrity. *International Journal of Software Informatics, 5*(3), 415−433.

Rule, P. G. (February 3, 2011). *What do we mean by "Lean"? Guest lecture for professional practice series.* UCL: Department of Computer Science, Faculty of Engineering Sciences.

Schwartz, M. (2016). *The art of business value.* Portland, Oregon: IT Revolution.

Shalloway, A., Beaver, G., & Trott, J. R. (2010). *Lean-agile software development: achieving enterprise agility.* Addison-Wesley.

Sharma, S., & CTO, DevOps Technical Sales and Adoption IBM. (June 30, 2016). *Innovate, learn and disrupt: A practical discussion on becoming agile.* Enterprise DevOps Summit London.

Siemens. (April 20, 2016). *Autonomous systems: Spider workers.* Available online http://www.siemens.com/innovation/en/home/pictures-of-the-future/digitalization-and-software/autonomous-systems-siemens-research-usa.html.

Skelton, M., & O'Dell, C. (2016). *Continuous delivery with windows and .NET.* O'Reilly Media.

Strode, D., Huff, S., Hope, B., & Link, S. (2012). Co-ordination of co-located agile software development projects. *Journal of Systems and Software, 85*, 1222−1238.

Sutherland, J. (2005). *Future of scrum: support for parallel pipelining of sprints in complex projects.* Denver, CO: Agile (Conference).

Trimble, J., & Webster, C. (Jan. 2013). From traditional, to lean, to agile development: finding the optimal software engineering cycle. In *46th Hawaii international conference on system sciences* (pp. 4826−4833).

Van Heesch, U., Eloranta, V.-P., Avgeriou, P., Koskimies, K., & Harrison, N. (2014). Decision centric architecture reviews. *IEEE Software, 31*(1), 69–76.

Weill, P. (June 14, 2007). *IT Portfolio management and it savvy: Rethinking it investments as a portfolio*. Centre for Systems Research (CISR) MIT Sloan School of Management.

Weill, P., & Woerner, S. (2015). Thriving in an increasingly digital Ecosystem. *MIT Sloan Management Review, 56*(4).

Appendix 2: Artificial Intelligence (AI) and Big Data

Graham Collins
Faculty of Engineering Sciences, University College London, London, United Kingdom

Introduction

Artificial intelligence (AI) and Big Data are inexorably linked. The phenomenal explosion in the availability of data, together with advances in AI algorithms to see patterns in this data, has created an opportunity for improvements to assist us in our work and everyday lives. The increase in computer speeds and the use of cloud computing have enabled vast amounts of data to be processed and analysed and have created a reality of an AI-driven and data-driven future. Although AI relies heavily on Big Data, the two topics are separated as Big Data is also utilized by many other disciplines in industry, commerce, financial services and medicine. However, an overlap between the coverage of the topics is inevitable.

AI

History of AI

During the Second World War, military radar operators in the United Kingdom would try to identify enemy aircraft as quickly as possible from their screens. Operators were skilled, but their rate of success in identifying the plane accurately varied. For each operator, the correct identification rate of enemy aircraft, the true positives and the false alarms or false positives was recorded. This information was plotted as a receiver operating characteristic curve to determine the accuracy of each operator. In a time of war, the most accurate prediction was necessary to secure defences. The operators performing more accurately remained, whereas operators who were less accurate at predictions had to leave.

This is, in effect, the same process in AI today. Data scientists select the best algorithms, the mathematical formula that provides the best prediction. The magnitude of data in the meantime has escalated. It is now beyond the capacity of even thousands of project professionals to assimilate. The volume of data is often too large for traditional computer systems to process: this is called Big Data. The volume of data is increasing exponentially. One is not looking at one dot on a screen, one is often searching millions of pixels and across millions of images, whether images on the internet or detailed medical scans. The ability to train these AI systems and their underlying algorithms has also vastly improved. In the last few years, a range of technological advances in cloud and network architecture, computer processing and sensors have made a future supported by AI systems not only a possibility but a reality. As one progresses from proof of concept (POC) to AI optimized systems, advanced medical diagnosis, smart manufacturing and smart cities are all creating projects that can improve lives.

One could argue that the accuracy of an AI system recommending a product when searching the internet does not matter too much. However, most would agree that obtaining an accurate prediction from a radar operator, or an AI system identifying a cancerous tumour from a medical scan certainly does matter, as both affect the chances of survival.

As an example, within images, one could train an AI system to recognize specific objects. One may wish to install such a system to check that workers in an environment such as a hospital or workplace are complying with safety measures during a pandemic. The system is then trained with images of people wearing face masks and other images showing people not wearing face masks: these images now comprise the training set. As more images are collected and labelled, this AI system becomes better at recognizing those wearing or not wearing face masks. The probability of a correct identification increases. Tryolabs (Tryolabs, 2020) uses deep learning to help identify from images such objects as face masks (Fig. B.1).

Some image training systems for AI may not seem particularly complex but invariably contribute to extensive databases. One may be asked as an identity check whether one is a human or a robot by completing a reCAPTCHA quiz to protect websites from fraudulent transactions or synthetic accounts generated by bots. As part of this, process one may have to identify from a set of images which contain an object such as a traffic light. If one is asked which image contains a traffic light, this will be invariably training a software system, which may be applied to autonomous vehicles (Fig. B.2).

Figure B.1 Images classified in real time with a percentage probability that people are wearing face masks.
Image courtesy of Tryolabs. Image from video content created by Tryolabs. *Face mask detection in street camera video camera streams*; 2020. Available at: https://tryolabs.com/blog/2020/07/09/face-mask-detection-in-street-camera-video-streams-using-ai-behind-the-curtain/. [Accessed 25 November 2020].

Figure B.2 Image of a reCAPTCHA quiz asking the user to verify that they are not a robot.

A short video showing how to train an AI system to identify an object is available at lobe.ai, https://lobe.ai (Lobe, 2020). This outlines the sequence needed to train the AI system to recognize someone drinking a glass of water. The system needs to be provided with images of someone drinking water at different angles. The system also needs images to show someone who is not drinking a glass of water. The system is then tested to see if it can correctly identify when someone is drinking a glass of water. The video shows that at first, the AI system may not identify actions correctly. The system can then be trained to improve accuracy. Each time the AI system incorrectly identifies what is happening it can be manually corrected. As the system is trained with more examples, the accuracy of prediction can increase substantially.

Using training data is one approach to machine learning. An AI application for vehicle identification may include labelled images of the different types of vehicles expected: trucks, passenger vehicles and so on. This is termed supervised learning. The system is trained to automatically recognize what sort of vehicle it is. It is not always necessary for a user to develop a training set as there may be pre-trained models already be available. This level of capability is available via Microsoft Cognitive

Services which have pre-trained algorithms already installed. When an image of a vehicle is uploaded, the category of the vehicle can be identified, and provided the appropriate data set is available, the location may be identified as well. This service can provide the identification as text, which can be voice-activated, improving accessibility. This system can also be trained to update continually, enabling the system to recognize specific vehicles entering a depot, or crossing an international border.

AI research and applications are expanding rapidly with areas such as machine learning, deep learning, vision systems, natural language processing. AI is generally considered a collective term for computing intelligence that mimics tasks that humans make. IBM in particular often uses the same abbreviation, AI, for augmented intelligence, which reminds us that the purpose is not to replace humans but to enhance their capability. McKinsey (McKinsey & Company, 2021) defines AI as the 'ability of a machine to perform cognitive functions we associate with human minds such as perceiving, reasoning, learning, interacting with the environment, problem-solving and even exercising creativity'. Machine learning is essentially a subset of AI where machines learn from data. Typically, one uses historical data, then feeds this through an algorithm or a combination of algorithms to try to predict a future scenario. Deep learning can be considered an extension of machine learning where vast amounts of data are passed through a network to identify patterns. Spotify and Google use this when one searches, or when Google Assistant helps users via speech recognition. AI recognizes patterns in the waveform of songs, spoken questions, or sorting through millions of internet images. The data is fed through a network and at each layer of the network complex features are extracted to provide a prediction.

Finding new data sets

AI systems can be trained on specific data sets, and with more data become better at recognizing patterns and making more accurate predictions. As data volumes increase, AI tools combined with technologies such as cloud computing are needed to cope with the scale of the data to find the relevant patterns. AI-driven analysis can even identify new data sets in many fields of work. Audit practices can use such methods to extract terms from contracts to reduce their auditor's workload, saving both time and money. This can also provide information for internal requests, to answer similar queries, or find previously resolved cases in insurance. Deloitte uses a smart tool that pools the knowledge within their organization. Auditors not only use their expertise, but they are also able to use the experience of their fellow professionals. Deloitte trains their systems and label insurance cases as fraudulent or non-fraudulent, and so can reduce the exposure for their clients. Their systems can also categorize the level of risk new customers may pose to an insurance company. This kind of AI analysis is applied within their online technical library which can be searched via speech recognition. Their chatbot system tries to emulate a human conversation so that it can recognize certain patterns and predict likely requests. This helps auditors search through their extensive resources and find legal, tax and regulatory documents. Each search adds value to the system. As the AI system learns to become more efficient at finding each document, it also learns the route the auditor has taken, potentially shortening the search for others in the future.

However, one of the problems in AI is that developing algorithms to deliver a range of competencies has been difficult. Recent attempts have used reinforcement learning with deep neural networks, layers of a network building abstract representations of the data sets and learning objects directly from input signals. The research teams at Google's AI unit DeepMind have been investigating whether these networks can generalize beyond their training experience. They have created an artificial agent able to take part in a diverse range of tasks (Hill et al., 2019). Reinforcement learning has been used in games such as when DeepMind's computer competed in a game of AlphaGo. Here, the reinforcement does not need to be trained or have prior knowledge of the environment. It learns by a process of trial and error. Typically, this is based on a decision process such as the Markov decision process so that software agents take actions that optimize the required outcomes. This has the potential to improve project and strategic level decisions in the industry.

Reducing bias and increasing accountability

Training data from biased data sets inevitably leads to biased results. For AI systems, there needs to be trust and accountability as to why certain predictions have been made. Although algorithms have been discredited and blamed for many events such as the examinations grading fiasco in the United Kingdom, algorithms are not good or bad, they are simply finding a pattern in the information. Algorithms are trying to ascertain what something may be, a likely grade, an object, or from smart city data an area of high pollution. Where things go wrong is when there are poorly designed algorithms or poor-quality data sets. It seems intrinsically unfair that an algorithm for exam grading should be heavily weighted on the school attended, or that an algorithm for recruitment for a tech company should be biased against female applicants because it is trained to identify characteristics of its current predominantly male workforce.

Understanding AI decisions can often be problematic. For example, with financial decisions, where regulators want to ascertain how a decision is made. Often decisions are considered as a 'black box', where it is difficult to determine how the decision has been made. However, there have been considerable improvements in understanding these processes, with some systems showing how different factors affect the weighting or bias to certain outcomes.

AI identification using neural networks can be applied to CVs in recruitment for a company or a project. For this to be effective there needs to be a recognition of the balance of power towards the organization, and that there must be adequate checks in place. Organizations using AI systems need to understand any inherent biases in the data. Were the algorithms based on historical data that favoured male candidates? Likewise, for facial recognition systems, there needs to be a check if the system is biased and can only identify certain ethnic groups accurately. In judicial systems that use AI, these are trained on data sets of people charged with crimes and the lengths of sentences they were given. Research has shown these systems are biased increasing the likelihood that certain groups are more likely to be sentenced and have higher custodial sentences imposed.

The Silicon Valley-based company Eightfold uses AI in recruitment. From millions of skill categories, this helps find staff for each role within their client's projects. AI is transforming talent management into a competitive advantage for organizations, avoiding the costs and time normally associated with losing high-performing staff and recruiting new employees. However, there can be issues with bias particularly if algorithms are based on historical data, mainly of men. Then these systems are likely to be biased against female applicants. Amazon developed such a project with their algorithms trained on their employment data over previous years. Most data came from men reflecting the male dominance of the software sector. Amazon eventually abandoned their project (Dastin, 2018).

The World Economic Forum has identified that many AI recruitment algorithms are biased, reproducing the same human biases demonstrated by recruiters (Schulte, 2019). If AI systems are taught which candidates are preferable, they will emulate humans and reinforce these biases. One advantage of AI systems is that they can also ignore aspects where humans may have a bias. AI systems can be trained to ignore gender and focus on competencies that are relevant to the role.

The algorithms selected should be examined for any potential bias. For example, an algorithm for a bank loan is often based on what has happened in the past. If this algorithm is trained on a set of customers that have been denied credit, and this sample is biased, then this unfairness will be perpetuated. More equitable decisions are needed. Not just on past financial decisions, which may be biased against certain groups, but what is right for society. The data scientists and developers need to ensure that they are not biased in their selection of data. One way to reduce this effect is to ensure diversity in the team developing these applications. Another approach is to follow the guidance of the Information Commissioner's Office regarding the principles of fairness, lack of bias and openness so that there is an increased level of account-ability. Other frameworks include IBM's AI Fairness 360, Google's Explainable AI BETA, and Rolls Royce's toolkit for AI ethics and trustworthiness: The Aletheia Framework (Rolls Royce, 2021).

Democratization of AI

Increasingly we are seeing the expansion of the democratization of AI, allowing those without many years of experience in machine learning and analytics to use these power-ful tools. There are popular visualization software suites, and increasingly there are applications where pre-trained machine models are available. However, one of several problems is the explainability of these models. There are increasing attempts to under-stand why specific models and machine learning solutions provide particular answers. One of these is the 'AI Explainability 360' toolkit and library project, created by IBM and open-sourced by the LINUX Foundation (The Linux Foundation, 2021), which helps developers better understand how machine learning makes predictions throughout the project life cycle.

AI projects need data scientists to select the right data sources and algorithms. Previously, in companies, this may have been the role of statisticians. However, software engineering skills are also needed, and good communication skills with colleagues to understand their project needs. It is unlikely all the necessary skills for a successful AI project will be in one person. Therefore, a team of skilled professionals with a variety of data skills is required. This enhances the role of the project manager, who will be needed to put a diverse team of engineers and business people together for an AI project.

AI can improve how projects are managed

Increasingly AI is being applied to areas of project management, optimizing the schedule, integrating data sets and rescheduling projects. In project management, these AI systems are particularly helpful in finding new resources, for instance, automatically searching through CVs and selecting appropriate candidates.

Some project management tools have the functionality to suggest who should be included in the team. These AI tools collect historical data on staff and the areas they have worked on and automatically use machine learning to optimize and reschedule, considering the availability and different skills of the project team. The digital agency Strømlin uses the AI software Forecast (Forecast, 2021), to reduce time on administrative work. This has provided an improved high-level view of how their projects are progressing. During development, project status is shared collaboratively with their clients. According to Peter Marius Stampe, Strømlin's Commercial Strategy Lead, using this software has helped the company strengthen relationships with their clients, as this software provides improved visibility which is 'key to building trust and better vendor-client relationships'. Their teams have noticed an improved precision in their time estimates for tasks and their schedules. Strømlin has noticed that the AI estimations help their teams more accurately determine how many hours a developer will take to complete a task. Another important feature for them is the functionality to synchronize with the tools their developers use, such as Gitlab, Google Drive and Calendar.

AI systems make budgets and schedules easier to monitor, as the embedded software algorithms learn from the data made available within the organization. Teams can be automatically provided with available staff, their skill sets, and predictions of how long tasks will take based on previous project data. As these algorithms use the estimates based on historical data they tend to improve at predicting costs and completion times for similar projects, as more project data is collected. If this software recognizes a delay it can automatically re-prioritize the schedule and the project manager can either accept this or manually adjust the schedule or the allocation of resources. The AI functionality helps reduce time and costs for routine tasks, preparing schedules, financial reporting and sourcing appropriate data for teams. These tools also provide visualizations and insights, enabling project managers to make better decisions. This type of software helps project staff by linking to resources across the organization. The project manager is then better able to find available staff with the experience for the current type of project and be able to ensure the completion of the project on time.

There are tools to provide the automatic tracking of resources, planning and forecasting projects. The more advanced tools centralize data for modelling, make predictions based on previous project data, reducing schedule variance and predict timings more accurately. Gartner (Costello, 2019) anticipates that 80% of project management work will be eliminated by 2030, as AI takes on project management functions and can analyse data faster than humans, raising project performance. These systems will collect data and make project professionals better planners. However, survey data collated in a report (PMI, 2019) by the Project Management Institute (PMI) suggest that AI will increase the numbers needed to manage projects and ensure these automated solutions are delivering effectively.

Atlassian is adding machine learning to their cloud products such as the software development tool Jira, and their team collaboration tool Confluence. Using data from more than 150,000 customers, supports teams using their products become more productive. The software provides suggestions to speed up searches and find relevant documents and other resources. Their team responsible for the product Jira, is developing its predictive capability, to make searches personalized and project teams more productive. Their integrated AI software recommends people for their projects. It suggests only those who are working on a similar stage of development within the project life cycle.

AI can, with Big Data and supporting technologies, help predict what will happen, optimize resources and automate processes so that humans can focus on more creative work. For projects, this is about gaining insight, reducing risk and automating processes as well as supporting humans to deliver higher quality solutions. Whether it be in project management, manufacturing, or healthcare, AI and Big Data have the potential to enhance work and life for everyone.

Security threats

Adversarial deep learning can be used to protect systems, but it can also be used by cybercriminals who can use the enhanced power of adversarial algorithms to attack websites. Low-quality sources of data, out-of-date data, or even data sets maliciously planted to bypass security systems and allow access to confidential files (Munoz-Gonzalez & Lupu, 2018). However, AI security systems can solve the problem of the limited number of cybersecurity experts available. It is not just about replacing humans, it is about solving the problem at scale in seconds so that threats can be identified, isolated and contained. AI systems also help lower the level of technical skills needed by providing either automated steps to counteract threats or suggest strategies for security teams using them.

A team supported by an AI security system analysing millions of data items can alert system users of any unusual behaviour, from a human or bot, trying to access a file. The AI software not only avoids continual alerts making security teams fatigued, but it also allows teams to focus on more creative work.

Organizations need to protect the valuable ideas they develop. Algorithms and the associated software and analytic processes developed are valuable assets and should be protected as intellectual property (IP). A recent (2020) online lecture at University College London (UCL) (UCL, 2020) discussed the dilemmas created with the advances in AI and algorithms. Who will own the IP is an ongoing question, as algorithms can generate their own algorithms. For the present, IP is likely to remain a human entity; a person who has ultimately designed or created the environment for the algorithm to create an enhanced version. With increasingly complex AI software systems using multi-objective optimization for simulation, design and manufacturing, this may have to be kept under review. Another challenge for the legal profession is deciding whom to blame in industrial accidents involving robots and accidents involving autonomous vehicles. Legislation will be needed to catch up with advances in technology. Society seems to expect a higher level of safety from autonomous systems than is expected from humans. Fortunately, organizations are creating frameworks to make AI systems safer.

Advances in AI optimization and simulation

In industrial design, for example, for the surface of an aircraft wing, surfaces need to be designed and optimized for many attributes. These may include aerodynamics, weight, strength and stress factors. Often, software simulations process these surfaces as a mesh, an array of points across a curved surface, to model an aircraft wing or the housing of the engine in a racing car. Deep-learning methods using convolutional networks work well with representing points on a grid-like structure in two dimensions as the pixels within an image. However, they work less well with complex data such as multi-dimensional points. Recently tools have been developed which can parse more complex data such as meshed surfaces in computer graphics for computer-aided design, where 3D shapes need to be optimized. Engineers also need to consider how these designs interact with the laws of physics with computer-aided engineering. Neural Concept collaborating with Bosch Research are using 3D geometric convolutional neural network techniques to understand how electric drive (E-drive) motor housing simulations can be optimized and learn how physical factors impact the final design. They are achieving these simulations in milliseconds, as opposed to classic simulators which may take days. This allows engineers to simulate in real time, accelerate the R&D cycles, enhance product performance and accelerate time to production (Neural Concept, 2020).

For advanced simulations and accelerated engineering, high-performance computing (HPC) is required. HPC will enable advances in AI-driven generative design, drug discovery and climate modelling. The European Union is funding an initiative to be at the forefront of supercomputing with LEONARDO which will be able to perform over 248 million billion calculations per second. Atos is a leading company behind this European project. Robert Viola, Director General of the Directorate-General for Communications Networks, Content and Technology of the European Commission stated that this will combine 'Artificial Intelligence and HPC technologies' and will help humanity respond to, climate change, natural disasters and pandemics (EuroHPC, 2020). Atos is also a partner in the AI4EU project to construct the first AI on-demand platform and ecosystem for the European community. These initiatives provide access to supercomputing, and the algorithms, data and services necessary to support commercial innovation and scientific research.

The societal impact of AI

News broadcasts covering AI systems beating humans in games such as AlphaGo, and the deployment of AI drones in war zones has fuelled concerns of man against machine. However, this is not an inevitable scenario. Research projects need to focus on aspects of intelligence so that humans can work together. AI systems are capable of so many work activities that many humans also fear being replaced by robots. Daniel Susskind in his book, *A World Without Work* (Suskin, 2020), argues that robots are taking our 'routine' jobs. It seems many jobs once considered secure, are being progressively being taken by AI systems. Within Amazon warehouses, where employees run to collect items for customer orders, their roles are increasingly replaced by robots. In a few years, some of the work of their delivery drivers is likely to be replaced by drones. It is not just warehouse work, but areas such as law, medicine and agriculture that are already being affected. It could be argued this is a dystopian view, or that these machines free us from arduous and repetitive work and allows workers to be re-skilled in other areas.

Increasingly AI is augmenting the capabilities of humans so that humans can avoid errors, have a safer work environment, and businesses can reap the rewards of automation. During the COVID-19 pandemic, communications and AI-assisted technologies have enabled many who are office-based, to transition to remote working. Although it may reduce the subtleties of face-to-face communication, technologies can allow workers to avoid contact and reduce infection during a pandemic. However, it needs to be recognized that many workers do not have this opportunity to lower their risk by working from home.

Streaming data is critical to detecting anomalies in industrial systems

Processing sensor data in real time is the most effective way to detect anomalies if remedial action needs to be taken. Streaming data can be used in conjunction with historical time series sensor data to compare the current patterns to expected and identify any deviations from those expected. The dynamic features within time series can also help understand the normal range. This is true of many scenarios. Heart activity during exercise shown in an electrocardiogram, or industrial processes, such as the vibrations of an oil drill bit entering a different stratum. Industrial sensor data can be displayed as visualizations and integrated into dashboards. AI and algorithms can identify with a high degree of accuracy anomalies, arrhythmia within heart conditions, or problems with machinery in an industrial system.

AI systems can, for example, be used to monitor pumps via sensors within a water supply network. Anomalies can then be detected from data streams and this information can be used for predictive maintenance so that the pump can be replaced before it fails. This analysis can be done for industrial plants, jet engines and fleets of delivery vehicles. Detecting unusual patterns, allows components to be replaced before failure, increasing safety for passengers and ensuring supply chain resilience. Airbus uses the application software Splunk to achieve greater visibility of their supply chains and location of maintenance parts globally. They also use this software to gain cognitive insights into their applications. Airbus has specifically set up a unit for digitization to allow greater visibility of their data across the enterprise, not just for IT operations but also in business use and security. Many companies have disparate monitoring tools used to manage IT performance. These tools can help identify and understand performance issues and anomalies. The software stack can be holistically monitored and managed. Visibility across platforms can be increased to identify problems and their root cause. ERT, a pharmaceutical company, has accelerated clinical trials with AI-assisted monitoring software (ERT, 2019). Their open telemetry improves visibility across their infrastructure, helping to identify and remedy issues. For pharmaceutical companies, AI systems tracking supply chains using blockchain technologies are becoming increasingly important, as the data cannot be altered or deleted. The immutable nature of blockchain assures medical staff that the data has not been tampered with. Hyperledger, an open-source Linux project, is a blockchain technology that can be used in conjunction with sensors to track and ensure authenticity and provenance and confirm vaccines are stored at the correct temperature.

AI provides the opportunity for improved efficiency

Artificial Intelligence is having a profound impact wherever automated analysis or decision-making is applied. It is affecting numerous areas of our lives: transport, work and consumer purchases. AI is improving the efficiency of energy, and transport systems, giving us a potential future of a net-zero carbon society. It is helping companies to recruit staff, find documents and better plan and resource their projects. Every day, AI helps us with our Google searches and suggests products we may want to buy. AI systems can process data via their algorithms far more quickly and accurately than people would ever be able to. It is, therefore, no surprise that Arvind Krishna, the CEO of IBM, stated during the keynote of their virtual Think Digital 2020 event (IBM, 2020c): 'every company will become an AI company, not because they can, but because they must'. It is the convergence of the advances in AI, cloud computing, processing power, and the explosion and availability of data that has given us this opportunity.

IBM is applying its AI-enabled hybrid cloud infrastructure to workflows within ServiceNow's intelligent workflow systems. Applying IBM Watson's AIOps solution, reduces resolution times by 65%, helping to reduce outages. Issues such as outages can cost over half a million dollars per hour (IBM, 2020a). Addressing these problems translates into considerable cost savings as well as improving experiences for customers, employees and partners.

The Internet of Things (IoT), sensors and other technologies, incorporated into industrial settings and workflows, provide data streams from production lines. These technologies enable machine learning decisions to be made within real time and can increase the quality of products in manufacturing. Intel has been working with Nebbiolo Technologies to improve the inspection of spot welds within Audi's production lines. This POC project is to improve the quality of spot weld inspections on the vehicle production line. The aim is to increase the number of vehicles inspected. Before the project, the spot welds for one vehicle, each day, were manually inspected. This took considerable time and also involved highly trained technicians. The intention is also to inspect every vehicle automatically using an AI system without any delay in the production line. Using machine learning, these systems are trained to identify high-quality images of welds and to identify images with minor flaws. By analysing all the welds, for all vehicles in production and not just a sample, Audi is aiming to increase quality and reduce costs. Currently, Audi is extending this project to other production roles, such as paint spraying, to assure quality. To fully realize this migration to Industry 4.0, large-scale automated manufacturing systems that have machine-to-machine communication, millions of items of data need to be analysed. This data needs to be integrated so that instead of only one application being enhanced, this capability can be scaled and used across the factory (Intel, 2020). AI-enhanced visual inspection systems can be applied to other areas. However, we need to know that the system is doing this reliably and we need assurance that there is no drift from what is required. This is particularly important for safety-critical systems and their constituent components.

AI-enhanced medical imaging systems

AI is now being increasingly applied to medical imaging systems. At UCL, magnetic resonance imaging (MRI) has been used for several years used in conjunction with machine learning, to provide life-saving identification of soft tissue tumours. Since 2005 the resolution of images of cancerous tissues has rapidly advanced. By 2015 the team at UCL led by Prof. Mark Emberton (National Institute for Health Research (NIHR), 2020) had pioneered the use of MRI using machine learning as a primary investigation tool for prostate cancer in several studies. Beforehand, biopsy the was standard way of diagnosing this type of cancer. Now, based on data from several studies, NICE, the guiding authority on clinical practice in the United Kingdom, recommends that an MRI should be the first investigation rather than an invasive biopsy. Using MRI results in higher survival rates, as all tumours can be identified via radiographers or with the assistance of machine learning. Whereas, with a biopsy, there is a chance that the tissue sample taken misses the malignant cells when they are present.

At another research centre, at the Mater Hospital University College Dublin, AI applications are being developed for surgery. Images of carcinogenic tissue at surgery are differentiated with the use of fluorescence, dyes that stay for a limited time within the patient. Using vision systems to interpret pictures and the probability that specific areas of tissue may be carcinogenic helps surgeons decide what tissue to remove.

Removing the cancerous tissue and retaining as much of the healthy tissue as possible considerably improves the recovery and the long-term survival of patients. AI systems are good at recognizing regular structures with straight edges and angles, such as buildings and vehicles, but struggle with irregular objects, such as organs and tissues. For surgical procedures, far more data and training images are required to accurately interpret images particularly those that are moving. As a biopsy during surgery would take too long, the team is aiming to obtain real-time AI diagnosis during the surgical operation. Algorithms classify the images and specifically assigns each area of the image, highlighting the boundaries of these tissues and assigning a probability of the tissue being healthy, cancerous, or benign. This prediction then helps the surgeon decide which area of tissue to remove. Professor of Surgery, Ronan Cahill outlined during the IBM's Cloud & AI Forum 2020 (IBM, 2020b), that current robotic surgery devices such as the DaVinci robot are in effect an extension of the surgeon's hands and that by incorporating a 'smart interpretation of the appearance of tissues during surgery' it would enhance these machines. The robotic device and the surgeon would then be able to work in synergy. The surgeon viewing the digital image with the tissues classified by the AI system could quickly identify and remove any cancerous tissue.

AI can radically advance imaging and simulations for healthcare and engineering. AI can deliver production processes and supply chains more efficiently. AI will provide the improved capability of organizing, planning and delivery of these projects. With these projects, there will also be a need for more technically trained staff and project professionals to interpret the decisions that these AI systems make. With advances in networks and communications technology, increased availability of open data and new supercomputer facilities, AI provides the opportunity for research teams to find solutions to some of humanities' most complex problems.

Big Data

The emergence of the term Big Data

During the late 1990s, the computer graphics scientist, John R. Mashey, working at SGI, was using the term 'Big Data', for data volumes that are too large to be processed by traditional computing systems. Big Data, became more widely used during the start of this century and the Economist (Sunmin, 2015) was outlining the ubiquitous nature of data and how science and business are using Big Data. Data is expanding exponentially. IDC, the market intelligence organization, is now predicting that the world will create more data in the next 3 years more than it did in the last 30 years (IDC, 2020). Just one connected vehicle creates four terabytes of data a day and 65 billion searches are made on WhatsApp. Collecting and understanding data on social media, what videos are viewed, location data and even keystroke patterns create a powerful resource for targeting potential customers. Using this data in conjunction with AI is now radically impacting numerous facets of our lives, including helping to diagnose illnesses.

The data being processed by many systems, particularly medical imaging systems, can be considered Big Data. The sheer volume of data that form these images requires deep learning methods to help identify patterns. Medical scanning technology is rapidly developing. There are orders of magnitude difference in the volume of data generated by computer tomography scans compared to more detailed MRI scans. In 2005, the boundary of soft tissue cancers was hardly perceptible to experienced radiologists with MRI. Now MRI scans provide high-resolution images at a cellular level. The consequent data requirements necessary to store and process these images have increased, in addition the demands of deep learning technologies have further increased memory requirements (Hering et al., 2019).

Empirical evidence is needed

Big Data can also transform business. Rather than guessing the next step, a data-driven approach using empirical data provides the evidence to support the direction a business should take. In addition to Big Data being too large in volume for traditional computer systems, often it is also too complex. Sometimes, data is delivered too fast and needs specialist technologies and software to process, such as streaming data from an aircraft jet engine or from medical sensors attached to a patient in an intensive therapy unit. Combining this with other technologies such as cloud computing and telecommunication technologies increases the effectiveness of these systems. The nexus of these technologies enables engineers to view historical data as well as see what is happening in real time. They allow emerging patterns to be seen and make far more accurate predictions. The scale of Big Data is increased exponentially by remote sensors and technologies. IoT can reduce the demands on a network by being able to process data at the edge of the network: edge computing. This also reduces the response time for smart devices such as industrial and environmental sensors. This brings the possibility of being able to create smart city infrastructures, and Industry 4.0. In these smart infrastructures and advanced industrial systems, Big Data becomes increasingly important, as sensors vastly increase the volume of data and the need for real-time analysis.

Although Big Data has sometimes been referred to as the 'new oil', this analogy can be unhelpful. Data is an exponentially expanding resource and becomes more valuable when combined with other vast data sets, especially when analysed with AI tools. The use of fossil fuels is increasingly harming the environment. With Big Data and AI, we have the potential to better understand our environment and reverse this trend and help bring about a net-zero carbon economy. For energy networks, both for supply and consumption, we have the data available to reduce our energy needs and create far less wasteful energy usage. We can apply AI to many other systems to reduce wastage. For instance, in agriculture, using AI vision systems on agricultural machinery can even identify individual crops, apply just the optimum amount of fertilizer and pesticide, reduce harmful run-offs to water supplies and reduce the costs of production.

Increasing the value of data by making it openly available

Open data is a valuable resource for citizens. As part of smart infrastructure, sensors can be placed in cities to monitor air pollution and can make pollution data available in real time. With this data, citizens can make decisions to improve their health, avoid running at certain times of the day, or stay at home if they are particularly at risk with a medical condition such as chronic asthma. It can even be integrated into smart city systems to divert traffic if there is a traffic jam, reducing pollution within an area. The deployment of 5G will bring further opportunities due to the speed that people, vehicles and the infrastructure are connected. This will help optimize of transport systems and track the supply of goods.

Within medicine, the sharing of information amongst medical professionals is vital to increase knowledge of illnesses. Combining data from different hospital researchers and doctors can quickly gain insight into the early diagnosis of the SARS-CoV-2 virus (COVID-19). Increasingly a particular type of image classification, a convolutional neural network is used to assist in the identification of illnesses from medical scans. UCL is developing AI automated diagnostic and prognostic tools for sonography in COVID-19 and related lung conditions (Centre for Medical Image Computing (CMIC) UCL, 2020). The aim is to collect as much data as possible from COVID-19 and non-COVID-19 patients around the world using lung ultrasound data to create a tool to identify illnesses accurately and as rapidly as possible. Each scan is annotated by medical experts, which contributes to their web-platform, hosting the tool and open database of images. The accuracy of this tool increases as more images are added and making this data available for all hospitals allows this project to benefit as many patients as possible.

Providing an open data resource to protect the planet

Oceans provide a vital resource. They also provide a vital sink for the increases in temperature and CO_2 levels to be absorbed. To improve the understanding of our oceans, the Mayflower Autonomous Ship (MAS) research project is aiming to collect ocean data and make it an openly available resource. Don Scott, the Chief Technology Officer of the MAS project explained during the IBM Forum 2020 (IBM, 2020b), that the project led by ProMare and IBM provides a sustainable way to collect vast amounts of data. He explained that it is a particularly challenging project due to the ocean's extreme conditions and continual changes in current and waves.

The project, inspired by the original *Mayflower* journey in 1620, aims to recreate the same journey from Plymouth, United Kingdom, to Massachusetts, United States. The autonomous ship is scheduled to set 'sail' during 2021, coinciding with the UN Decade of Ocean Science Research for Sustainable Development.

Navigation decisions are made by the 'AI captain'. Due to the challenges of communications and time delays with the team on land, decisions must be made on the autonomous vessel. These are based on real-time environmental sensor data, combined with radar detection and collision analysis of objects such as ships. These systems are based on IBM's technology using models trained before being installed on the graphics processing unit (GPU) Nvidia system. This GPU is designed to cope with the processing demands of machine learning and simultaneous data from sensors. The sensors' telemetry includes the ocean current, depth of water and sea temperature. Radar systems with weather information create a hazard map and the onboard models decide appropriate actions to follow, adjusting the route and the ship's speed accordingly. Also, the weather forecast is incorporated in the decisions to feed into the route planner. This provides an autonomous way to gather data continuously and at the same time lower the cost of data acquisition. The vessel will create an open platform for oceanographic data to help scientists to understand climate change.

Other projects are providing environmental data such as the monitoring of the thickness of ice within the polar regions. The rate of ice melting needs to be understood to protect coastal communities around the world. Not only are AI systems used to analyse the data sets, but AI is also used to improve the precision of the images from underneath the ice shelves, providing more accurate measurements of ice thickness and better predictions of ice melting (Rahnemoonfar, Johnson, & Paden, 2019).

Data has become ever more important for the survival of enterprises. However, having this data available is no guarantee of success. Unless the analysis of this data and the insights are harnessed, these efforts will be wasted. Project teams need to work in conjunction with these resources to make better-informed decisions. With advances in AI, computing power, cloud computing and the vast amount of data available, we can allow our AI systems to find the optimum path to achieve the desired outcomes, with potential increases in speed and quality. Organizations need to be data-driven as well as AI-driven, to enable the continuous cycle of data analysis, insights and predictions to exploit these opportunities.

Case Study: Open data and AI solutions

It is problematic to use medical data for training research staff. Even anonymous data can potentially identify someone if they have a rare genetic illness. Within UCL open data sets are used to increase the skills of their research teams. This accelerates their learning curves of how to analyse Big Data empirically and the techniques in machine learning.

During 2015 numerous press reports were outlining the poor air quality in major cities. One seminal scientific study (Howard, 2015) highlighted that over 9000 people died in London each year due to air pollution. Research students in the Department of Computer Science at UCL were keen to create AI solutions to help reduce this problem. They collaborated with researchers from the National Institute of Informatics, Tokyo to produce apps aimed at helping users avoid areas of high pollution. The teams had access to open data sets: data that was freely available, including Copernicus, the European Union's programme for earth sciences data, and from London Air, a research group at King's College London. The research teams used machine learning to create apps, to enable citizens, including those exercising via cycling or running, to select routes that would lower their exposure to air pollutants (Collins, Varilly, & Yoshinori, 2015). The goal of machine learning is to create a mathematical model that can make an accurate prediction. The algorithms within the models were initially provided with training data so that the model could learn the desired outcomes and could be tuned to give the correct outcome, for this case, suggesting a route with the lowest exposure to air pollutants. The research groups faced many challenges: data being in different formats and which data sets to use. They needed to ensure that a suitable route could be displayed on mobile phones without any noticeable delay. Another challenge was providing a route that did not add significantly to the journey time. Alternatively, to ask the user if they would accept a trade-off of additional journey time for an improvement in air quality. Finally, researchers estimated the improvement in air quality for each journey the user made and cumulatively with regular use of the app.

The research groups embarked on different aspects of the project, but all relating to improving air quality for users. Some of the data sets were from only a limited number of sensors from London Air, but each sensor was providing streams of data based on different pollutants, including particulate matter, PM2.5 and PM 10.0, which has been shown to exacerbate lung conditions. The problems the research students encountered were numerous, how to deal with gaps in the data, how to transfer data for processing and which algorithms to select. Creating the visual display of the apps was the relatively easy part. Assimilating data on a variety of mobile platforms and guiding users was a far from trivial problem, as can be seen by the time taken to create the test and trace apps for the United Kingdom during the COVID-19 pandemic. The algorithms and the code were open-sourced so that organizations such as the London Sustainability Exchange and other researchers could incorporate or refine these solutions further.

Bibliography

Centre for Medical Image Computing (CMIC) UCL. (2020). *Lung ultrasound in COVID-19 data collection website*. Available at http://niftyweb.cs.ucl.ac.uk/index.php. (Accessed 11 December 2020).

Collins, G., Varilly, H., & Yoshinori, T. (2015). Pedagogical lessons from an international collaborative Big Data undergraduate research project, ECSAW'15. In *Proceedings of the 2015 European Conference on Software Architecture Workshops* (pp. 1—6). September 2015, Article. No.:32 https://dl.acm.org/doi/10.1145/2797433.2797467.

Costello, K. (2019). *Gartner says 80 percent of today's project management tasks will be eliminated by 2030 as artificial intelligence takes over, Gartner*. Available at https://www.gartner.com/en/newsroom/press-releases/2019-03-20-gartner-says-80-percent-of-today-s-project-management. (Accessed 15 December 2020).

Dastin, J. (2018). *Amazon scraps secret AI tool that showed bias against women, Reuters*. Available at https://www.reuters.com/article/us-amazon-com-jobs-automation-insight-idUSK CN1MK08G. (Accessed 11 December 2020).

ERT. (2019). *Transforming clinical development with AI*. Available at https://www.ert.com/news/transforming-clinical-development-with-ai-applications-imaging/. (Accessed 12 December 2020).

EuroHPC. (2020). *Leonardo a new EuroHPC world class pre-exascale supercomputer.* Available at https://eurohpc-ju.europa.eu/news/leonardo-new-eurohpc-world-class-pre-exascale-supercomputer-italy. (Accessed 15 December 2020).

Forecast. (2021). Available at https://www.forecast.app. (Accessed 15 December 2020).

Hering, A. D., et al. (2019). Memory-efficient 2.5D convolutional transformer networks for multi-modal deformable registration with weak label supervision applied to whole-heart CT and MRI scans. *International Journal for Computer Assisted Radiology and Surgery, 14*(11), 1901−1912. https://doi.org/10.1007/s11548-019-02068-z

Hill, F., et al. (2019). *Environmental drivers of systematicity and generalization in a situated agent.* Cornell University. Available at https://arxiv.org/abs/1910.00571v4. (Accessed 25 January 2021).

Howard, R. (2015). *Up in the air: How to solve London's air quality crisis, Policy Exchange Unit.* Available at https://policyexchange.org.uk/wp-content/uploads/2016/09/up-in-the-air.pdf. (Accessed 20 November 2020).

IBM. (2020a). *IBM and ServiceNow.* Available at https://www.ibm.com/cloud/blog/announcements/ibm-and-servicenow. (Accessed 10 December 2020).

IBM. (2020b). *IBM Cloud and AI Forum 2020.* Available at https://www.ibm.com/uk-en/events/cloud-and-ai-virtual-forum. (Accessed 26 November 2020).

IBM. (2020c). *IBM Think Digital 2020.* Available at https://www.ibm.com/events/think/watch (Accessed 25th January 2021).

IDC. (2020). Available at https://www.idc.com/getdoc.jsp?containerId=prUS46286020. (Accessed 15 December 2020).

Intel. (2020). *Audi's automated factory moves closer to industry 4.0.* Available at https://www.intel.com/content/dam/www/public/us/en/documents/case-studies/audis-automated-factory-closer-to-industry-case-study.pdf. (Accessed 12 December 2020).

Lobe. (2020). *Microsoft.* Available at https://lobe.ai. (Accessed 21 January 2021).

McKinsey & Company. (2021). *An executive guide to AI.* available at https://www.mckinsey.com/business-functions/mckinsey-analytics/our-insights/an-executives-guide-to-ai. (Accessed 15 January 2021).

Munoz-Gonzalez, L., & Lupu, E. C. (2018). The Secret of Machine Learning. *ITNow* (Volume 60,(Issue 1), 38−39. Spring 2018 https://academic.oup.com/itnow/article/60/1/38/4858513. (Accessed 25 November 2020).

National Institute for Health Research (NIHR). (2020). *Transforming prostate cancer diagnosis for men around the world.* https://www.nihr.ac.uk/case-studies/transforming-prostate-cancer-diagnosis-for-men-around-the-world/26274. (Accessed 12 January 2021).

Neural Concept. (2020). *3D learning software for enhanced engineering.* Available at https://neuralconcept.com/wp-content/uploads/2020/11/Bosch-customer-stories.pdf. (Accessed 15 December 2020).

PMI. (2019). *AI@Work: New Projects New Thinking (2019).* Available at https://www.pmi.org/learning/thought-leadership/pulse/ai-at-work-new-projects-new-thinking. (Accessed 25 November 2020).

Rahnemoonfar, M., Johnson, J., & Paden, J. (2019). AI radar sensor: creating radar depth sounder images based on generative adversarial network. *Sensors, 19*(24), 5479. https://doi.org/10.3390/s19245479. . (Accessed 18 January 2021)

Rolls Royce. (2021). *The Aletheia Framework.* Available at https://www.rolls-royce.com/sustainability/ethics-and-compliance/the-aletheia-framework.aspx. (Accessed 25 January 2021).

Schulte, J. (2019). *AI assisted recruitment is biased heres how to beat it.* World Economic Forum. Available at https://www.weforum.org/agenda/2019/05/ai-assisted-recruitment-is-biased-heres-how-to-beat-it/. (Accessed 25 November 2020).

Sunmin, K. (2015). *Big data evolution, the economist intelligence unit, Economist.com.* Available at https://eiuperspectives.economist.com/technology-innovation/big-data-evolution. (Accessed 10 December 2021).

Suskin, D. (2020). *A world without work: technology, automation, and how we should respond.* Henry Holt & Company.

The Linux Foundation. (2021). *AI explainability 360.* Available at https://ai-explainability-360.org. (Accessed 15 December 2020).

Tryolabs. (2020). *Face mask detection in street camera video camera streams.* Available at https://tryolabs.com/blog/2020/07/09/face-mask-detection-in-street-camera-video-streams-using-ai-behind-the-curtain/. (Accessed 25 November 2020).

UCL. (2020). *UCL Sir Hugh Laddie Lecture 2020.* Available at https://www.ucl.ac.uk/laws/events/2020/nov/online-2020-sir-hugh-laddie-lecture.

Appendix 3: Case study 1: Cement plant

Cement Factory, Rugby Portland Cement Co. Ltd. South Ferriby, Lincolnshire.

Introduction

For years, numerous articles and books have maintained that the earlier the designated project manager is associated with the project the more likely it is that it will be successful. This has certainly been borne out in the case of the project described in this chapter for which I was the project manager. My involvement in this project, a 1000 ton/day Portland cement plant, started almost as soon as Tarmac Civil Engineering (Tarmac) the company I worked for, was approached in January 1965 by the client organization, Rugby Portland Cement Ltd (RPC) to build a new semi-dry process cement production plant in South Ferriby, Lincolnshire.

The client specified that the suppliers of the kiln and associated equipment would be the German firm of Polysius and my first job was to accompany the Tarmac technical director to a visit to the Polysius factory in Neu Beckum in Germany for technical discussions which included a visit to one of their operating plants.

Contract

The original contract was for Tarmac to design and build the civil works only. This comprised the plant and building foundations, piling, roads, drains, silos, chimneys, steel structures and buildings. The design of this work was subcontracted by Tarmac to a firm of Consulting Engineers who also carried out soil surveys and soil tests. The contract was based on the ICE Conditions of Contract 4th edition.

As Tarmac also had a process engineering division in London, RPC placed a separate contract for the project management, design, procurement and installation of the mechanical, electrical and instrumentation works with this division.

RPC had already decided on the semi-dry cement process using the Polysius Lepol grate re-heater, kiln and cooler. This meant that this equipment together with the raw meal and cement mills became nominated subcontracts. The suppliers of this specialist plant and machinery would also supervise the erection and commissioning of their plant.

All the remaining equipment such as feeders, conveyors, mixers, compressors, fans, aeration pads, precipitators, switchgear, control panels, etc. together with pipes ducts and cables was specified and procured using conventional competitive tendering based on the IMechE Model Form of General Conditions of Contract MF1.

All equipment, materials and plant were charged to RPC at cost while all project management, engineering, procurement and related head office support costs were charged in accordance with an agreed schedule of man-hour rates.

Because there were in effect two separate contracts between RPC and Tarmac, it was decided that once work had started on site, the civil engineering and the mechanical/ electrical works would be organized as two separate profit centres, each with its own site manager and support staff. This meant that if the civil works were held up due to late information from the process engineering division, they could claim damages for delay. Similarly, if the mechanical erection was delayed due to a foundation not being ready, a claim could be made by the process engineering against the civil engineering division. The start and finish dates of all the design, procurement and site activities were set out on the critical path programme drawn up by the process engineering division and agreed to by the civil engineering division before work started on site.

The progress control for the civil engineering work was by measurement of completed work and comparing it to the amounts given in the Bills of Quantities produced at the tender stage. The man-hours of the mechanical and electrical works were recorded on weekly time sheets and invoiced to the client monthly at the agreed man-hour rates. All extra work requested by the client was recorded and confirmed back in writing within 2 days. Design and construction changes proposed by Tarmac were submitted to RPC for approval and could not be executed until this was received in writing.

'Verbal Instructions' had to be confirmed within 24 hours with the standard wording: 'Cost and programme implications to be advised'. It was agreed with the client that any unresolved costs due to extra work would be submitted as part of the final account.

Process

A flow diagram was prepared which clearly showed the manufacturing process starting with the limestone quarry and clay pit which are the fundamental ingredients for cement, and ending with the bagging plant and bulk dispatch bay.

Fig. C.1 shows the process flow diagram

The limestone is delivered to the site from the limestone quarry by a 1-mile long belt conveyor. The clay from the clay pit is excavated and loaded into narrow gauge rail tipper wagons to be hauled to the site by a diesel engine. After the clay has been shredded both materials are weighed, mixed and ground to a powder in a raw-meal mill known as a double rotator.

The raw meal is stored in a silo from which it is fed by a belt conveyor into a blending silo and from there via belt conveyors into two nodulizers which, with the addition of water, produce pellets. These are heated by the kiln exit gas in a travelling grate pre-heater before being fed into the refractory lined rotating kiln.

In the kiln, the pellets are heated to about 1450°C (calcined) which produces a cement clinker which is passed through a clinker cooler and conveyed to a clinker bunker. The clinker is then fed into a rotating cement mill to produce the cement which is either fed from a cement silo into bulk-cement lorries for site deliveries or passed through a bagging plant to produce the familiar 1 cwt cement bags.

The kiln is fired by pulverized coal blown through burners by a combined pulverizer-blower. An FD fan supplies the combustion air. The hot exit gas from the kiln passes through the pellet preheater and a bank of electro-static precipitators before being exhausted by an ID fan to the chimney.

Site layout

When preparing the site layout drawings, the Tarmac design team, which included the construction and planning engineers in London worked closely with RPC. Because the new plant shared the site with the existing cement-making plant, it was essential that during the construction and commissioning period the operation of the existing plant would not be adversely affected. As the existing plant was in fact a wet process plant, the two were completely separate in every way. The early involvement of Polysius and the relationship created during our initial visit helped us to obtain the necessary layout drawings and design data at an early stage. In addition, we received useful information on access and servicing requirements as well as advice on erection and operating procedures.

All discussions and negotiations were in English as all the staff who were employed by Polysius and the other German plant manufacturers dealing with this project spoke very good English.

Project structure

The project structure was a functional organization. The only project management team members were myself, an assistant project manager (APM) and a secretary. The APM was recruited from one of the technical departments and was seconded to the project as part of his/her training programme. At the end of the project, these APMs returned to their departments or became project managers.

The design engineers allocated to the project carried out their work geographically within the specialist departments of the process division in London and were responsible to their departmental managers for quality, performance and design cost (efficiency). However, they were answerable to the project manager for adherence to the project schedule.

Project management plan (co-ordination procedure)

Once the tender was accepted by the client and the official order had been received, one of my first tasks after being appointed project manager was to write the Co-ordination Procedure or what is now known as the project management plan (PMP). This meant in effect shutting myself away and not participating in any other company activity for a week while studying all the drawings, specifications and conditions of the two contract documents between Tarmac and RPC. The main purpose of the PMP is to inform all the stakeholders of the 'What, Why, When, How, Where and Who' of the project, but it is also an excellent way for the writer to obtain a good understanding of what was required to be delivered.

The draft was distributed to all the department managers for comment, which had to be returned to me within 3 days. There were inevitably many suggestions, additions and queries and once these had been addressed or incorporated, the document was issued as Rev.0 in accordance with a previously agreed distribution schedule.

From then on, this document became the 'Bible' of the project, which was amended when major changes to scope, specifications or programme became known. As a constant reminder of what had to be 'delivered', it was always on my desk and accompanied me to all progress meetings, whether in the office or on site.

One of the main items in the PMP was the family tree, that is, who reports to whom, which also became the basis of the distribution schedule. It was stressed that all communications between Tarmac and RPC regarding contract conditions, progress and costs would only be between the Tarmac PM and his opposite number at RPC − the Chief Engineer of RPC. Specialist technical department managers could communicate with their opposite number working for the client or a supplier on technical matters, but any decisions on technical, financial and programme matters had to be put in writing and sent to the PM for confirmation. This rule was strictly enforced.

Project schedule

The project scheduling was carried out using critical path analysis. This was the first time that this method would be used by this division of Tarmac and I was keen to try it out. Because of its size, we opted to use the ICL Computer Centre in central London to process the data. Every month during the design phase, the activities and durations were entered on input forms and sent off to the Computer Centre for analysis. A day later, we picked up reams of concertinad printouts giving the starting and end dates, critical paths and floats of all the activities. It only required one small error in the input data for all these sheets to be useless so that many hours were expended in trying to find the error.

Even after correcting the print-outs, sending this mass of sheets to the department heads or site proved to be unpopular with the recipients. I decided therefore to cancel the computerized analysis and do it manually on the Activity on Arrow (AoA) network diagram. This was both quicker and cheaper and any unrealistic results could quickly be identified and corrected. We discovered that this manual analysis on the networks took only a fraction of the man-hours required for filling in the input sheets, delivering them to the computer centre, collecting them and updating the network.

Progress was recorded on the network sheets by simply thickening the arrows on the activities to about the same proportion as the percentage complete. Every month prints of these updated networks were sent to all the heads of department and the site.

The AoA networks were divided into six main sections:

Overall network:	Hammock of activities showing milestones.
Civil site work network:	Roads, drains, foundations, concrete silos, chimney, security.
Structural construction network:	Buildings, sheds, steel towers, steel bunkers.
Main plant network:	Nodulizer, grate pre-heater, kiln, cooler, raw-meal mill, cement mill.
Mechanical network:	fans, coal grinder, piping, ducts, air-pads, precipitator bagging plant.
Electrical network:	Cables, instruments, switchgear, control panels, lighting communication.

The mechanical and electrical networks were each sub-divided into two parts:

Part 1: Design, specification, procurement, inspection, delivery to site
Part 2: Offloading, storage, site installation (erection), testing, commissioning

This method of programming was so successful during the design stage, that it was later used on site where all the network programs for both the civil and mechanical works were pasted on A-0 size hardboard panels and placed in racks for easy access. It was the site planning engineer's function to record progress daily by marking up the completed or partially completed activities (as was done during the design stage) with a red pencil. In this way, everyone could see where we were and what still had to be done.

Apart from its use as a scheduling system, network analysis also turned out to be an excellent tool for solving specific problems. While the cement mill was being erected, it became apparent that some civil engineering work had to be carried out directly beneath the steel structure which was still being completed by the steelwork erectors. The program, which had been recently revised due to a modification to the mill foundations, clearly showed that both these activities were due to be performed at the same time and both were on the critical path. Aware of the danger of work being done on two levels simultaneously, both site managers came into my office to discuss the problem. When the civil engineering site manager suggested that we draw a network to sort it out it was music to my ears. I then realized that a man brought up on bar charts appreciated the use of a network to solve individual problems. After a short discussion, the solution was to stagger the work of both teams into a series of 3-hour day and night shifts. Although this increased the costs slightly, the programme was maintained.

Site management

Because of the size of the contracts and because the civil engineering work and mechanical/electrical work were measured and costed differently, it was decided to appoint a 'Site Supremo' to whom both site managers reported for progress, costs, quality, safety and discipline. His other function was to ensure that there would not be any spurious or unreasonable claims or counterclaims.

As the design and procurement stage was almost finished, and as none of the other site construction managers were available, I was asked to go to the site and take over the role of Supremo as well as being the project manager.

After discussion with my wife, we agreed to spend the next year on site until the plant was commissioned and handed over. Fortunately, both my children were too young to go to school, so we let our house near London and moved to Lincolnshire for the construction and commissioning phases of the project. Tarmac provided us with a house near the site in Winterton and a company car.

Both the civil and mechanical site managers had their own support team accommodated in their own suite of offices. My own team as overall site manager only consisted of a deputy site manager, who was recruited from the London mechanical design engineering department, a planning engineer, a secretary and a telex operator/typist who were both recruited locally. While each site manager had his own telephone line, the telex facility was shared by all the managers. I was extremely fortunate in having a very efficient secretary who could almost read my mind. When I had to return to London for regular or ad-hoc meetings, all I had to do was to tell her the purpose of the meeting and a folder of all the documents I needed was on my desk on the evening before my departure.

It was my habit to visit the site every Sunday morning while only a skeleton maintenance crew was at work. These visits enabled me to see what physical progress had been achieved in the various areas during the week. The following Monday morning I chaired a site progress meeting which was attended by both the mechanical and civil site managers and the site planning engineer. My secretary took notes and by the afternoon the instructions for further action (decided at the progress meeting) were handed to the site foremen. If, at the progress meeting it became apparent that some of the activities were delayed, the cause was discussed in detail and an appropriate action plan agreed to bring the work back on programme. Following the meeting, the planning engineer updated or amended the various construction networks which were then photographically reduced to A3 size for use by the site foremen. All the planning was done on networks. The only bar charts produced were manually drafted charts based on the networks to facilitate resource (manpower) smoothing in specific work areas.

It was on one of my Sunday morning visits that I came face to face with inter-union demarcation. It was a very windy day, and as I walked through one of the sheds, I noticed that one of the large asbestos-cement cladding sheets was flapping in the wind due to a dislodged hook bolt. I grabbed the hook bolt and held the sheet against the sheeting rail. The nut for the hook bolt was missing, but I noticed a piece of string on the ground about 3 or 4 metres away. The wind seemed to be getting stronger and I was afraid to let go in case the sheet flew off across the yard. Just then a workman came into the shed, and I called him over to pick up the string so that I could lash the hook bolt to the rail.

"Sorry Sir, I can't do that", he said, 'I'm a fitter, you need a rigger'. Fortunately, 5 minutes later another man turned up and helped me secure the sheet. I assumed he must have been a rigger.

Piling problem

I was on vacation with my family on the French Riviera when I received a phone call from the Manager of Operations, who happened to be on holiday in Liguala on the Italian Riviera. The day he got there, he invited us to lunch after which he gave me news that almost ruined my holiday. Apparently, after driving about 100 piles for the combined raw meal storage and blending silos, three of the completed piles started to lift out of the ground in a phenomenon known as pile heave. All further piling work was immediately suspended and after investigating the problem more closely we discovered that the piles, which were spaced in accordance with BS recommendations and had only been reinforced for the top half of their length, had snapped below the reinforcement. This meant that the lower half of the piles which were still in place could not be recovered. All the other piles had therefore to be abandoned because we could not know whether or not they were fractured below the reinforcement level. The whole area covered by the pile cluster was therefore in effect sterilized as it could not be relied upon to support heavy loading. The raw meal silos which we originally designed as a double silo (one on top of the other) had to be completely redesigned as two separate silos so that the whole conveying system which was based on the use of bucket elevators had to be redesigned to a conveying system using belt conveyors. However, with the help of network analysis, the subsequent design-work and site construction effort enabled these silos and conveying system to be completed in time to meet the overall programme milestones.

Silo and chimney construction

Unfortunately, on a large project, it is not always possible to plan that all concreting work is carried out between March and November. In our case, because of the delay caused by the previous piling problems, the revised programme demanded that the raw meal silos and chimney had to be built in February. The construction involved sliding formwork which was a 24-hour operation. For the roads and foundations, all concrete was supplied by mixer trucks, but because there was a risk of the ready-mixed concrete lorries not being able to reach the site regularly due to of adverse (ice or fog) road conditions, it was decided to mix the concrete for the silos and chimney on site, requiring the erection of a mixing plant and stockpiling enough sand and aggregate to build the silos. In addition, a small boiler and heaters had to be procured to produce hot water and stop the aggregate from freezing.

Once the sliding operation started, it could not be stopped until the last lift was completed, and when visiting the site at midnight, I always found it exhilarating to watch heated concrete being poured into these slowly moving shutters under an array of floodlights by a dedicated team of engineers and operatives.

Safety

During the whole construction period, we only had one serious accident, which almost fatally injured an electrician who was not wearing a hard hat. Although hard hats were issued to all operatives and site staff, in 1956, when the plant was being built, wearing a helmet was not obligatory. It was not until the Personal Protective Equipment at Work Regulations came into force in 1992, that the wearing of hard hats became mandatory. All the site staff wore hard hats as an example to the operatives, but if management took disciplinary action against a worker for not wearing a hard hat, the rest of the men would come out on strike.

The accident was caused by a welder pulling on his snaked welding cable while making a welded pipe connection on a platform at the top of an access tower. The kicking angles for the platform were lying on the deck waiting to be welded into position and as the welder pulled the cable, it straightened and pushed one of the kicking angles over the edge hitting an electrician who was fixing a junction box on the base of the tower. The angle iron penetrated his skull and knocked him unconscious. The first aid team was quickly at the scene, but it took about half an hour for the ambulance to arrive and take him to the hospital. I am not a religious man, but that night I prayed and spent two sleepless nights worrying whether the man would survive.

When we investigated the cause of the accident, we found that there were three safety violations, which were all preventable.

1. The kicking angles should have been fitted around the platform perimeter before further work was carried out on the platform.

2. The electrician should not have been working at the base of the tower while work was being carried out on the platform above him.
3. The electrician should have been wearing the hard hat (helmet) issued to him and all other operatives.

By a miracle, the man survived but never returned to site. For the next few days, all the men wore their hard hats, but soon the macho riggers reverted to wearing commando-style knitted caps and only a few of the other trades wore their safety helmets.

As a result of this accident:

We instructed the site safety officer to carry out more frequent inspections to ensure that safety regulations and practices were adhered to.

We issued new hard hats with emphasized recommendations to wear them at all times.

We taped off areas at the bases of towers, structures, etc. when operatives were working above.

We arranged for an ambulance to be stationed permanently on site.

We issued instructions that kicking angles were to be erected on stairs and platforms as the erection work on these sections proceeded.

Security

Because we shared the site with the existing wet process plant which operated 24 hours a day, the security of the site was the responsibility of the client.

One morning I arrived on site to be met by the civil site manager who reported that our complete stock of over 100 corrugated-asbestos cement roofing sheets had been stolen during the night. Apparently, despite the fact that the existing kiln was in operation, the thieves cooly drove right to the stockyard with a big lorry and loaded up what was there.

When I discussed the problem with the client, I suggested we employ a security firm to supply a guard and a dog to keep watch at night. The client agreed to the night watchman but refused permission to use a dog on the grounds that all the dog did was to awaken the guard from his nap. A week later, we had a guard and there were no further thefts, but the stolen sheets were never recovered.

End of project

Over the life of the project, there were numerous extra work orders issued by the client. These were all meticulously recorded, logged and costed but both parties agreed that payment would be deferred until project completion. At the final account meeting attended by the Directors of both companies, I submitted these claims which amounted to a total of £90,000, expecting the usual objections and denials. To my surprise and delight, our claim was accepted without argument. However, my elation was short-lived, because after the coffee break, the client tabled a counterclaim for loss of useful buildable site area due to the broken buried piles, which came to exactly £90,000! The Directors then had lunch and the matter was settled.

The end of a project is always a mixture of elation and sadness. Although the work of the project manager is not finished when the formal handover has taken place and the plant is fully operational, there is inevitably a great sense of achievement and satisfaction as the all the required documents are handed over to the client. To celebrate the event, we organized a dinner party attended by all the Tarmac and RPC site staff and Directors. The Tarmac permanent employees were looking forward to returning home, but many of the locally recruited staff were in tears.

Project statistics, South Ferriby cement works

Capacity	1000 tons of Portland cement per day
Contract value	£4M (todays value £80M)
Contract date	July 1965
Site start date	February 1966
Completion date	October 1967
Rotating Kiln	Polysius Lepol
Cement Mill	F.L Smidth
Double rotator	Polysisus

Lessons learnt

Appoint project manager during contract negotiation stage

Open early discussions with subcontractors and major suppliers

Use Critical Path Methods for scheduling and control

Construct concrete site roads before excavations and foundations

Enforce safety procedures

Set up site security early

Ensure pile spacing exceeds minimum spacing recommended by BS

Key to process flow diagram cement plant

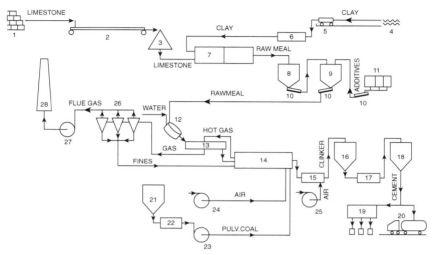

Figure C.1 Process flow diagram.

1. Limestone Quarry
2. Limestone Belt Conveyor
3. Limestone Stockpile
4. Clay Pit
5. Clay Wagons
6. Clay Shredder
7. Double Rotator
8. Raw Meal Silo
9. Blending Silo
10. Feeder
11. Additives Store
12. Nodulizers
13. Grate Pre-Heater
14. Rotary Kiln
15. Clinker Cooler
16. Clinker Silo
17. Cement Mill
18. Cement Silo
19. Bagging Plant
20. Bulk Cement Tankers
21. Coal Bunker

22.   Coal Mill
23.   Coal Blower
24.   F.D. Fan
25.   Cooling Fan
26.   Multi-Cyclones or E.S. Precipitator
27.   I.D. Fan
28.   Chimney

Appendix 4: Case study 2: Teesside oil terminal

Introduction

The Teesside oil terminal at Seal Sands near Seaton Carew, Cleveland, was the first receiving facility in Britain for North Sea Oil. The engineering design for the receiving, processing, storing and distributing the crude from the Norwegian Ekofisk complex was started in the second half of 1973. The client was a consortium of oil companies which was led by Phillips Petroleum of Bartlesville, Oklahoma, and included Petrofina, Agip, Norsk Hydro, Elf, Total, Aquitaine, and three smaller oil companies. The unstabilized crude was brought ashore at Seaton Carew via an undersea Norpipe pipeline from the Ekofisk North Sea oil platform 220 miles off shore. The main process unit, including the Natural Gas Liquid (NGL) plant was located at Seal Sands, which together with the tank farm at Greatham, 2 miles to the north, was at the time, the largest petrochemical construction project in Europe.

Contract

The main contractor responsible for project management, design, procurement and construction was the Sim Chem division of Simon Engineering, who established a bespoke project management, design and procurement office in London for the total delivery of the project. However, the NGL plant was designed at Sim Chem's offices in Cheadle Heath Cheshire and the design of initial civil works and jetties was placed by the client with independent consultants.

The overall plant lay-out and design of the roads, tanks, control buildings, steel structures, columns, piping, electrical network and computerized instrumentation were carried out by the different specialist engineering sections of the (task force) project team in London. Also, part of the team was the client's project manager who was party to all major decisions and approved every major contract document and purchase order.

The plant was designed to be commissioned in three phases.

Phase 1, the least complex but most important phase, was the completion of the receiving spheres, tank farm, four desalting and stabilization units, metering station, deballasting facility, effluent treatment plant and six tanker jetties. This was scheduled to be completed in the summer of 1975 to coincide with the completion of the undersea pipeline from the Ekofisk field and was officially opened by representatives of the British and Norwegian government exactly on schedule, despite the 3-day week crisis in 1974, which adversely affected some of the equipment and materials suppliers.

Phase 2, the NGL plant at Seal sands was completed in August 1979. and

Phase 3, the commissioning of the last three stabilizing units, was completed in mid-1980.

Process

Unstabilized crude was pumped from the Ekofisk field via a 34-inch pipeline to Seaton Carew and from there, via an underwater pipe and cable bundle under the creek to the reception and storage spheres at Seal Sands. After stabilization, the crude was pumped back to the tank farm at Greatham where it was stored in ten 75,000 barrel floating roof storage tanks. From these tanks, the stabilized crude was pumped back under the creek to be metered and loaded into the crude oil tankers berthed at the jetties at Seal Sands. Because, unusually for the United Kingdom, the tankers arrived empty and under ballast, they first had to be de-ballasted. The ballast water was then passed through the effluent treatment plant before it was discharged into the sea.

The gas from the stabilizers was piped to the NGL plant for the production of ethane, propane, butane, isobutane and methane fuel gas. However, until the NGL plant, which was part of Phase 2, was able to process the gas from the stabilizers, it was flared.

Site layout

The whole of the Seal Sands site is built on sand dredged from the shore to form the deepwater harbour for four 150,000 tons crude oil tankers and four smaller NGL tankers. The dredged sand was pumped inshore and consolidated mechanically by massive concrete blocks dropped from cranes. Apart from the reception spheres, and the ethane, propane, butane and isobutane tanks which were piled, all the buildings, stabilizers, fractionation columns and other tanks on Seal Sands had reinforced concrete footings, as were the crude oil storage tanks at the Greatham tank farm. This storage facility was separated from the process unit by a navigable creek which had to be crossed by the pipes and cables between the two areas. This crossing was achieved by constructing a 417 m long pipe and cable bundle in a dry dock on the south bank of the creek protected by a coffer dam. After each of the nine 48 inches diameter concrete coated pipes in the bundle had been pressure tested, the coffer dam was removed and the 4800 tons bundle was floated out and sunk to the bottom of a previously dredged trench crossing the creek.

The process unit on Seal Sands contained the reception and stabilization units, NGL plant and tanks, boiler house, control buildings, fire station, offices, stores, workshops, raw and potable water tanks and flares.

The area on the east side of the facility comprised the dredged harbour, six jetties for the crude and NGL tankers, the metering station, deballasting tanks and the effluent treatment plant.

For project control purposes, the site facility was divided into three main areas.

Area 1 comprised the reception and storage spheres, under-creek pipe and cable crossing, the admin and control buildings, fire station, metering station, and deballasting facility, effluent treatment plant and ground flare.

Area 2 comprised the stabilization and desalting plant, which was part of phase 1, and the NGL plant, tankage on Seal Sands, flare stack, gas turbines, propane compressors, and boiler house which were phase 2 of the project.

Area 3 comprised the stabilized crude tanks and crude pumps at Greatham.

As all these areas were of course connected by roads, sewers, pipes and cables, contract boundaries had to be drawn between them and agreed by all the contractors.

Project structure

Project management, design, procurement services and site supervision were covered by a fixed fee calculated on the estimated total number of engineering and procurement hours at agreed individual rates. Pre-determined percentages of the fixed fee were paid by the client on the completion of mutually agreed design, procurement and construction milestones. The size and scope of the project dictated the use of a task force type of project organization headed by a project director and three project managers, each with an assistant (deputy) project manager.

Project manager A was responsible for the design and construction stages of Areas 1 and 3.

Project manager B was responsible for the design and construction stages of Area 2.

Project manager C was responsible for establishing and maintaining the project schedule, writing the terms and conditions of the subcontracts, as well as being accountable for the letting, monitoring and delivering of all subcontracts.

After most of the equipment orders and major subcontracts had been placed, project managers A and B left the project and the focus of responsibility was changed from area completion to Phase completion. As a result, project manager C was appointed to ensure the timely delivery of Phase 1 and the deputy of project manager B became responsible for the completion of Phases 2 and 3.

To meet the client's break-down of costs, the site had to be divided into nine investment units reflecting the main areas of activity. The nine investment units which each had its own cost code, were as follows:

1. Unstabilized crude reception and storage spheres at Seal Sands
2. Desalting and stabilization plant at Seal Sands
3. Stabilized crude storage (tank farm) and crude pumps at Greatham
4. Crude loading terminal and jetties and ballast tanks at Seal Sands
5. NGL processing plant at Seal Sands
6. Boiler house including the gas turbines at Seal Sands
7. NGL storage facilities including refrigerated tankage at Seal Sands
8. Terminal bunkering facilities at Seal Sands
9. NGL loading terminal and jetties at Seal Sands

This meant that the cost of all equipment, materials and site work had to be recorded against one of the nine cost codes, depending on which area the work took place. This required the setting up of a highly sophisticated cost coding, logging, monitoring and reporting procedure.

In addition, because three members of the consortium had a special interest in particular parts of the facility, certain specified drawings and documents, including the monthly cost and progress reports, had to be sent to the three companies as well as Phillips Petroleum, the lead operator.

Project management plan (co-ordination procedure)

One of the first functions of the project managers was to write and distribute the project co-ordination procedure or project management plan. This document contained or referenced all the systems and procedures to be used on the project, including the site description, programme, list of specifications, applicable standards, senior personnel, responsibility chart, distribution schedule and test procedures. A separate co-ordination manual was written for operations on site. Both documents were amended and re-issued to reflect any major changes to the project team, procedures, specification or contractual obligations

Change control

A project of this size was invariably subject to numerous changes and improvements, most of which were generated by the client. A strict change control procedure was instituted which required every change to be confirmed in writing, both at the design and construction stage. The variation request was then sent to all relevant design sections and site to assess the cost and time implications. This was then returned by the project manager to the originator for agreement or modification, but before it was implemented the final change had to be approved by the project manager and the client.

Project schedule

All scheduling was carried out by the planning section using Activity on Arrow Critical Path Analysis. The client specified a computerized scheduling system which was processed by a large out-sourced computer centre in London. However, although the voluminous print-outs were regularly delivered monthly to the client, it was found that the manual networks (which were the basis of the computer print-outs), were the preferred scheduling and monitoring medium by both the design office and site construction teams. After receiving regular feedback from the design sections and site, the planning section based in London, monitored, updated and distributed the revised networks monthly in accordance with the distribution schedule given in the co-ordination procedure.

There were two basic network programme domains:
1. Programmes covering the mechanical, electrical and instrumentation design stages, the procurement of equipment and materials and the letting of subcontracts.
2. Programmes covering each of the three site construction areas. These were subdivided into civil works, structural and mechanical erection including pipe installation and testing, and electrical and instrumentation installation and testing.

Detailed sub-networks were produced for a number of design functions and selected construction areas which included the crude tanks, stabilization streams, reception spheres, fractionation columns, boiler house and crude pumps.

A master network containing the interfaces of the different construction area networks linked up with the appropriate design networks. This network also included final commissioning. All the major subcontractors used the relevant site networks and augmented them with their own more detailed schedules. Manual bar charts based on the networks were initially used by the design departments and subcontractors for resource allocation, but for re-programming and monitoring purposes the CP networks themselves were found to be the best method. Erection progress was monitored by the area site engineers against set milestones on the site construction networks and was reported monthly to the client and the project manager at head office.

Engineering design

During the design phase, regular progress meetings were held monthly, most of which were attended by the client's resident project manager. It became apparent early in the design phase that the design process was being delayed by contractually required data or drawings from equipment suppliers not being received by the specified time.

To overcome this problem, every purchase order included a table of equipment items normally used on a refinery project, which showed what drawings or data were required by the design departments for incorporation in the layout or assembly drawings. The supplier was therefore made aware exactly what information had to be submitted. The dates when these data were required were stated in the purchase order and their delivery was vigorously expedited.

At a certain point of the design stage, a major review was called by the client, at which every layout drawing was checked by specialist engineers flown in from the client's operations worldwide, whose task was to ensure that the design met all the specified design and safety standards, operating procedures and spares provisions. Particular attention was paid to operability and unrestricted safe access to valves, instruments and routine maintenance points. This resulted in a smooth and relatively trouble free commissioning stage.

The design of the NGL plant was carried out by a process engineering team in Simon Engineering's head office in Cheadle Heath Cheshire, which at one stage was augmented by 50 engineers and technicians flown in from their Indian office. The design associated with the jetties, ground consolidation, vibro-compaction, piles, and foundations was carried out by civil engineering consultants employed directly by the client.

Procurement

The procurement department was responsible for preparing and issuing the tender documents for equipment, materials and subcontracts, which, because of the size of the project, were reimbursable. Equipment and materials had to comply with the client's and Sim Chem's specifications and were then selected on price, delivery, sustainability and vendor track record. The client was then invoiced for the fully discounted costs including delivery, and, where applicable, setting to work and commissioning.

With the exception of domestic site contracts, all subcontracts were let by head office in London under the control of project manager C and were based on one specially written common comprehensive set of General Conditions of Contract. Where considered necessary, a set of Special Conditions of Contract were added to reflect the discipline or special requirements of the subcontractor. This procedure covered every subcontract such as civil engineering work, steelwork erection, mechanical installation, boilers, tankage, electrical works, instrumentation, non-destructive testing, insulation and painting. The number of subcontracts let using this one set of General Conditions of Contract was over 120 with a combined contract value of about £150M (£850M present value).

The boiler house, reception spheres, crude oil and NGL tanks and a number of specialist items were let as lump-sum contracts. All other site construction work was measured monthly by a firm of quantity surveyors, using an agreed schedule of rates prepared by the firm and based on unit rates generally used in the petrochemical industry. This included the civil, mechanical, electrical and instrumentation and insulation work.

Inspection and expediting

As well as witnessing all the intermediate and final performance tests of all mechanical and electrical items, an inspection of the equipment was carried out at the vendor's premises at predetermined stages of the manufacturing process.

While the inspectors of equipment, which were part of the procurement department, were located at the two head offices, expediters were based at strategic locations around the country and visited the equipment suppliers weekly to ensure that the required information, as well as the actual equipment or material, was delivered on time. This vigorous expediting gave an early warning to the project manager. On some occasions, when a potential late delivery threatened the end date, visits by the project manager to top management of the offending supplier invariably had the desired effect.

Site organization

Each of the three site areas had their own site office run by a construction superintendent responsible to the construction manager, who reported to each of the project managers. Each of the three area offices was staffed to administer and control its own civil, mechanical, piping, electrical and instrumentation subcontractor, but there were only two insulation subcontractors for the whole site, one for piping and one for the cold tanks, that is, ethane, butane and isobutane. There was one painting subcontractor.

The decision to appoint a separate civil, mechanical and electrical subcontractor for each site area was made primarily in order not to overload the resources of any one company. In the event, when halfway through the project, the contract of the mechanical subcontractor of one area was terminated for not meeting the agreed milestone dates, it was relatively easy for the mechanical subcontractor of one of the other areas to take over the work, being already familiar with all the contractual site conditions, procedures and technical specifications.

A chief field engineer, reporting to the construction manager, was responsible for quality control and planning and the client's resident construction manager on site ensured compliance with Phillips Petroleum's procedures and standards.

All major contractors in each site area supplied their own messing facilities for their own men and those of their own subcontractors. A bespoke site agreement, which set out the role and code of conduct of the subcontractors, also included the standards for the site welfare facilities, safety and security as well as a clause that forbade contractors poaching each others workers.

It became apparent well before the site work was due to start, that there were insufficient welders in the north of England qualified to petrochemical plant welding quality standards. To overcome this problem, a welding school was established at which welders from the closing shipyards could be trained to produce welds to the required quality standards. However, most of these welders belonged to the Union of Shipbuilders, Blacksmiths and Boilermakers, whilst traditionally all the pipe welders on petrochemical plants were members of the Construction Engineering Union (CEU), who now felt their jobs threatened. This inter-union dispute was resolved before work started on site by a mutual agreement, whereby the boilermakers welded the joints of all the electric resistance welded (ERW) pipes whilst the CEU members welded the joints of the seamless pipes. Pipe welding was carried out during normal working hours to allow all joints to be radiographed during a night shift by an independent NDT company.

Another early decision was to construct the roads and sewers immediately after the soil consolidation was completed, that is, before any foundations or piling work. Constructing the roads to base course level, allowed the construction traffic including crawler cranes to run 'in the dry', so that subsequent excavation and foundation work was not delayed by equipment bogged down due to bad weather. At the end of the construction period, the road base course was repaired and levelled to receive the top coat.

Handover documents

As the site work proceeded, the procurement department collected and collated all the many operating instruction and maintenance manuals, equipment guarantees, vendor's drawings, parts lists, priced spares recommendations, test certificates and lubrication schedules. These were indexed and bound in hard covers for storage and use in the operation control building.

As-built drawings of the whole facility were updated and handed over after take-over by the client.

Lessons learnt

Carry out a design review by operating staff and process specialists.

Use critical path methods for scheduling

Establish and operate strict change control procedures.

Carry out vigorous expediting of documentation as well as the hardware from suppliers.

Combine regular stage inspection with expediting of delivery.

Include and impose liquidated damages on late delivery of documentation as well as the hardware.

Ensure test certificates are available for every stage of manufacture from steel mill to performance test.

Draw up a site agreement with trade unions before the start date.

Hold back construction work until all drawings and materials are available.

Build roads to base course level early to keep plant moving in bad weather.

Do not skimp on the size of the soil survey grid.

Minimize the number of site let subcontracts.

Establish good relations with public bodies and neighbouring companies as early as possible.

Key to process block flow process diagram

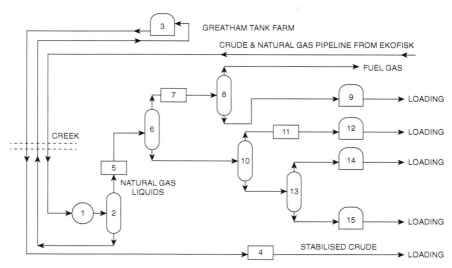

PROCESS BLOCK FLOW DIAGRAM

1.	High Pressure Reception of Pipeline Crude and Natural Gas	Seal Sands	
2.	Crude Stabilization	Seal Sands	
3.	Stabilized Crude Storage	Greatham Tank Farm	
4.	Crude Meters and Loading	Seal Sands	
5.	Compression and Dehydration	Seal Sands NGL Plant	
6.	Deethanizer	"	"
7.	Treating and Drying	"	"
8.	Demethanizer	"	"
9.	Ethane Storage and Loading	"	"
10.	Depropanizer	"	"
11.	Treating and Drying	"	"
12.	Propane Storage and Loading	"	"
13.	Butane Splitter	"	"
14.	ISO Butane Storage and Loading	"	"
15.	Butane Storage and Loading	"	"

Facility Statistics

The following statistics of the terminal give some idea of the scale of this project.

Contract Value: £320M (equivalent present day value £2.3 Billion)

Designed throughput of stabilised crude oil: 1M barrels per day

Throughput of NGL products, i.e. ethane, propane, butane, isobutane: 67,000 barrels/day.

Total site area including tank farm; 676 acres.

The complex required:

1,500,000 engineering and procurement manhours.

20,000,000 construction manhours

12,000 purchase orders

At peak, the construction force numbered over 5,000 operatives.

The combined engineering departments of Sim Chem produced over 40,000 drawings and over 7,000 piping isometrics.

The facility required:

Over 100,000 cubic metres of concrete,

7000 piles,

10 miles of roads,

100,000 tonnes of steel,

300 miles of piping,

20,000 values,

1000 miles of cables,

80,000 panel & field mounted Instruments, and

3 state-of the-art computer control centres.

The complex included:

4 unstabilised crude reception spheres, each of 45,000 barrels capacity

10 floating roof crude storage tanks, each with a storage capacity of 750,000 barrels and a roof area the size of an American football field

7 stabiliser & de-salter trains

Effluent treatment plant with 4 ballast tanks.

6 jetties for crude and NGL tankers and export metering station.

3 gas turbine driven propane compressors

3,113 Tonnes/ hour steam boilers

Appendix 5: Abbreviations and acronyms

Abbreviation	Meaning	Usage
ACC	Annual Capital Charge	Finance
ACWP	Actual Cost of Work Performed	EVA
ADR		Construction
AI	Artificial Intelligence	General and AI
ANB	Adjudicator Nominating Body	Construction
AoA	Activity on Arrow	CPA
AoN	Activity on Node	CPA
APM	Association for Project Management	PM
ARM	Availability, Reliability, Maintainability	MOD
BC	Business Case	PM
BCWP	Budgeted Cost of Work Performed	EVA
BCWS	Budgeted Cost of Work Scheduled	EVA
BoK	Body of Knowledge	PM
BOQ	Bill of Quantities	Construct
BrIM	Bridge Information Management	VDC
BS	British Standard	General
BSI	British Standards Institution	General
CAD	Computer-aided Design	General
CAM	Computer-aided Manufacture	General
CAR	Contractor's All Risk	Construct
CBS	Cost Breakdown Structure	PM
CDE	Common Data Environment	AI
CDM	Construction, Design and Management	Construction
CEO	Chief Executive Officer	Business
CEN	Comité Européen de Normalisation	General
CIF	Carriage, Insurance, Freight	Procurement
CIM	Civil Information Management	VDC
CM	Configuration Management	PM
COBie	Construction Operation Building Information Exchange	
CPA	Critical Path Analysis	PM
CPA	Contract Price Adjustment	Procurement
CPD	Continuing Professional Development	General
CPI	Cost Performance Index	EVA
CPM	Critical Path Methods	CPA
CSCS	Cost and Schedule Control System	EVA
CSP	Communications Service Provider	AI
CWA	Construction Work Areas	VDC
CV	Cost Variance	EVA
CV	Curriculum Vitae	General
DCF	Discounted Cash Flow	Finance

Continued

—Cont'd

Abbreviation	Meaning	Usage
DDP	Delivery Duty Paid	Procurement
DE	Digital Engineering	AI
DIN	Deutsche Industrie Normen	General
EAC	Estimated cost at Completion	EVA
ECC	Estimated Cost to Complete	EVA
ECHR	European Convention of Human Rights	Governance and Law
EU	European Union	General and Governance
EV	Earned Value	EVA
EVA	Earned Value Analysis	PM
EVMS	Earned Value Management System	EVA
FCC	Forecast Cost to Complete	EVA
FF	Free Float	CPA
FLAC	Four Letter Acronym	General
FMEA	Failure Mode and Effect Analysis	MOD
FOB	Free on Board	Procurement
FOR	Free on Rail	Procurement
FRC	Financial Recording Council	AI
GCNN	Geometric Convolutional Neural Networks	AI
GML	Graphic Machine Learning	AI
GPU	Graphics Processing Unit	AI
HAZOP	Hazard and operability	General
HPC	High Performance Computing	AI
HR	Human Resources	General
H&S	Health and Safety	General
HSE	Health and Safety Executive	General
IA	Investment Appraisal	Finance
IDC	Intelligent Design and Construction	BIM
IFC	Industry Foundation Class	AI
IoT	Internet of Things	General and AI
IP	Intellectual Property	General
IPMA	International Project Management Association	PM
IPMT	Integrated Project Management Team	PM
IPR	Intellectual Property Rights	General
IRR	Internal Rate of Return	Finance
IS	Information Systems	General
ISEB	Information Systems Examination Board	General
ISO	International Organization for Standardization	General
IT	Information Technology	General
ITT	Invitation to tender	Procurement
ITU	Intensive Care Unit	Medicine
JIT	Just in time	General

—Cont'd

Abbreviation	Meaning	Usage
IWP	Information Work Pack	VDC
KPI	Key Performance Indicator	PM
LAD	Liquidated Ascertainable Damages	Construct
LCC	Life Cycle Costing	PM
LD	Liquidated Damages	Construct
LOB	Line of Balance	Construct
LOD	Level of Design	VDC
LRM	Liner Responsibility Matrix	PM
LSX	London Sustainability Exchange	Sustainability and AI
MDP	Markov Decision Process	AI
MOD	Ministry of Defence	General
MRI	Magnetic Resonance Imaging	Medicine
MTO	Material take-off	Construct
NDT	Non-destructive Testing	Construct
NII	National Institute of Information	AI
NLP	Natural Language Processing	AI
NOSCOS	Needs, Objectives, Strategy and Organizations Control System	MOD
NPV	Net Present Value	Finance
OBS	Organization Breakdown Structure	PM
OD	Original duration	EVA
OECD	Organisation for Economic Cooperation and Development	Sustainability
OGC	Office of Government Commerce	General
ORC	Optimal Replacement Chart	Finance
ORM	Optimal Replacement Method	Finance
P3	Projects, Programmes and Portfolios	PM
PBS	Product Breakdown Structure	PM
PDM	Precedence Diagram Method	CPA
PEP	Project Execution Plan	PM
PID	Project Initiation Document	PM
PERT	Program Evaluation and Review Technique	CPA
PFI	Private Finance Initiative	Finance
PIM	Project Information Management	BIM and VDC
POC	Proof of Concept	VDC
PM	Project Management	PM
PM	Project manager	PM
PMI	Project Management Institute	PM
PMP	Project Management Plan	PM
PPE	Post-project Evaluation	PM
PPP	Public—Private Partnership	Finance
PRD	Project Definition	PM
QA	Quality Assurance	General
QC	Quality Control	General

Continued

—Cont'd

Abbreviation	Meaning	Usage
QMS	Quality Management System	General
QP	Quality lan	General
R&D	Research and Development	General
RFQ	Request for Quotation	Procurement
ROC	Receiver Operating Characteristics	AI
RR	Rate of Return	Finance
SFR	Sinking Fund Return	Finance
SMART	Specific, Measurable, Achievable, Realistic, Time-bound	MOD
SOR	Schedule of Rates	Construct
SOW	Statement of Work	PM
SPI	Schedule Performance Index	EVA
SRD	Sponsor's Requirement Definition	PM
SV	Schedule Variance	EVA
TCP	Time, Cost and Performance	PM
TF	Total Float	CPA
TQM	Total Quality Management	General
TOR	Terms of Reference	General
UCD	University of Dublin	General
UCL	University College London	General and PM
VA	Value Analysis	General
VCD	Virtual Design and Construction	BIM
VE	Value engineering	General
VM	Value Management	General
WBS	Work Breakdown Structure	PM

Acronyms used in project management

AIM	Asset Information Management
ARM	Availability, Reliability, Maintainability
BIM	Building Information Modelling
CAD/CAM	Computer-aided Design/Computer-aided Manufacture
CADMID	Concept, Assessment, Demonstration, Monitoring, In-service, Disposal
CFIOT	Concept, Feasibility, In-service, Operation, Termination
CS^2 (CSCS)	Cost and Schedule Control System

—Cont'd

AIM	Asset Information Management
EMAC	Engineering Man-hours and Cost
FLAC	Four Letter Acronym
HASAWA	Health and Safety at Work Act
IPMA	International Project Management Association
NAPNOC	No Agreed Price, No Contract
NEDO	National Economic Development Office
NIMBY	Not In My Backyard
NOSCOS	Needs, Objectives, Strategy and Organization Control System
NOSOCS&R	Needs, Objectives, Strategy, Organization Control, System and Risk
PAYE	Pay As You Earn
PERT	Program Evaluation and Review Technique
PESTLE	Political, Economic, Sociological, Technological, Legal, Environmental
PRAM	Project Risk Analysis and Management
PRINCE	Projects In a Controlled Environment
RAMP	Risk Analysis and Management for Projects
RIDDOR	Reporting of Injuries, Diseases and Dangerous Occurrences Regulations
RIRO	Rubbish In — Rubbish Out
SAPETICO	Safety, Performance, Time, Cost
SMAC	Site Man-hours and Cost
SMART	Specific, Measurable, Achievable, Realistic and Time-bound
SOW	Statement of Work
SWOT	Strengths, Weaknesses, Opportunities and Threats

Appendix 6: Glossary

Activity An operation on a network which takes time (or other resources) and is indicated by an arrow.

Actual cost of work performed (ACWP) Cumulative actual cost (in money or man-hours) of work booked in a specific period.

Actual hours The man-hours actually expended on an activity or contract over a defined period.

Adjudication Procedure for resolving a dispute by appointing an independent adjudicator.

Advance payment bond Bond given in return for advanced payment by client.

Analytical estimates Accurate estimate based on build-up of all material and labour requirements of the project.

AoA Activity on arrow.

AoN Activity on node.

Arbitration Dispute resolution by asking an arbitrator to make a decision.

Arithmetical analysis A method for calculating floats arithmetically.

Arrow A symbol on a network to represent an activity or dummy.

Arrow diagram A diagram showing the interrelationships of activities.

Back end The fabrication, construction and commissioning stage of a project.

Backward pass A process for subtracting durations from previous events, working backwards from the last event.

Banding The subdivision of a network into horizontal and vertical sections or bands to aid the identification of activities and responsibilities.

Bar chart See Gantt chart.

Belbin type One of nine characteristics of a project team member as identified by Meredith Belbin's research programme.

Beta (β) distribution Standard distribution giving the expected time te = (a + 4m + b)/6.

Bid bond Bond required with quotation to discourage withdrawal of bid.

Bond Guarantee given (for a premium) by a bank or building society as a surety.

Budget Quantified resources to achieve an objective, task or project by a set time.

Budget hours The hours allocated to an activity or contract at the estimate or proposal stage.

Budgeted cost of work performed (BCWP) See Earned value.

Budgeted cost of work scheduled (BCWS) Quantified cost (in money or man-hours) of work scheduled (planned) in a set time.

Business case The document setting out the information and financial plan to enable decision-makers to approve and authorize the project.

Calendar Time scale of programme using dates.

Capital cost The project cost as shown in the balance sheet.

Cash flow Inward and outward movement of money of a contract or company.

Change control The process of recording, evaluating and authorizing project changes.

Change management The management of project variations (changes) in time, cost and scope.

Circle and link method See Precedence diagram.

Close-out procedure The actions implemented and documents produced at the end of a project.

Close-out report The report prepared by the project manager after project close-out.

Comparative estimates Estimates based on similar past project costs.

Computer analysis The method for calculating floats, etc., using a computer.

Conciliation The first stage of dispute resolution using a conciliator to improve communications and understanding.

Configuration management The management of the creation, maintenance and distribution of documents and standards.

Conflict management Management of disputes and disagreements using a number of accepted procedures.

Contingency plan Alternative action plan to be implemented when a perceived risk materializes.

Cost breakdown structure (CBS) The hierarchical breakdown of costs when allocated to the work packages of a WBS.

Cost code Identity code given to a work element for cost control purposes.

Cost control The ability to monitor, compare and adjust expenditures and costs at regular and sufficiently frequent intervals to keep the costs within budget.

Cost performance index The ratio of the earned value (useful) cost and the actual cost.

Cost reporting The act of recording and reporting commitments and costs on a regular basis.

Cost variance The arithmetical difference between the earned value cost and the actual cost. This could be positive or negative.

Cost/benefit analysis Analysis of the relationship between the cost and anticipated benefit of a task or project.

Counter-trade Payment made of goods or services with materials or products that can be sold to pay for the items supplied.

CPA Critical path analysis. The technique for finding the critical path and hence the minimum project duration.

CPM Critical path method. See CPA

CPS Critical path scheduling. See CPA.

Critical activity An activity on the critical path which has zero float.

Critical path A chain of critical activities, i.e., the longest path of a project.

Dangle An activity that has a beginning node but is not connected at its end to a node that is part of the network.

Deliverable The end product of a project or defined stage.

Dependency The restriction on an activity by one or more preceding activities.

Direct cost The measurable cost directly attributed to the project.

Discounted cash flow (DCF) Technique for comparing future cash flows by discounting by a specific rate.

Distribution schedule A tabular record showing by whom and to whom the documents of a project are distributed.

Dummy activity A timeless activity used as a logical link or restraint between real activities in a network.

Duration The time taken by an activity.

Earliest finish The earliest time at which an activity can be finished.

Earliest start The earliest time at which an activity can be started.

Earned value hours See Value hours.

End event The last event of a project.

Estimating Assessment of costs of a project.

EVA Earned value analysis.

Event The beginning and end node of an activity, forming the intersection point with other activities.

Expediting Action taken to ensure ordered goods are delivered on time. Also known a progress chasing.

Feasibility study Analysis of one or more courses of action to establish their feasibility or viability.

Feedback The flow of information to a planner for updating the network.

Float The period by which a non-critical activity can be delayed.

Forward pass A process for adding durations to previous event times starting at the beginning of a project.

Free float The time by which an activity can be delayed without affecting a following activity.

Front end The design and procurement stage of a project. This may or may not include the manufacturing period of equipment.

Functional organization Management structure of specialist groups carrying out specific functions or services.

Gantt chart A programming technique in which activities are represented by bars drawn to a time scale and against a time base.

Graphical analysis A method for calculating the critical path and floats using a linked bar chart technique.

Graphics Computer-generated diagrams.

Grid Lines drawn on a network sheet to act as coordinates of the nodes.

Hammock An activity covering a number of activities between its starting and end node.

Hardware The name given to a computer and its accessories.

Herzberg's theory The hygiene factors and motivators that drive human beings.

Histogram A series of vertical columns whose height is proportional to a particular resource or number of resources in any time period.

Incoterms International trade terms for shipping and insurance of freight.

Independent float The difference between free float and the slack of a beginning event.

Indirect cost Cost attributable to a project, but not directly related to an activity or group within the project.

Input The information and data fed into a computer.

Interface The meeting point of two or more networks or strings.

Interfering float The difference between the total float and the free float. Also the slack of the end event.

Internal rate of return (IRR) The discount rate at which the net present value is zero.

Investment appraisal Procedure for analysing the viability of an investment.

Key performance indicators (KPI) Major criteria against which the project performance is measured.

Ladder A string of activities which repeat themselves in a number of stages.

Lag The delay period between the end of one activity and the start of another.

Latest finish The latest time at which an activity can be finished without affecting subsequent activities.

Latest start The latest time at which an activity can be started without delaying the project.

Lead The time between the start of one activity and the start of another.

Leadership The ability to inspire and motivate others to follow a course of action.

Lester diagram Network diagram that combines the advantages of arrow and precedence diagrams.

Letter of intent Document expressing intention by client to place an order.

Line of balance Planning technique used for repetitive projects, subprojects or operations.

Litigation Act of taking a dispute to a court of law for a hearing before a judge.

Logic The realistic interrelationship of the activities on a network.

Logic links The link line connecting the activities of a precedence diagram.

Loop A cycle of activities that returns to its origin.

Manual analysis The method for calculating floats and the critical path without the use of a computer.

Maslow's hierarchy of needs The five stages of needs of an individual.

Master network Coordinating network of subnetworks.

Matrix The table of activities, durations and floats used in arithmetical analysis.

Matrix organization Management structure where functional departments allocate selected resources to a project.

Mediation Attempt to settle a dispute by joint discussions with a mediator.

Menu Screen listing of software functions.

Method statement Narrative or graphical description of the methods envisaged to construct or carry out selected operations.

Milestone Key event in a project which takes zero time.

Milestone slip chart Graph showing and predicting the slippage of milestones over the project period.

Negative float The time by which an activity is late in relation to its required time for meeting the programme.

Negotiation Attempt to reach a result by discussion which is acceptable to all sides.

Net present value (NPV) Aggregate of discounted future cash flows.

Network A diagram showing the logical interrelationships of activities.

Network analysis The method used for calculating the floats and critical path of a network.

Network logic The interrelationship of activities of a planning network.

Node The intersection point of activities. An event.

Organization breakdown structure (OBS) Diagrammatic representation of the hierarchical breakdown of management levels for a project.

Organogram Family tree of an organization showing levels of management.

Output The information and data produced by a computer.

P3 Primavera Project Planner.

Parametric estimates Estimates based on empirical formulae or ratios from historical data.

Pareto's law Doctrine which shows that approx. 20% of causes create 80% of problems. Also known as 80/20 rule.

Path The unbroken sequence of activities of a network.

Performance bond Bond that can be called by client if contractor fails to perform.

PERT Program evaluation and review technique. Another name for CPA.

PESTEL Political, economic, sociological, technical, legal, environmental.

Phase A division of the project life cycle.

Planned cost The estimated (anticipated) cost of a project.

Portfolio management Management of a group of projects not necessarily related.

Post project review History and analysis of successes and failures of project.

Precedence network A method of network programming in which the activities are written in the node boxes and connected by lines to show their interrelationship.

Preceding event The beginning event of an activity.

Printout See Output.

Procurement Operation covering tender preparation, bidder selection, purchasing, expediting, inspection, shipping and storage of goods.

Product breakdown structure (PBS) Hierarchical decomposition of a project into various levels of products.

Program The set of instructions given to a computer.

Programme A group of related projects.

Programme management Management of a group of related projects.

Programme manager Manager of a group of related projects.

Progress report A report that shows the time and cost status of a project, giving explanations for any deviations from the programme or cost plan.

Project A unique set of co-ordinated and controlled activities to introduce change within defined time, cost, and quality/performance parameters.

Project close-out The shutting down of project operations after completion.

Project context See Project environment.

Project environment The internal and external influences of a project.

Project life cycle All the processes and phases between the conception and termination of a project.

Project management The planning, monitoring and controlling of all aspects of a project.

Project management plan (PMP) A document which summarizes of all the main features encapsulating the why, what, when, how, where and who of a project.

Project manager The individual who has the authority, responsibility and accountability to achieve the project objectives.

Project organization Organization structure in which the project manager has full authority and responsibility of the project team.

Project task force See Task force.

Quality assurance Systematic actions required to provide confidence of quality being met.

Quality audit Periodic check that quality procedures have been carried out.

Quality control Actions to control and measure the quality requirements.

Quality management The management of all aspects of quality criteria, control, documentation and assurance.

Quality manual Document containing all the procedures and quality requirements.

Quality plan A plan that sets out the quality standards and criteria of the various tasks of a project.

Quality policy Quality intentions and directions set out by top management.

Quality programme Project-specific document that defines the requirements and procedures for the various stages.

Quality review Periodic review of standards and procedures to ensure applicability.

Quality systems Procedures and processes and resources required to implement quality management.

Random numbering The numbering method used to identify events (or nodes) in which the numbers follow no set sequence.

Requirements management Capture and collation of the client's or stakeholders' perceived requirements.

Resource The physical means necessary to carry out an activity.

Resource levelling See Resource smoothing.

Resource smoothing The act of spreading the resources over a project to use the minimum resources at any one time and yet not delay the project.

Responsibility code Computer coding for sorting data by department.

Responsibility matrix A tabular presentation showing who or which department is responsible for set work items or packages.

Retention bond Bond given in return for early payment of retention monies.

Retentions Moneys held by employer for period of maintenance (guarantee) period.

Return on capital employed Profit (before interest and tax) divided by the capital employed given as a percentage.

Return on investment (ROI) Average return over a specified period divided by the investment given as a percentage.

Risk The combination of the consequences and likelihood of occurrence of an adverse event or threat.

Risk analysis The systematic procedures used to determine the consequences or assess the likelihood of occurrence of an adverse event or threat.

Risk identification Process for finding and determining what could pose a risk.

Risk management Structured application of policies, procedures and practices for evaluating, monitoring and mitigating risks.

Risk management plan Document setting out strategic requirements for risk assessment and procedures.

Risk register Table showing the all identified risks, their owners, degree of P/I and mitigation strategy.

Schedule See Programme.

Schedule performance index The ratio of earned value cost (or time) and the planned cost (or time).

Schedule variance The arithmetical difference between the earned value cost (or time) and the planned cost (or time).

Sequential numbering The numbering method in which the numbers follow a pattern to assist in identifying the activities.

Situational leadership Adaptation of management style to suit the actual situation the leader finds him/herself in.

Slack The period between the earliest and latest times of an event.

Slip chart See Milestone slip chart.

SMAC Site man-hour and cost. The name of the computer program developed by Foster Wheeler Power Products Limited for controlling man-hours in the field.

Software The programs used by a computer.

Sponsor The individual or body who has primary responsibility for the project and is the primary risk taker.

Stakeholder Person or organization who has a vested interest in the project. This interest can be positive or negative.

Start event The first event of a project or activity.

Statement of work (SOW) Description of a work package that defines the project performance criteria and resources.

Sub-network A small network that shows a part of the activities of a main network in greater detail.

Subcontract Contract between a main contractor and specialist subsidiary contractor (subcontractor).

Subjective estimates Approximate estimates based on 'feel' or 'hunch'.

Succeeding event The end event of an activity.

Task The smallest work unit shown on a network programme (see also Activity).

Task data The attributes of a task such as duration, start and end date, and resource requirement.

Task force Project organization consisting of a project team that includes all the disciplines and support services under the direction of a project manager.

Teamwork The act of working harmoniously together in a team to produce a desired result.

Time estimate The time or duration of an activity.

Toolbar The list of function icons on a computer screen.

Topological numbering A numbering system where the beginning event of an activity must always have a higher number than the events of any activity preceding it.

Total float The spare time between the earliest and latest times of an activity.

Total quality management (TQM) Company-wide approach to quality beyond prescriptive requirements.

Updating The process of changing a network or programme to take into account progress and logic variations.

Value engineering The systems used to ensure the functional requirements of value management are met.

Value hours The useful work hours spent on an activity. This figure is the product of the budget hours and the percentage complete of an activity or the whole contract.

Value management Structured means aimed at maximizing the performance of an organization.

Variance Amount by which a parameter varies from its specified value.

Weightings The percentage of an activity in terms of man-hours or cost of an activity in relation to the contract as a whole, based on the budget values.

Work breakdown structure (WBS) Hierarchical decomposition of a project into various levels of management and work packages.

Work package Group of activities within a specified level of a work breakdown structure.

Appendix 7: Sample examination; questions 1: Questions

The following pages show 60 typical examination questions that may appear in the APM PMQ examination. The answers are given in Appendix 8. The numbers in brackets refer to the relevant chapter numbers in the book.

1. List 12 items (subjects) which should be set out in a PMP. (14)
2. Explain the purpose and structure of a WBS. (12)
3. Describe the most usual risk identification techniques. (15)
4. Explain the risk management procedure. (15)
5. Set out the risks associated with travelling from Bath to London by road. Draw a risk register (log) and populate it with at least four perceived risks. (15)
6. Describe a change management procedure. Draw up two forms relating to change management. (17)
7. Draw a bar chart for the following activities:

Activity	Duration (days)	Preceding activity
A	5	–
B	7	A
C	9	A
D	7	B
E	2	C
F	6	C
G	2	E and D
H	3	F
J	4	G and H

What is the end date?
What is the effect of B slipping by 3 days. (19)

8. Explain the difference between project management and programme management. (2 and 3)
9. Explain the purpose of stakeholder management and describe the difference (with examples of positive and negative stakeholders). (7)
10. Explain what is meant by configuration management. (18)
11. Describe a risk management plan and give its contents. (15)
12. State four risk mitigation strategies excluding contingencies. (15)
13. Explain the main tools used in quality management. (16)
14. Explain the main topics of a quality management system. (16)
15. Describe the purpose of milestones and draw a milestone slip chart showing how slippage is recorded. (25)
16. Explain the advantages of EVA over other forms of progress monitoring. (32)
17. Explain the purpose of a project life cycle and draw a typical life cycle diagram. Explain what is meant by product life cycle and expanded life cycle. (11)

18. Describe what documents are produced at the various stages of a life cycle.   (11)
19. Explain the difference between the three main types of project organization.   (9)
20. Explain what is meant by communication management. Give eight barriers to
         good communication and explain how to overcome them.   (38)
21. Explain the advantages of a project team; list six features and give four barriers to
         team building.   (39)
22. Describe what is meant by conflict management and list five techniques.   (42)
23. Describe the purpose of Belbin test and explain the characteristics of the eight
         Belbin types.   (39)
24. Explain Herzberg's motivation theory and Maslow's theory of needs.   (39)
25. Explain the main constituents of a business case. Who owns the business case.   (5)
26. Explain what are the qualities which make a leader.   (40)
27. Describe the main stages of a negotiation process.   (41)
28. Explain what is meant by cash flow. Draw the format of a cash flow chart.   (31)
29. Describe a close-out procedure. List six documents that must be prepared and
         handed over to the client on close-out.   (45)
30. Describe six topics to be considered as part of a procurement strategy. List four
         types of contract.   (34)
31. Describe the selection process for employing contractors or subcontractors.   (34)
32. List and explain the phases of a project as suggested by Tuckman.   (39)
33. Describe what is meant by the project environment.   (4)
34. List 10 reasons why a project may fail and suggest ways to rectify these
         failures.   (15, 16, 36)
35. Explain the role of a programme manager and show the advantages to the
         organization of such a position.   (3)
36. What is meant by requirement management.   (5)
37. Describe the principal reasons for an investment appraisal and give the
         constituents of such an appraisal.   (6)
38. What are the four main types of estimating techniques and what is their
         approximate degree of accuracy.   (13)
39. What is meant by resource levelling and resource smoothing.   (30)
40. Describe the roles of the client, sponsor, project manager and a supplier.   (10)
41. List at least six documents that have to be handed over at the end of a project. (44)
42. Describe six common generic causes of accidents in industry.   (36)
43. Draw a work breakdown structure for the manufacture of a bicycle. Limit the
         size to four levels of detail.   (12)
44. What is meant by internal rate of return (IRR)? Show how this can be obtained
         graphically.   (6)
45. Describe the functions carried out by a project office.   (2)
46. What is the purpose of a post project review.   (45)
47. Describe three methods of conflict resolution when mediation has failed.   (42)
48. State four main characteristics of a good project manager.   (2)
49. Describe two pros and cons of an AoA network and an AoN network.   (19)
50. Describe two pros and cons of DCF and payback.   (6)
51. What is meant by portfolio management.   (3)
52. What are success criteria.   (37)
53. Describe a change register.   (17)
54. What is meant by information management.   (37)
55. Explain what is meant by value management.   (35)

56. Explain the difference between project management and line management.   (1)
57. What is meant by matrix project management.   (9)
58. What is benefit analysis.   (6)
59. What is meant by project governance.   (43)
60. Give eight topics of an information management plan.   (37)

Appendix 8: Sample examination; questions 2: Bullet point answers

1. **Project-management plan (PMP)**
Reflects the PM's understanding of the project
Should be written by the PM
This enables him/her to fully understand the sponsor's requirements
Must be updated regularly with good configuration management
Is in fact the bible of the project
Covers: Why, What, When, How, Where, Who and How much (Kipling poem)
Not to be confused with a time schedule (plan)
Can be known by other names by different companies
Contents well set out in BS 6079
Contents to include
Reasons for planning, baseline programme, responsibility matrix, document distribution schedule, procedures for monitoring and control, resource allocation, network preparation, estimating, risk management, health and safety, change control, configuration management, etc.

2. **Work breakdown structure (WBS)**
Next stage after conception from project life cycle
Divides the project into manageable packages
Can be product-based or work-based
Shows hierarchy of work packages
Leads to assignment of work packages and resources
Task-oriented family tree
Leads to PBS, CBS, OBS and RBS
Helps in creating a responsibility matrix
Foundation for planning and CP network
WBS called PBS in PRINCE methodology
Gives better definition of work
Acts as check list to find missing stages and areas of work
Good basis for risk identification process
Does **not** show interdependencies
Team members have good picture of main stages
Top down and bottom up estimating leads to network, bar chart, histogram S curve

3. **Risk identification techniques**
Assumption analysis
    Specific, quick
    Subjective, related to assumptions, restricted
Brainstorming
    Cost-effective, wide ranging, involves participants, assists communication and understanding
    Throws up wild ideas, requires good facilitation, risks may be exaggerated, group must be *restricted, participants must be chosen*
Check lists or prompt lists
    Past experience tapped, good coverage, focused on right areas, rapid response
    Restricted to items on list, lacks originality, could miss new problems

Delphi technique

>   Can be conducted by phone, post, e-mail or fax, taps expert opinion, possibility of agreement

>   Requires the relevant experts, takes time, requires good coordination

Document reviews

>   Simple to do, authoritative, very useful if document is a close-out report

>   Could take a long time, identification of risks may be difficult

Interviews

>   Confidential, specific, enables risks to be analysed, allows discussion

>   Requires good interviewing technique, time-consuming, requires subsequent comparison of results

Questionnaires

>   Structured questions and answers, encourages originality, can be answered at any time, allows for additional ideas if requested

>   Repetitions have to be compared, responses could be fatuous, could take time

SWOT (strengths, weaknesses, opportunities, threats) analysis

>   Highly focused, lists opportunities as well as threats, structured

>   Needs good control, parts not necessarily risk

WBS (work breakdown structure)

>   Each stage or task can be investigated for risk content

4. **Risk management procedure**

    Types of risk, political, economic, technical, security, environmental

    Risk management plan, diagrams, P—I 3 × 3 matrix, risk log, techniques

    Risk management process, software tools, Monte Carlo, @Risk, Predict, 3-point estimating

    Identification, techniques, brain storming, check list, prompt list, interviews, WBS, Delphi

    Assessment, priorities, impact, probability, SWOT, decision trees, Ishikawa diagrams

    Qualitative analysis, quantitative analysis, risk owner, risk register

    Risk management should be initiated at start of project

    All projects have risk

    Risk identification

    Risk history or diary

    Risk owner

    Risk log or register

    Qualitative analysis

    Quantitative analysis, Monte Carlo, @Risk, Predict

    3-point estimating

    Impact/probability matrix

    H, M and L assessment

    Managing risk

5. **Travel risks**

    Motorway congestion

    Bad weather (ice, snow, storm)

    Car breakdown

    Roadworks

    Tyre puncture

    Feeling ill or sick

    No parking space at destination

    For risk register (log) see 53

6. **Change management**

Customer responsibility: impact assessment, evaluation, agreement

Sponsor responsibility: authorization, review, benefit assessment

Project manager responsibility: control procedure, monitoring, evaluation, implementation

Difference between external and internal change, effect on budget

Change control process, stakeholder input

All departments must assess time and cost effect of change on them

Key documents

Change request form

Change record: this records the effects of the change on the project

Change control register

Change authorization: this gives the effect of the change on cost, time and performance

Feedback to customer (who has right to cancel) after examining the consequences

Change register

7. **Bar chart**

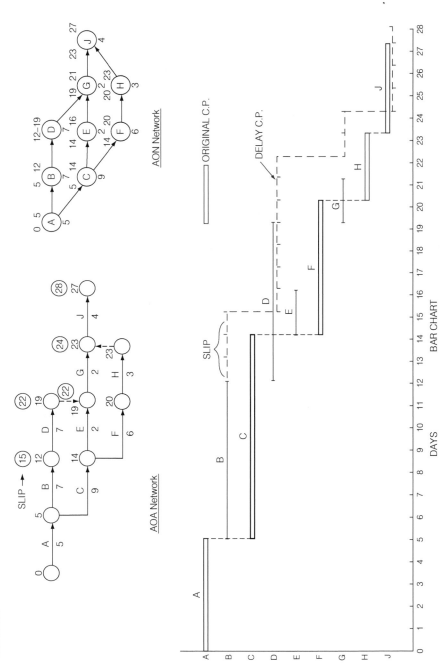

8. **Project and programme management**
   Project is unique process to achieve a set objective
   Has a defined start and finish
   Constraints are time, cost and resources
   Project managers change
   Organization can be task force, matrix or functional
   Definition of a programme
   A collection of *related* projects
   Advantages and role of programme manager
   Coordinate resources
   Assign priorities
   Oversee project managers
   Reduce risk
   Prepare overall milestone plan
   Resolve conflicts between project managers

9. **Stakeholder management**
   Positive stakeholders: project team, client, contractors
   Negative stakeholders: pressure groups, environmental groups, planning authorities
   Influence of stakeholders to be assessed, for example
   Power to affect project
   Financial muscle
   Malleability
   Personal involvement
   Vested interest
   Political bias or affinity
   Stakeholder prioritization

10. **Configuration management (CM)**
    Covers documents, drawings and components
    CM planning
    Item identification
    Control of configuration items
    Status
    Verification audit

11. **Risk management plan**
    Introduction: why risk management, company policy
    Purpose of process: client requirement, environmental, health and safety, project viability
    Scope of process, Political, Economic, Sociological, Technical, Environmental, Legal
    Description of project
    Specific aspects to be covered: innovations, erection, finance, sabotage, security
    Type of organization: functional matrix, task force
    Roles and responsibilities: managers, risk owners
    Tools and techniques required: Monte Carlo, @Risk, Predict, etc.
    Deliverables: risk register, matrix, action plans, contingency plans, etc.
    Mitigation

12. **Risk mitigation**
    Transference
    Insurance
    Reduction
    Elimination
    Deference
    Acceptance
13. **Quality tools**

Quality Objectives	Specifications
Quality systems	Company systems, ISO 9000
Quality plan	Analysis, inspection, recording, degree of testing
Quality assurance	Procedures, processes
Quality control	Checking
Quality manual	Contains all the above
Checking and testing	Verification, sampling, validation, certificates
Quality audit	Check that procedures are being adhered to
Quality reviews	
Pareto analysis 80/20 rule	
Cause and effect analysis, Ishikawa or fishbone diagram, decision trees	
Brainstorming (could lead to risk assessment)	
Check lists	
Process flow sheets	
Concentration diagrams	
Activity sampling	
Ranking and rating	
Bathtub curves	
Poka Yoka	
Qualification certificates, for example, welders qualification tests	

14. **Quality management**
    Quality planning
    Quality assurance
    Quality control
    Quality standards
    Fitness for purpose
    Meets acceptance criteria
    Total quality management (TQM)
    Right first time and zero defect
    Quality circles
    ISO 9000
    Quality audits

15. **Milestones**
    Important stages
    Payment points (expenditures or receipts)
    Achievement points
    High-level programme
    Milestone slip charts show record and predictions
    Milestone schedule
    Marked on Gantt charts
    Target on networks
    Often written into a contract
    Approval for change and date
    For milestone slip chart see Chapter 25

16. **Earned value management (EVA)**
    Reasons for EVA:
    Shows difference between work planned, work booked (on time sheets) and work performed
    Abbreviations: BCWS, ACWP and BCWP. Note: BCWS is planned (not budget)
    These abbreviations are now being phased out
    $EV = BCWP = budget \times percent\ complete$
    Control curves: budget, planned, actual, earned value, percent complete, efficiency (CPI)
    Interpretation and solution if CPI and SPI are negative
    Calculations: $CPI = BCWP/ACWP$, $SPI\ (cost) = BCWP/BCWS$
    Cost and time variances: cost variance $= BCWP - ACWP$
    Time variance $= BCWP - BCWS$; Final cost $=$ Original budget/CPI
    Final time $=$ Planned time/SPI (time) or SPI (cost)
    Should be geared to CP network, which can be updated from EVA feedback
    If returns of percent complete are on time sheets, regular feedback is guaranteed
    Computer programs have been developed to do the calculations
    Advantages:
    Shows trends, shows estimating errors, predicts final cost and completion time using curves and approx. calculations
    Percent complete of multi-discipline, multi-contractor project possible
    Trend can be seen early on in project and corrective action can be taken
    Shows cost and time position on one report
    Shows percent complete and efficiency on one graphical report
    Weakness:
    Difficulty in assessing percent complete
    Wrong budget estimates
    Poor feedback system
    Time sheets essential

17. **Project life cycle**
    No. of phases
    Standard life cycles
    BS 6079: conception, feasibility, implementation, operation, termination
    Common phases:
    Initiation, conception, feasibility, definition, design, development, production, manufacture, installation, implementation, commissioning, operation, disposal
    Extended life cycle includes disposal. Different industries have different life cycles

Advantages:
Ensure no unnecessary expenditure
Review possible after each phase. Go or no-go gates
Trigger for further funding
Each phase a control stage with checkpoints
Basis for WBS
Phases can be split again into stages
Phases called stages in PRINCE used in IT
Different phases can have different project managers
Gives top management an overall picture of project
Gives a rough progress position (crude bar chart)
Phases can be costed for control purposes
Can show up resource and continuity problems
Shows milestones
Developed by sponsor and project manager

18. **Stage documents**

Conception	**Business case, DCF**
Feasibility	WBS, OBS, CBS, risk register
Implementation	PMP, CP network, resource chart, milestone chart, EVA
Operation	Operating manuals
Termination	Handover certificate, maintenance manuals, spares list

19. **Organizations**
    The three project organizations are
    a. Functional
    b. Project (task force)
    c. Matrix [combination of (a) and (b)]

20. *Functional organization*
    Each department responsible for its own work
    Suitable for business as usual
    Ideal for routine operations and mass production
    Project manager usually appointed from one department for small change projects
    Project organization (task force)
    Used on large projects
    Team created especially for the project
    Team located in one office or building for good interaction
    Good communication
    Project manager has full responsibility for cost, time and performance
    Team includes design, procurement and construction specialists
    Matrix organization
    Most common type of organization
    Personnel allocated to project by departments
    PM responsible for time cost and quality
    Department manager responsible for the work by his/her department
    See also 56 and 57

21. **Communication management**
    Ensures good communication systems
    Clear instructions and messages
    All external communications via project manager
    Importance of documents to be in writing
    Communication by walk-about
    Communications can be formal or informal
    Barriers to communication
    > Cultural
    > Geographical location
    > Language, pronunciation
    > Bad translation
    > Technical jargon
    > Perception, attitude, lack of trust
    > Poor leadership, unclear objectives
    > Misunderstanding
    > Personality clash
    > Dislike of sender, selective listening
    > Group think
    > Bad equipment or equipment failure
    > Poor working environment, noisy office
    > Poor document distribution system, lost files
    > Unnecessarily long messages
    > Withholding information
    > Assumptions
    > Hidden agenda
    > Poor knowledge retention
    > Poor distribution and selective listening
    Overcoming barriers to communication
    > Simple messages
    > Follow-up by testing
    > Confirm in writing
    > Improve office facilities
    > Bring team together
    > Improve motivation
    > Maintain equipment
    > Can all be overcome by good planning, good equipment maintenance and ensuring the instructions are clear and unambiguous?
    > It is helpful to give the reason for an instruction as this will help to ensure the cooperation.
    > Initial lectures on the etiquette and customs of the different cultures reduce misunderstandings. Translations should always be made by a native speaker of the language into which the document is translated.

22. **Team building and teamwork**
    Advantages of project team
    Complementary skills
    Increased productivity
    Project manager support
    Informal communications
    Strong identity with project
    Common objectives

Motivation for project
Will to succeed
Focus
Good team spirit
Other methods used to build teams
Organized events, training, away days, discussions
The main characteristics of a successful team are
Mutual trust
Firm but fair leadership
Mutual support and loyalty
Open communications
Cooperation and full participation
Pride in the project and a sense of belonging
Good mix of technical skills and talent
Clear understanding of goals and objectives
Enthusiasm and confidence in success
Good support from top management
Apart from technical skills, different personality traits can be utilized to build up a balanced team and ensure that 'round pegs' are not driven into square holes.
For example, whilst some people like checking details, others prefer to be concerned with the wider picture.
By assembling the team with the different Belbin types, a balanced and competent workforce can be created.
Barriers to team building
 High staff turnover
 Poor leadership
 Geographical separation
 Internal conflict and unresolved conflict
 Low morale
 Lack of motivation
 Ill-defined objectives
 Poor environment
 Lack of trust
 Too many changes
 Poorly defined roles and responsibilities
 Poor communications

23.   Conflict management

Withdrawing
Smoothing
Confronting
Forcing
Compromising
Thomas and Kilman
Mediation
Conciliation
Adjudication
Arbitration
Mitigation
Litigation

24. **Belbin team roles**
    Balanced team, individuals in the correct slot. The roles are
    Plant
    Shaper
    Resource investigator
    Coordinator
    Monitor evaluator
    Implementer
    Team worker
    Completer finisher
    Specialist
    Not always possible to find the correct type
    Project manager must use resources available
25. **Motivation**
    Herzberg motivational theory
            Recognition of achievement
            Advancement and growth prospects
            Pleasant working conditions
            Increased responsibility
            Job freedom
            Interesting work
            Perks
    Maslow hierarchy of needs
            Physiological needs
            Security
            Safety
            Social
            Esteem
            Self-actualization
26. **Business case**
    Defines 'Why' and 'What' the requirements are
    Outlines objectives
    Outlines cost, time and performance/quality criteria
    Might include success/failure criteria
    Should include major risks
    Can have other names such as brief, scheme, statement of work (SOW) statement of requirements (SOR)
    Owned by sponsor or client
    Could include investment appraisal and possibly DCF/NPV, etc.
    Should have assessed other options
    Should have identified and considered all other stakeholders
    Submitted to board for approval
27. **Leadership and leadership qualities**
    Ability to influence rather than direct
    Negotiation skill
    Motivation
    Initiative
    Communication skills
    Ability to listen
    Fairness
    Firmness

28. **Negotiation**
    Planning and case preparation
    Set minimum acceptances
    Build up relationship
    Exchange information
    Bargaining
    Concessions
    Agreement
    Documentation of settlement
29. **Cash flow**
    Costing and monitoring methods
    Cash flow forecast (outflow and inflow)
    Cash flow curves (see Chapter 31 for diagram)
    Cost breakdown structure
    EVA for monitoring costs
    Budget preparation
    Commitments
    Accruals
    See 36 'estimating'
30. **Handover and close-out**
    Acceptance by client
    Acceptance certificate
    Transfer of responsibility to sponsor or operator
    Transfer of ownership to client
    Deliverables completed and handed over
    Acceptance certificate received
    Operating instruction handed over
    Spares lists handed over
    Transfer responsibility to user/client
    Material disposal arranged. This covers
    Sell surplus material to client if possible
    Return surplus to contractor's stores
    Return redundant office equipment and materials
    Sell unusable materials as scrap
    Clear site, remove huts and temporary fences
    Make good roads and other areas (stockyards)
    Complete all contract documentation
    Complete all audit trails and file documents
31. **Procurement strategy**
    Based on value for money
    Includes all feasible options
    Single source supply or competition
    Partnering or not
    Is construction included?
    Is delivery to site included?
    Standard or special contract conditions
    Types of pricing, firm, fixed, target, cost plus, reimbursable, schedule of rates
    Contractor selection criteria

Contractor bid meetings and attendees by contractors
Minimum and maximum number of bidders
Tender opening policy
Letter of intent possibilities
Limitations of procurement areas — UK, EC, Western countries, country of final user
Long lead items
Signatures for different levels of contracts
Types of contract
lump sum
remeasured
reimbursable
target, design
build and operate

32. **Contractor selection**
Expediting and inspection requirements
Procurement and delivery schedule
Liquidated damages
Incentives
Discounts required
Cash flow and payment terms
Shipping restrictions INCO terms
Guarantees and liabilities
Packaging and storage requirements
Spares lists
Operating and maintenance manuals
After-sales service

33. **Tuckman team phases**
Team development may change to reflect phase changes
Tuckman found teams undergo stages

Forming	Start of project
Storming	Jockeying for position; potential staff conflict
Norming	Smooth running
Performing	Fully integrated working and high morale
Mourning	End of project, breakup of team

Project manager must take care staff does not leave too soon as project nears the end.

34. **Project context environment**
Context within which project is undertaken
Takes account of internal and external forces
Client, company, contractors, suppliers, consultants, public, end users, etc.
PESTLE (political, economic, sociological, technical, legal, environmental)

35. **Why do projects fail?**
Poor planning and cost control
Incompetent project management
Sponsor support lacking
Client makes too many changes
Changes too large or extensive
Late safety requirements
Costs underestimated

Lack of change control

Lax configuration management

Environmental changes

Unforeseen climatic conditions

Integration problems

Technical and teething problems

Inadequate resources of staff and equipment

Poor working conditions

Uncooperative line management

Sabotage and political upheaval

Cost, time and performance requirements not met

Inadequate support from top management

Inadequate specification

Loose agreement with client

Loose contractual conditions

Ill-trained staff or operatives

Fundamental design faults

Late deliveries of information and equipment

Insufficient inspection and expediting

Poor cash flow

Insolvency of subcontractor

Unclear original brief causing misunderstanding

Commissioning problems stakeholder management

36. **See answer to question 8.**

37. **Estimating**

Sets out what information and data are required from:

Clients, end-users, other stakeholders

How information is captured, collated, assessed

How data are analysed and tested

Changes must be logged, reviewed and tested

38. **Investment appraisal**

DFC, NPV

> Advantages: time value of money, compares competing projects
>
> Disadvantages: complex, wrong estimates, relies on accurate discount rate

Payback

> Advantages: simple, can be discounted
>
> Disadvantages: time value ignored, cash flow after payback period ignored
>
> IRR, graphical solution, average return per annum, return on investment percent
>
> *Intangible benefits:* marketing, impact on business, prestige, social benefits, environmental benefits
>
> Consider other options, risks, accounting practices, stakeholder views

39. **See answer to question 36.**

40. **Resource management**

Resource allocation

Resource histogram

Resource scheduling

Resource levelling due to constraints, that is, lack of labour, materials, plant, etc.

Resource smoothing uses up float to reduce labour usage

Replenishable resources and renewable resources

41. **Organizational roles**
    Steering group or steering committee
    Function of project office (if established)
    Role of client and end user: specifies and operates finished plant
    Role of sponsor: owns business case, monitors performance, ensures resources, ensures benefits are realized, agrees variations
    Role of project manager: controls cost, time and performance of project, responsible for quality and safety, reports to sponsor and stakeholders
    Project manager's charter
    Stakeholders
    Contractors and suppliers responsibilities; supply specified equipment, materials and plant and manpower to meet set programme and performance requirements.

42. **Handover documents**
    Handover certificate
    Equipment list
    Operating manuals
    Maintenance manuals
    Spares list
    Guarantees
    Lubrication schedule

43. **Health and safety**
    H and S Standards
    Laws: HSWA 1974 (Health and Safety at Work Act)
    Management of Health and Safety
    Regulations 1992 covers
    Safe plant and equipment
    Dangerous substances
    Safe workplace and access
    Protective clothing
    Safe environment
    Information and training
    Reporting
    CDM (construction, design and management) Regulations
    Consumer Protection Act 1987
    Common law duties (negligence) apply
    Accident causes
        Lack of safety training
        Badly maintained equipment
        Inadequate protective clothing and headgear
        Unsecured ladders and scaffolding
        Poor unsafe design
        Unprotected pits and trenches
        Unsafe construction of trench supports
        Ignorance of safety procedures
        Inadequate protection when using toxic substances
        Inadequate safety warning signs

44. **WBS of bicycle**
    Assembly drawing
    Frame
    Handlebars

Wheels and tyres
Pedals
Gears and chain
Saddle and carrier
Mudguards
Lights and bell

45. **See answer to question 37.**

For graphical solution, see Chapter 6, Fig. 6.2

46. **Project office**

Collection and collation of reports and time sheets, etc.
Administrative support to project manager
Dissemination of instructions
Operation of configuration management system
Administration of change control
Library for standards and procedures
Records progress information
Produces curves and tabular information for use by project manager
Writes up and distributes minutes of meetings

47. **Post-project review and evaluation**

Reviews should be held throughout the project
Close-out meeting with project team
Evaluation against success criteria
Prepare post-project appraisal and write close-out report
Include abstracts from project diary and site diary where necessary
Highlight problems encountered, cost and time overruns, special delays
Report on project team performance
Recommendations for future projects, learning from experience
Evaluate project-management process
Evaluate techniques employed

48. **See answer to question 22.**

49. **Project manager**

Characteristics:
Knowledge of planning tools
Ability to control costs
Ability to get on with people
Does not panic when things go wrong
Is fair but firm
Ability to negotiate
Appreciation of importance of safety
Ability to write clear, concise reports
Good communicator and listener
Can motivate the team he/she leads
Is aware of corporate priorities and policies
Leads by example

50. **AoA and AoN networks (CPA)**

AoA network pro
As activities and durations are on the link lines (arrows), they can be more fully described
Can be drawn rapidly by hand without instruments
Progress can be updated by 'redding' up the lines in proportion of percent complete

As activities take time, they are more logically represented by an arrow

Streams of activities can be interlinked with dummy activities which can cross other activities

Can be drawn on a grid with links being able to cross activities

AoN (precedence) networks pro

Activities and durations are shown in or above the nodes

There is no requirement for dummy activities

Activities are defined by one number instead of two as in AoA

Can be drawn on a grid, but links cannot cross activity nodes

Lester diagram is a combination of both, gives best of both worlds

51. **See answer to question 37.**

52. **Portfolio management**

If projects are *not directly* related, it is portfolio management

Prioritize projects against needs of organization

Allocate resources and eliminate bottlenecks

Assess risks of projects in portfolio

Decide on timing of project starts

Monitor performance of all projects

Ensure good cash flow and profitability

Carry out cost—benefit analysis of projects

Portfolio management is a subset of corporate management

It is concerned with the organization, planning and control of a number of projects which are not necessarily connected with each other.

Its function is to ensure that the projects in the portfolio are aligned with the organization's strategy and meet the cost, programme, quality and performance criteria.

53. **Project success criteria**

The success of a project can be measured by the following criteria

Completion on time

Performance in accordance with the specified requirements

Meeting the quality standards

Completion within cost budget

Compliance with Health & Safety regulations

54. **Change register**

Essential headings on change register

Project title

Date of request to change

Name of instigator

Description of change

Reason for change

Time and cost estimates from each department

PM's summary of effects

Approval of change and date

55. **Information management**

Information management is concerned with the following functions

Acquisition of information

Data collection

Preparation of data

Dissemination to interested parties (stakeholders)

Storage, retrieval and archiving

Collection and capturing of information

Collation and distribution

Procurement and maintenance of equipment

Confidentiality

Modern IT systems permit huge amount of data to be collected.

This requires processing for both quantity and quality.

Timing is also essential, as late dissemination can seriously affect decisions.

All this has to be done efficiently by using an information system which is enshrined in an information policy plan.

56. **Value management**

In a constantly changing environment, systems, procedures and products must be regularly examined and updated to meet the needs of the stakeholders' value management and its subset value engineering is concerned with the strategic question of what should be done to improve performance without affecting the quality. Its aim is to improve the functionality of a product or system whilst reducing the overall cost.

Techniques include the following:

Functional analysis

Investigate alternative solutions to meet function requirements

Verbs and nouns technique

Evaluation

Acceptance, implementation and audit

57. **Project management versus functional line management**

PM advantages:

Greater efficiency and effectiveness in employing tools and techniques used in project management

One person, the project manager, is responsible

Sponsor knows at start what the deliverables will be

Sponsor can see the PMP and be assured the correct procedures are being used

Dedicated manager in charge of cost, expenditure and programme, strong commitment

More line organizations are moving towards project management for change

Monitoring and control through life of project

Single line of communication with all parties, especially external parties

Single line of reporting

Project manager is trained to handle the stresses of change

Learning from experience through post-project reviews

Project-management good training for top management due to wider vision

Line managers may be distracted by having to deal with line management functions

Spirit of dedicated project team not present in functional organization, high motivation

No competition for resources from other departments

Disadvantages:

Less job security than in functional organization

At times could be inefficient due to delays in information

Could develop parochialism and arrogance

Project objectives could eclipse company objectives

Not as strong specialist skills as functional department

Not as efficient in resources as functional department

Reporting line not as clear as in functional department

The difference between project management and line management is that the project has a defined start and end date with a set cost budget and performance criteria.

Line management on the other hand deals with the day-to-day business-as-usual operations of the company.

For this reason the matrix type of project management is arguably the preferred project organization, as the normal business is not disrupted whilst the changes such as new systems or structures, which require project management, can be accommodated.

58. **Matrix project management**

Advantages:

Fast response to resource change, flexible

More economical that task force

Stronger specialist base, knowledge not lost

Existing resources use state-of-the-art technology

Common facilities (computer programs) shared

Career prospects unchanged contract labour more easily taught and absorbed

Disadvantages:

More executive input required, possible conflict between PM and functional manager

Possible resource priority disputes between projects and function

Not as integrated as a task force

Less commitment to project than department

Personnel have two bosses, conflicting priorities

Requires more interdepartmental coordination which could be complex

PM has not the same authority to commit resources

PM not responsible for pay and rations

Of the three main types of project organization, that is, functional, matrix and task force, matrix is probably the most common.

Diagram of project organization.

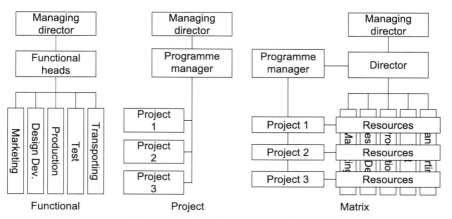

Diagram of project organization.

It utilizes the existing functional departments to supply the human resources, without disrupting the day-to-day operation of the various departments.

The personnel allocated to the project are responsible to the project manager for meeting the time and cost criteria, but report to the functional manager who is responsible for quality and technical competence.

As the functional manager is in charge for 'pay and rations' and recommendation for promotion, there may be a conflict of interest which has to be managed and can be a real challenge for the project manager.

59. **Project success and benefit analysis**
    Define benefits and agree on method to measure benefits
    Set success criteria
    Set and monitor key performance indicators (KP)
    Resource histogram, resource smoothing, peak reduction and 'S' curve
    PERT and use of three-point estimates for durations
    Maximizing parallel activities to reduce overall time
    Options to be examined as part of business case
    Non-financial benefits may be difficult to quantify
    Sydney opera house was a success, but a failure in terms of cost and time

60. **Project governance**
    Project governance covers the following:
    Structure to obtain objectives
    Monitoring performance
    Standard procedures
    Quality assurance
    Corporate policy and transparency
    Delegation of authority
    Reporting procedures
    Ethics

61. **Information management**
    Topics included in the information management plan are the following:
    Types of document
    Authority and standards
    Distribution
    Retrieval
    Acquisition/modification permits
    Acknowledgements of receipts
    Security of data
    Disaster recovery
    Configuration control
    Distribution of documents
    All these have to be done efficiently by using an information system that is embedded in an information policy plan

Appendix 9: Syllabus summary bullet points

Selected topics from *APM BoK*, sixth edition.

1. Context

Environment
External influences
Governance
Host organization
Internal influences
Legislation
P3
PESTLE
Setting
Stakeholders
Structured approach

2. Governance

Approval gates
Assurance scheme in force
Coherent business strategy
Compliance with external regulations
Continual improvement policy
Corporate ethics
Defined roles
Fostering trust
Frank disclosure and reporting
Good communications
Portfolio direction
Project management capability
Project sponsorship
Risk awareness
Stakeholder engagement

3. Project management
Definition of project management
Defined start and finish
Motivation of all involved
Planning, monitoring and controlling all aspects of the project
Project charter
Time, cost and performance requirement

4. Programme management
Assign priorities
Collection of related projects
Coordinate resources
Oversee project managers
Prepare overall milestone plan
Reduce risk
Resolve conflicts between project managers

5. Portfolio management
Allocate resources and eliminate bottlenecks
Assess risks of projects in portfolio
Carry out cost-benefit analysis of projects
Decide on timing of project starts
Ensure good cash flow and profitability
Monitor performance of all projects
Prioritize projects against the needs of the organization
Projects can be related or not related

6. Infrastructure
Access to all tools and techniques
Assurance of governance
Maintenance of infrastructure
Permanent organizational structure
Project office
Secretariat
Specialist skill supply when required
Technical support
Training and coaching programmes

7. Knowledge management

Corporate function
Experience recorded for future use
External and internal sources
Ownership of knowledge
Implementation mechanism
Knowledge capture
Lessons learned log
Maintenance of knowledge repository
Storage of knowledge

8. Life cycle

BS 6079: Conception, feasibility, implementation, operation, termination
Ensures unnecessary expenditure
Go or No-Go gates
Number of phases
Review after each phase
Trigger for further funding

9. Success factors

Define benefits and agree measure
Ensure senior management commitment
Identify and assess benefits
Implement proper governance
Provide good communications
Set and monitor KPIs (key performance indicators)
Set success criteria

10. Sponsorship

Business case owner
Communication skills
Governance
Oversees delivery of benefit
Responsible to Board or client
Steering group member
Supporting role to project or programme

11. Environment
Context in which project is undertaken
Client, company, contractor, supplier, consultant, end user, public, utilities
PESTLE (Political, Economic, Sociological, Technical, Legal, Environmental)
Takes account of internal and external forces

12. Operations management
Applies to product-based and service industries
For routine activities
Inputs are resources, for example, capital, people, materials and equipment
Outputs are products, services, solutions, information and distribution
Must work with P3 projects

13. Strategic management
Choice of strategy
External influences and demands
Mission statements
Opportunities
Stakeholders' requirements
Strategic analysis
Strategic implementation

14. Communication
Assumptions
Barriers to communication:
Bad translation
Cultural
Dislike of sender
Equipment failure
Group think
Hidden agenda
Jargon
Language and pronunciation
Misunderstandings
Perception, attitude, lack of trust

Personality clash
Poor document distribution, lost files
Poor knowledge retention
Poor working environment
Unclear objectives
Unnecessary long messages
Withholding information

15. Conflict management

Adjudication
Arbitration
Compromise
Conciliation
Confronting
Forcing
Mediation
Mitigation
Smoothing
Thomas and Kilman
Withdrawing

16. Delegation

Clear lines of authority
Clear specification of instruction
Communication skills
Ensure person delegated is trained
Knowledge of person delegated
Language problem
Monitor action taken
Motivation
Plan what can be delegated
Work done by others

17. Influencing

Cultural and contextual awareness
Communication skills
Compliance with commitments
Influencing techniques

Negotiation skills
Sensory awareness
Trust and understanding
Understand own attitude

18. Leadership

Ability to adapt to change
Ability to influence rather than direct
Ability to listen
Communication skills
Decision making
Does not panic
Fairness
Initiative
Integrity
Keeps a cool head
Motivation
Negotiation skills
Situational leadership
Wider perspective

19. Negotiation

Agreement
Bargaining
Build up relationship
Concessions
Documentation of settlement
Exchange information
Planning and case preparation
Set minimum acceptances

20. Teamwork

Advantages of project team:
Away days, discussions training events
Belbin team roles
Common objectives
Complementary skills
Focus

Herzberg's motivation theory
Increased productivity
Informal communications
Maslow's hierarch of needs
Motivation for project
Strong identity with project
Team spirit
Tuckman team phases
Will to succeed
Work satisfaction

21. Professionalism

Accountability
Breadth and depth of knowledge
Certification
Code of ethics
Code of practice
Commitment
CPD
Professional institutions
Public interest
Responsibility
Specialized knowledge

22. Communities of practice (CoP)

Best practice
Collective competence
Common interest in project management
Dissemination of knowledge
Motivation
Mutual support
New knowledge generated
Repertoire of resources
Shared experience

23. Competence

Combined skill, knowledge and experience
Competence frameworks

Competency assessments
Defined expectations
Defines roles
Defines statements of work
Practical experience
Technical expertise
Theoretical understanding

24. Ethics framework
Awareness of cultural differences
Knowledge of legal boundaries
Moral leadership
Personal code of conduct
Professionalism
Transparency of actions
UK Bribery Act

25. Learning and development
Ability levels
APMP qualification
Continual improvement
CPD Certificates
Learning and development programme
Learning objectives
Mentoring
Performance management
Performance reviews
PRINCE two qualification
Support for development

26. Business case
Could include investment appraisal
Defines 'Why' and 'What'
Includes major risks
Might include success/failure criteria
Outline cost, time and performance/quality criteria
Outline objectives
Owned by sponsor or client

Requirements (SOR)
Should have assessed other options
Should have identified and considered all stakeholders
Submitted to Board for approval

27. Control

ACWP, BCWP, BCWS
Achievement points
Bar chart
Baseline
Cost performance index (CPI)
Cybernetic control
Decision points
Earned value management (EVM)
Earned schedule
Feedback requirement
Gantt chart
Milestone slip chart
Payment point
Programme
Resource smoothing
Schedule performance index (SPI)
Time variance
Time sheets
Tracking

28. Information management

Big data
Confidentiality
Collation and distribution
Collection and capture
Dissemination
Procurement and maintenance of equipment
Storing and archiving

29. Organization

Functional organization
Generic organization structure

Matrix organization
PRINCE two
Portfolio organization structure
Programme organization structure
Project office
Project organization structure
Steering group
Task force

30. Planning
Bible of project owned by project manager
Covers the Why, What, When, Where, How, Who and How much (Kipling poem)
Concept, Definition, Development, Delivery, Closure
Contents set out in BS 6079
Milestones
Project management plan
Sets out baseline of project
Updated regularly with good configuration management

31. Stakeholder management
Financial muscle
Influence of stakeholders to be assessed
Malleability
Negative stakeholders, pressure groups, environmental groups, planning authorities
Personal involvement
Political bias or affinity
Positive stakeholders, project team, client, contractors, sponsor
Power to affect project AoA
Stakeholder prioritization
Vested interest

32. Scope management (WBS)
Can be product based or work based
Divides project into manageable packages
Does not show interdependencies
Foundation for planning, CP network and bar chart
Gives better definition of work
Good basis for risk identification process

Helps to create responsibility matrix
Leads to PBS, CBS, OBS and RBS
Shows hierarchy of work packages
Task-orientated family tree
Top down and bottom up estimating

33. Benefits management

Benefits management plan
Benefits modelling
Capture baseline measurements
Covers projects, programmes and portfolios
Implementation of change
Measurable impact
Monitoring performance
Opportunities capture
Performance indicators
Realization of business benefits
Tangible benefits quantifiable

34. Change control

Authorization document giving change effect on cost, time and performance
Change control process
Change control register
Change request form
Customer responsibility: impact assessment, evaluation agreement
Departments must assess time and cost effect on them
Difference between external and internal change, effect on budget
Feedback to customer who has the right to cancel after assessment

35. Configuration management (CM)

Configuration audit
Control of configuration items
CM planning
Covers documents, drawings and components
Item identification
Status
Verification

36.   Change management
Assess, prepare, plan, implement, sustain
Carnall change management model
Generic change management
Kotter change management model
Lewin change management model

37.   Requirements management
Analyse requirements
Collect requirements from stakeholders
Ensure no contradictions
Ensure no misinterpretation
Justify requirements
Prioritize
Set up baseline requirements
Specify requirement
Update
Verify by inspection, tests, demonstration

38.   Solutions development
Alternative functionality
Baseline requirements
Evaluation and selection
Implementation (possibly phased)
Modelling and simulation
Progressive testing
Reduce capital cost
Validation against requirement
Value improvement
Verification of solution

39.   Resource scheduling
Aggregation
Availability limits
Bar chart
Histogram
Network analysis
Performance curves

Resource levelling
Resource scheduling
Resource smoothing
Reusable resources

40. Time scheduling
AoA or AoN (precedence) networks
Critical path (CP) has zero float
Estimate durations
Gantt chart
Lester diagram
Maximize parallel activities
PERT
Resource histogram
Resource smoothing, peak reduction
S curve
3-point estimating for durations

41. Financial and cost management
Commit funds
Cost/benefit analysis
Control expenditures
Estimating costs
Evaluate outcome
Financial governance
Justify costs
Reviews
Secure funding

42. Budgeting and cost control
Accruals
Analytical estimating
Bill of quantities
Bottom-up estimating
Budget preparation
Cash-flow forecast
Comparative estimating
Contingencies

Costing and monitoring methods
CBS (cost breakdown structure)
Commitments
Estimating
Parametric estimating
Schedule of rates
Subjective estimating
Top-down estimating

43. Funding

Build, own, operate and transfer (BOOT)
Budget holder
Capital expenditure (CAPEX)
Combined funding
Credit guarantees
Grants
Internal and external funding
Loans
Operational expenditure (OPEX)
Overdrafts
Private finance initiative (PIF)
Public private partnership (PPP)
Share issue
Valuations
Venture capital (VC)

44. Investment appraisal

Average return per annum
Advantages of DCF, NPV:
Compares competing projects
Time value of money
Advantaged of payback:
Can be discounted
Simple
Intangible benefits:
Impact on business
Marketing
Prestige
Social benefits

IRR, graphical solution
Return of investment percent
Risks to be considered
Stakeholder views to be considered

45. Risk management

Assessment, priorities, probability, SWOT, decision trees, Ishikawa diagram Delphi, WBS
Identification techniques, brainstorming, checklist, prompt list, interviews
Risk management plan, diagrams, P–I matrix, risk log, techniques
Risk management process, software tools, Monte Carlo, predict, 3-point estimating
Qualitative analysis
Quantitative analysis
Risk owner
Risk register
Types of risk, political, economic, technical, security, environmental

46. Risk context

Client requirement
Company risk policy
PESTLE
Positive and negative risk
Risk averseness, risk seeking, risk neutral, risk appetite
Risk management plan

47. Risk techniques

Assumption analysis
Brainstorming
Checklists
Delphi techniques
Document reviews
Interviews
Prompt lists
Questionnaires
SWOT analysis
WBS

48.   Quality management
Acceptance criteria
Fitness for purpose
ISO 900
Quality assurance
Quality audits
Quality circles
Quality control
Quality manual
Quality plan
Quality standards
TQM (total quality management)
Zero defect

49.   P3 assurance
Assurance process
Assurance providers
Audits
External assurance providers
Governance processes
Integrated assurance
Quality management plan
Regulatory compliance
Sponsor led
Support functions reviews
Risk register
Security
Stakeholder confidence

50.   Reviews
Agenda for review meetings
Business case review
Deliverables review
Frequency set out in quality management plan
Gate reviews
Management processes review
P3 assurance
Post project review
Report to owners
Triggered by events

51. Resource management

Acquired through procurement
Obtained internally or externally
Replenishable resources
Renewable resources
Resource allocation
Resource levelling
Resource scheduling
Resource smoothing
Service agreement
Shared infrastructure

52. Contract

Agreement between two parties
Bespoke contract
Close-out
Consideration
Competitive contract
Contracts: NEC, ICE, IMech.E, IChem.E, JCT
Cost plus contract
Disputes procedure
Final account
General conditions of contract
Hand over
Incentives
Joint Contract Tribunal (JCT)
Legal document
Liquidated penalties
Maintenance period
Offer and acceptance
Special conditions of contract
Standard contract
Subcontracts
Target contract
Time, cost and performance requirements

53. Mobilization

Accommodation for labour
Demobilization strategy
Facilities on site
Internal or external resources

Plant and machinery requirements
Premises and sites
Recruitment
Resource availability
Software availability
Telecommunications
Temporary facilities

54. Procurement

Based on value for money
Bid meetings and attendees by contractor
Cash flow and payment terms
Construction included or not
Contractor selection criteria
Discounts required
Expediting and inspection requirements I
Guarantees and liabilities
Incentives
Includes all feasible options
Letter of intent
Liquidated penalties
Long lead items
Min. and max. number of bidders
Operating and maintenance manual
Packaging and storage requirements
Partnering or not
Procurement areas, UK, EU, USA, Country of user, etc.
Shipping restrictions, INCO terms
Single source of supply or competition
Spares list
Tender opening policy
Types of contract

55. Provider selection and management

Contractual terms and conditions
Identifying providers
Interviews
Managing providers
Monitoring progress

Part of procurement function
Pre-qualification process
Questionnaires
Selecting and appointing providers
Selection process
Tendering policy
Vetting providers

56. Accounting

Balance sheet
Business management requirements
Collecting financial information
Communicating financial information
Corporate accounting
Financial accounting
Legal requirements
Profit and loss statement
Project accounting
Stakeholder needs

57. Health and safety

CDM (Construction, Design and Management) Regulations
Common law duties (negligence)
Consumer Protection Act 1987
Dangerous substances
H&S standards
Information and training
Management of Health and Safety Regulations (1992)
Laws: HSWA 1974 (Health and Safety at Work Act)
Protective clothing
Reporting
Safe environment
Safe plant and equipment, safe workplace and access

58. Human resource management (HRM)

Database of staff
Development of people
Dispute resolution

Employee relations
HRM policy
Legal requirements
Managing people
Pay reviews
People related activities
Personal development
Promotion
Recruitment
Redundancy

59. Law

Awareness of laws
European Convention of Human Rights (ECHR)
Civil law
Compensation
Criminal law
Intellectual property law
Jurisdiction
Legal duties
Legal system
Precedents
Processes
Rights
Statutory instruments
Working time directives

60. Security

Asset protection
Confidential data protection
Guards
Minimize disruption
Personnel security
Resource requirements
Risk mitigation
Security assessment
Security policy
Site security

61. Sustainability

Biological systems
Ecosystem
Environmental impact assessment
Minimum contamination
Zero pollution aim

Appendix 10: Words of Wisdom

Cash flow	More businesses go bust because of poor cash flow than low profitability
Cash flow	A bird in the hand is worth two in the bush
Claims	You need three things for a successful claim: 1. Good backup documentation 2. Good backup documentation 3. Good backup documentation
Communication	Listen carefully before talking; you have two ears and one mouth
Communication	Read twice, write once (for examinations)
Contract	A verbal contract is not worth the paper it is (not) written on
Contract	Do not use two if one will do
Control	The wheel that squeals gets the grease
Delegation	Do not keep a dog and bark yourself
General	If it looks wrong, it probably is wrong
General	If it looks too good to be true, it probably is
General	A wise man learns from experience, a fool does not
General	If in doubt, say nowt
General	Every problem has a solution
General	The early bird gets the worm
General	To take off, go against the wind, not with it
Planning	The shortest distance between two points is a straight line
Planning	The longest distance between two points is a shortcut
Planning	Forewarned is forearmed
Planning	It is later than you think
Procurement	If you do not inspect it arrives wrong
Procurement	If you do not expedite it arrives late
Quality	Quality is remembered long after the price is forgotten
Quality	A good product goes out
Quality	A bad product comes back
Risk	Nothing ventured, nothing gained
Safety	It is better to be old than bold
Safety	Look before you leap
Safety	The most important bolt is the one that is loose

Appendix 11: Bibliography

Adair, J. (2010). *Decision making and problem solving strategies*. Kogan Page.

Adams, J., & Adams, J. R. (1997). *Principles of project management*. PMI.

Adkins, L. (2010). *Coaching agile teams*. Pearson Education Inc.

Alvesson, M. (2002). *Understanding organisational culture*. Sage.

Andersen, E. (2008). *Rethinking project management*. Pearson.

Andrews, N. (2011). *Contract law*. Cambridge University Press.

APM. (2004). *Risk analysis and management guide*.

APM. (2008). *Interfacing risk and earned value management*.

APM. (2008). *Earned value management guidelines* (2nd edition).

APM. (2008). *Introduction to project planning*.

APM. (2010). *Introduction to project control*.

APM. (2013). *Earned value management handbook*.

APM. (2014). *Guide to integrated assurance*.

APM. (2015). *Planning, monitoring, scheduling and control*.

APM. (2016). *Guide to conducting integrated baseline reviews*.

APM. (2016). *Introduction to programme management* (2nd. edition).

APM. (2016). *Directing agile change*.

APM. (2017). *Guide to contracts and procurement for project, programme and portfolio managers*.

APM. (2017). *Introduction to managing change*.

APM. (2018). *Guide to project auditing*.

APM. (2018). *Directing change* (3rd ed.).

APM. (2018). *Sponsoring change* (2nd ed.).

APM. (2019). *Body of knowledge* (7th ed.).

APM. (2019). *Guide to cost estimating*.

APM. (2019). *Introduction to Portfolio Management*.

APM. (2019). *Guide to using a benefits management framework*.

APM. (2020). *Engaging stakeholders on projects*.

Atkins, L. (2010). *Coaching agile teams*. Addison-Wesley.

Bagueley, P. (2010). *Improve your project management: Teach yourself*. Hodder & Stoughton.

Bahsoon, R., & Emmerich, W. (2004). Evaluating architectural stability with real options theory. In *Proceedings of the 20th IEEE International Conference on Software Maintenance (ICSM'04)*.

Balogun, J., et al. (2008). *Exploring strategic change*. Prentice-Hall.

Barker, S. (2009). *Brilliant project management*. Prentice Hall.

Barkley. (2004). *Project risk management*. McGraw-Hill.

Barkley. (2006). *Integrated project management*. McGraw-Hill.

Barnes, P., & Davies, N. (2015). *BIM in principle and practice*. Thomas Telford.

Bartlett, J. (2005). *Right first and every time*. Project Manager Today.

Bartlett, J. (2010). *Managing programmes of business change*. Project Manager Today.

Bass, L., Clements, P., & Kazman, R. (2003). Software architecture in practice *SEI series in software engineering* (2nd ed.). Addison-Wesley.

Basu, R. (2011). *Managing project supply chains*. Gower.

Benko, C., & McFarlane, W. (2003). *Connecting the dots*. Harvard Business School.

Scott, Berkun. (2005). *Making things happen.* O'Reilly.

Bittner, E., & Gregorc, W. (2010). *Experiencing Project management.* John Wiley.

Boddy, D., & Buchannan, D. (2002). *Take the lead.* Prentice-Hall.

Boddy, D. (2002). *Managing projects.* Prentice-Hall.

Bosch, J. (May 16—20, 2011). *Keynote abstract Saturn Conference San Francisco.* SEI.

Boundy, C. (2010). *Business contracts handbook.* Gower.

Bourne, L. (2009). *Stakeholder relationship management.* Gower.

Bradley, G. (2010). *Benefit realization management.* Gower.

Bradley, G. (2010). *Fundamentals of benefit realization.* The Stationary Office.

Briscoe, D. (2015). *Beyond BIM.* Routledge.

Broome, J. C. (2002). *Procurement routes for partnering: A practical guide.* Thomas Telford.

Brown, N., Nord, R., & Ozkaya, I. (November/December 2010). *Enabling agility through architecture crosstalk.*

Brulin, G., & Svensson, L. (2012). *Managing sustainable development programmes.* Gower.

BSI, BS EN ISO 9000:2000. (2000). *Quality management systems, fundamentals and vocabulary.* BSI.

BSI, BS EN ISO 9000:2000. (2000). *Quality management systems, guidelines for performance improvement.* BSI.

BSI, PAS 2001:2001. (2001). *Knowledge management.* BSI.

BSI, BS EN ISO 9000:2003. (2003). *Quality management systems, guidelines for quality management in projects.* BSI.

BSI, PD 7501:2003. (2003). *Managing culture and knowledge.* BSI.

BSI, PD 7502:2003. (2003). *Guide to measurements in knowledge management.* BSI.

BSI, PD 7506:2005. (2005). *Linking knowledge management with other organizational functions and disciplines.* BSI.

BSI, BS EN ISO 900012008. (2008). *Quality management systems, requirements.* BSI.

BSI, BS 6079-1:2010. (2010). *Guide to project management.* BSI.

BSI, BS31100:2011. (2011). *Risk management.* BSI.

Burke, R., & Barron, S. (2007). *Project management leadership.* Burke Publishing.

Burke, R. (2011). *Advanced project management.* Burke Publishing.

Camilleri, E. (2011). *Project success.* Gower.

Cappels, T. (2003). Financially focused project management. *J. Ross.*

Carroll, J., & Morris, D. (2015). *Agile project management in easy steps.*

Carroll, T. (2006). *Project delivery in business-as-usual organizations.* Gower.

Carver, J., & Carver, M. M. (2009). *The policy governance model & role of the board member.* Jossey Boss.

Cavanagh, M. (2012). *Second order project management.* Gower.

Chapman, C. B., & Ward, S. C. (2003). *Project risk management* (2nd ed.). Wiley.

Chitkara, K. (2019). *Construct. Proj. management. planning, sched. control.* McGraw-Hill.

Cialdini, R. B. (2008). *Influence, science and practice* (5th ed.). Pearson.

CIOB. (2009). *Code of practice for project management for construction and development* (4th ed.). Wiley.

Clark Graig, Juana (2012). *Project management Lite.* Create Space Independent Publ.

Cleden, D. (2009). *Managing project uncertainty.* Gower.

Cleden, D. (2011). *Bid writing for project managers.* Gower.

Cleland, D. I. (2006). *Global project management handbook.* McGraw-Hill.

Cleland, D. I. (2007). *Project management: Strategic design and implementation.* McGraw-Hill.

CMMI.. (November 2010). CMMI® for development, version 1.3 CMMI-DEV, V1.3. SEI.

Cockburn. (2002). *Alistair, Agile software development.* Addison-Wesley.

Cohn, M. (2006). *Agile estimating and planning*. Addison-Wesley.

Cohn, M. (2009). *Succeeding with agile: Software development using scrum*. Addison-Wesley.

Cohn, M. (2020). *Agile estimating & planning*. Pearson Education.

Collett, P. (2003). *The book of tells*. Doubleday.

Collins, G. (June12−15, 2006). Experience in developing metrics for Agile projects compatible with CMMI best practice. In *SEI SEPG Conference, Amsterdam*.

Collins, G. (June 2011). Developing Agile software architecture using real-option analysis and value engineering. In *SEI SEPG Conference, Dublin*.

Costin, A. A. (2008). *Managing difficult projects*. Butterworth-Heinemann.

Covey, S. (2004). *7 Habits of highly effective people*. Simon & Schuster.

Crane, A., & Matten, D. (2010). *Business ethics* (3rd ed.). Oxford University Press.

Andy, Crowe. (2006). *Alpha project manager*. Velociteach.

Andy, Crowe. (2010). *The PMP exam*. Velociteach.

Edward, Dansker. (1992). *Integrated engineering construction projects*. Elsevier.

Davies, R. H., & Davies, A. J. (2011). *Value management*. Gower.

Davis, T., & Pharro, R. (2003). *The relationship manager*. Gower.

De Mascia, S. (2012). *Project psychology*. Gower.

De Vito, J. A. (2011). *Human communications* (12th ed.). Allyn & Bacon.

De Vito, J. A. (2012). *The interpersonal communications book*. Pearson.

Dent, F. E., & Brent, M. (2006). *Influencing skills and techniques for business success*. Palgrave Macmillan.

Diab, P. R. (2011). *Sidestep complexity*. PMI.

Egeland. B Project Smart. (2014). 7 deadly sins of project management. *Project Smart*.

Egeland, B. (2019). Ingredients for project management office success. *Project Smart*.

El-Reedy, M. A. (2011). *Construction management for industrial projects*. Wiley.

Elssamadisy, A. (2009). *Agile adoption patterns: A roadmap to organisational success*. Addison-Wesley.

European Committee for Standardisation. (2004). CWA 14924-4:2004, European guide to good practice in knowledge management. *Guidelines for Measuring KM*. CEN.

European Committee for Standardisation. (2004). CWA 14924-5:2004, European guide to good practice in knowledge management. *KM Terminology*. CEN.

European Committee for Standardisation. (2012). *FprEN 16271:2012 (E), value management*. CEN.

Ferraro, J. (2012). *Project management for the non-project manager*. Amacom.

Ferraro, M. (2012). *Venice, history of the floating city*. Cambridge University Press.

Field, M., & Keller, L. (1998). *Project management*. Cengage Learning EMEA.

Fisher, R., & Shapiro, D. (2007). *Building agreement*. Random House.

Fisher, R., Ury, W., & Patton, B. (2003). *Getting to Yes*. Random House.

Forsberg, K., Mooz, H., & Cotterman, H. (2000). *Visualising project management*. Wiley.

Frigenti, E., & Comninos, D. (2002). *The practice of project management*. Kogan Page.

Jeff, Furman. (2014). *The project management answer book*. Berrett-Koehler.

Gambles, I. (2009). *Making the business case*. Gower.

Gardiner, P. D. (2005). *Project management, a strategic planning approach*. Palgrave Macmillan.

Garlick, A. (2007). *Estimating risk*. Gower.

Dave, Garnett. (2011). *Project pain reliever*. J. Ross Publishing.

Gatti, S. (2007). *Project finance in theory and practice*. Academic Press.

Goel, B. B. (2002). *Project management principles & techniques*. Deep & Deep Publ.

Goldsmith, L. (2005). *Project management accounting*. Wiley.

Goleman, D., Boyatzis, R., & McKee, A. (2002). *Primal leadership*. Harvard Business School Press.

Goleman, D. (2007). *Social intelligence*. Hutchinson.

Goodpasture, J. (2010). *Project management, the Agile way*. Ross Publishing Inc.

Gordon, J., & Lockyer, K. (2005). *Project management and project network techniques* (7th ed.). Prentice-Hall.

Graham, N. (2010). *Project management for dummies*. Wiley.

Grimsey, D., & Lewis, M. K. (2007). *Public private partnerships*. Edward Elgar.

Roel, Grit. (2019). *Project management*. Routledge.

Brett, Hamed. (2017). *Project management for Humans*. Rosenfeld Media.

Hancock, D. (2010). *Tame, Messy and Wicked risk leadership*. Gower.

Harrison, F., & Lock, D. (2004). *Advanced project management*. Gower.

Haugan, G. T. (2002). *Effective work breakdown structure*. Kogan Page.

Haughty, D. (2014). *21 Ways to excel at project management*. Project Smart.

Heldman, K. (2011). *Project management jump start*. Syber.

Heldman, Kim (2018). *Project management jump start*. John Wiley.

Hersey, P. H., & Blanchard, K. H. (2012). *Management of organizational behaviour*. Prentice-Hall.

Highsmith, J. (2002). *Agile software development ecosystems*. Addison-Wesley.

Highsmith, J. (2009). *Agile project management* (2nd ed.). Addison-Westley Professional.

Highsmith, J. (2010). *Agile project management* (2nd ed.). Addison-Wesley.

Hillson, D. A., & Murray-Webster, R. (2005). *Understanding and managing risk attitude*. Gower.

Hillson, D. A. (2003). *Effective opportunity management for projects*. Marcel Dekker.

Hillson, D. (2009). *Managing risk in projects*. Gower.

Holzer, D. (2016). *BIM manager's handbook*. John Wiley.

Hopkinson, M. (2010). *The project risk maturity model*. Gower.

Horine, G. (2017). *Project management*. One Publishing.

Hossain, F. (2018). *Sustainable design and build*. Elsevier.

Hulett, D. (2009). *Practical schedule risk analysis*. Gower.

Hulett, D. (2011). *Integrated cost-schedule risk analysis*. Gower.

Humble, J., Molesky, J., & O'Reilly, B. (2015). *Lean enterprise: How high performance organisations innovate at scale*. O'Reilly Media.

ISO, ISO 31000:2009. (2009). *Risk management — principles and guidelines*. ISO.

ISO, ISO/FDIS 21500. (2012). *Guidance on project management*. ISO.

Jackson, B. J. (2006). *Construction management jump start*. John Wiley.

Johnson, G., & Scholes, K. (2004). *Exploring corporate strategy*. Prentice-Hall.

Jones, S. (2016). *Agile project management*. quick start guide (Create Space IPP).

Katzenbach, J. R., & Smith, D. K. (2005). *The wisdom of teams*. Harper Business.

Kemp. (2004). *Project management demystified*. McGraw-Hill.

Kerzner, H. (2004). *Advanced project management*. Wiley.

Kerzner, H. (2009). *Project management*. Wiley.

Khan, F., & Parra, R. (2003). *Financing large projects*. Pearson Education Asia.

Knaster, R., & Leffingwell, D. (2016). *SAFe® 4.0 Distilled: Applying the scaled Agile framework® for lean software and systems engineering*. Addison-Wesley.

Kor, R., & Wijnen, G. (2007). *59 Checklists for project and programme managers*. Gower.

Sam, Kubba. (2010). *Green construction project management & cost oversight*. Elsevier.

Kumar, B. (2016). *A practical guide adopting BIM in construction projects*. Whittles.

Larman, C. (2004). *Agile and iterative development: A Manager's guide*. Addison-Wesley.

Laudon, K. C. (2008). *Management information systems* (11th ed.). Prentice-Hall.

Layton, M. C. (2012). *Agile project management for dummies*. John Wiley.

Leach, L. P. (2005). *Critical chain management*. Artech House.

Leblanc, R. (2016). *A handbook of board governance*. John Wiley.

Leffingwell, D. (2007). *Scaling software agility: Best practices for large enterprises*. Addison-Wesley (Chapter 21 with Schwaber, K.).

Sidney, Levy. (2009). *Construction process planning & management*. Elsevier.

Lewis, H. (2005). *Bids, tenders and proposals*. Kogan Page.

Lewis, J. P. (2003). *Project leadership*. McGraw-Hill.

Lewis, J. P. (2005). *Project planning, scheduling and control*. McGraw-Hill.

Lewis, J. P. (2016). *Fundamentals of project management*. Amacom.

Linstead, S., Fulop, L., & Lilley, S. (2009). *Management and organization*. Palgrave Macmillan.

Lock, D. (2007). *Project management* (9th ed.). Gower.

Lock, D. (2013). *Gower book of people in project management*. Gower.

Longdin, I. (2009). *Legal aspects of purchasing and supply chain management* (3rd ed.). Liverpool Academic.

Ludovino, E. M. (2016). *Change management*. EM Press Ltd.

Mantel, S. J., et al. (2011). *Project management in practice* (4th ed.). Wiley.

Marchewka, J. T. (2012). *Information technology project management with CD-ROM* (4th ed.). Wiley.

Margerison, C., & McKann, D. (2000). *Team management: Practical new approach management books*.

Martin, N. (2010). *Project politics*. Gower.

Maylor, H. (2005). *Project management*. Prentice-Hall.

Meredith, J. R., & Mantel, S. J. (2010). *Project management: A managerial approach*. Wiley.

Minter, M., & Szczepanek, T. (2009). *Images of projects*. Gower.

Morris, P., Pinto, J. K., & Soderlund, J. (2011). *The Oxford handbook on project management*. Oxford University Press.

Mulcahy, Rita (2002). *PMP exam prep*. RMC Publication.

Muller, R., & Turner, R. (2010). *Project-oriented leadership*. Gower.

Muller, R. (2009). *Project governance*. Gower.

Nagarajan, K. (2004). *Project management*. New Age International.

Neil, J., & Harpham, A. (2012). *Spirituality and project management*. Gower.

Newton, R. (2006). *Project management, step by step*. Pearson.

Newton, R. (2009). *The project manager*. Pearson.

Nickson, D. (2008). *The bid manager's handbook*. Gower.

Nieto-Rodriguez, A. (2012). *The focused organization*. Gower.

Nokes, S. (2007). *The definitive guide to project management* (2nd ed.). Pearson.

O'Connell, F. (2010). *What you need to know about project management*. Wiley.

O'Connell, F. (2011). *All you need to know about project management*. Capstone Publishing.

Oakes, G. (2008). *Project reviews. Assurance and governance*. Gower.

Obeng, E. (2002). *Perfect projects*. Pentacle Works.

OGC. (2007). *Managing successful programmes*. The Stationary Office.

OGC. (2008). *Portfolio, programme and project offices*. The Stationary Office.

OGC. (2009). *Directing successful projects with PRINCE 2*. The Stationary Office.

OGC. (2009). *Managing successful projects with PRINCE 2*. The Stationary Office.

OGC. (2010). *An executive guide to portfolio management*. The Stationary Office.

OGC. (2011). *Management of portfolios*. The Stationary Office.

Paterson, G. (2015). *Getting to grips with BIM*. Routledge.

Pennypacker, J. S., & Dye, L. D. (2002). *Managing multiple projects*. Marcel Dekker.

Pennypecker, J. S., & Retna, S. (2009). *Project portfolio management*. Wiley.

Pidd, M. (2004). *Systems modelling: Theory and practice*. Wiley.

Pinto, J. (2010). *Project management* (2nd ed.). Pearson.

Frederick, Plummer. (2007). *Project engineering*. Elsevier.

PMI. (2013). *Project management body of knowledge*. PMI.

PMI. (2001). *PMI practice standard for work breakdown structures*. PMI.

PMI. (2003). *Organizational project management maturity model (OPM3): Overview*. PMI.

PMI. (2004). *Practice standard for configuration management*. PMI.

PMI. (2006). *Government extension to the PMBOK guide*. PMI.

PMI. (2007). *Project manager competency development framework*. PMI.

PMI. (2008). *A guide to the project management body of knowledge*. PMI.

PMI. (2008). *Organizational project management maturity model (OPM3): Knowledge foundation*. PMI.

PMI. (2008). *The standard for portfolio management*. PMI.

PMI. (2008). *The standard for program management*. PMI.

PMI. (2009). *Practice standard for project risk management*. PMI.

Rad, P. F., & Levin, G. (2002). *The advanced project management office*. St Lucie Press Project.

Rad, P. F. (2001). *Project estimating and cost management VA: Management concepts*.

Reiss, G., et al. (2006). *Gower handbook of programme management*. Gower.

Reiss, G. (2006). *The Gower handbook of programme management*. Gower.

Remington, K., & Pollack, J. (2007). *Tools for complex projects*. Gower.

Remington, K., & Pollack, J. (2008). *Tools for complex projects*. Gower.

Remington, K. (2011). *Leading complex projects*. Gower.

Reschke, H., & Schelle, H. (2013). *Dimensions of project management*. Springer Science.

Rigby, D., Elk, S., & Berez, S. (2020). *Doing Agile right*. Harvard Business Press.

Roberts, Paul (2007). *Guide to project management*. John Wiley.

Robertson, S., & Robertson, J. (2006). *Mastering the requirements process*. Addison-Wesley.

Rodriguez, A. (2012). *Earned value management for projects*. Gower.

Rogers, M. (2001). *Engineering project appraisal*. Blackwell Science.

Rose, K. (2005). Project quality management. *J. Ross.*

Ross, J. W., Weill, P., & Robertson, D. C. (2006). *Enterprise architecture as strategy: Creating a foundation for business execution*. Harvard Business Review Press.

Royce, W. (2011). Measuring agility and architectural integrity. *International Journal of Software Informatics, 5*(3), 415−433.

Rubin, K. S. (2012). *Essential Scrum*. Addison-Wesley.

Sadhan, C. (1988). *Project management*. Tata McGraw-Hill.

Sant, T. (2004). *Persuasive business proposals*. Amacom.

Sanwal, A. (2007). *Optimising corporate portfolio management*. Wiley.

Sauchez, A. X. (2016). *Delivering value with BIM*. Routledge.

Saxon, R. (2016). *BIM for construction clients*. NBS.

Terry, Schmidt. (2021). *Strategic project management made simple*. John Wiley.

Schwaber, K. (2004). *Agile project management*. Microsoft.

Ken, Schwaber. (2004). *Agile with Scrum*. Microsoft Press.

Schwaber, K. (2008). *Introduction to project management*. Course Technology.

Schwartz, M. (2016). *The art of business value*. Portland, Oregon: IT Revolution.

Schwindt, C. (2005). *Resource allocation in project management.* Springer.

Senaratne, S., & Sexton, M. (2011). *Managing change in construction projects.* Wiley.

Shalloway, A., Beaver, G., & Trott, J. R. (2010). *Lean-agile software development: Achieveing enterprise agility.* Addison-Wesley.

Shermon, D. (Ed.). (2009). *Systems cost engineering.* Gower.

Shore, J. (2007). *The art of agile development.* O'Reilly.

Shtub, A., Bard, J., & Globerson, S. (2004). *Project management.* Pearson.

Skelton, M., & O'Dell, C. (2016). *Continuous delivery with Windows and. NET.* O'Reilly Media.

Sleeper. (2006). *Design for six sigma statistics.* McGraw-Hill.

Smith, C. (2007). *Making sense of project realities.* Gower.

Spencer, L. M., & Spencer, S. M. (1993). *Competence at work.* Wiley.

Srevens, R., et al. (1998). *Systems engineering.* Addison-Wesley.

Stenzel, C., & Stenzel, J. (2002). *Essentials of cost management.* Wiley.

Stutzke, R. (2005). *Software project estimation.* Addison-Wesley.

Sutherland, J. (2005). *Future of scrum: Support for parallel pipelining of sprints in complex projects.* Denver, CO: Agile. Conference.

Sutherland, J. (2014). *The art of doing work in half the time.* Random House.

Sutt, J. (2011). *Manual of construction project management.* Wiley.

Taylor, J. C. (2005). *Project cost estimating tools, techniques and perspectives.* St. Lucie Press.

Taylor, P. (2011). *Leading successful PMOs.* Gower.

Peter, Taylor. (2015). *The lazy project manager.* Infinite ideas.

Taylor, P. (2016). *The social project manager.* Routledge.

Thacker, N. (2012). *Winning your bid.* Gower.

Thiry, M. (2010). *Programme management.* Gower.

Thiry, M. (2012). *Project based organizations.* Gower.

Trevino, L., & Nelsom, K. (2010). *Managing business ethics* (5th ed.). Wiley.

Turner, R., & Wright, D. (2011). *The commercial management of projects.* Ashgate.

Turner, J. R. (2003). *People in project management.* Gower.

Turner, R. (2003). *Contracting for project management.* Gower.

Turner, J. R. (2008). *The Gower handbook of project management* (4th ed.). Gower.

Ursiny, T. (2003). *The coward's guide to conflict.* Sourcebooks.

Venning, C. (2007). *Managing portfolios of change with MSP for programmes and PRINCE 2 for projects.* The Stationary Office.

Eric, Verzuh. (2005). *Fast forward MBA project management.* John Wiley.

Ward, S., & Chapmen, C. (2011). *How to manage project opportunity and risk.* Wiley.

Ward, G. (2008). *The project manager's guide to purchasing.* Gower.

Wearne, S. H. (1993). *Principles of engineering organizations.* Thomas Telford Publications.

Weaver, R. G., & Farrell, J. D. (1997). *Managers as facilitators.* Berret-Koehler.

Webb, A. (2003). *The project manager's guide to handling risk.* Gower.

Wenger, E., McDermott, R., & Snyder, W. (2002). *Cultivating communities of practice.* Harvard Business School Press.

West, D. (2010). *Project sponsorship.* Gower.

Williams, D., & Parr, T. (2006). *Enterprise programme management.* Palgrave Macmillan.

Williams, & Todd, C. (2011). *Rescue the problem project.* American Mgt. Association.

Wills, K. R. (2010). *Essential project management skills.* CRC Press.

Winch, G. M. (2010). *Managing construction projects.* Blackwell.

Wright, D. (2004). *Law for project managers*. Gower.

Wysocki, R. K., & McGary, R. (2003). *Effective project management*. John Wiley.

Yang. (2005). *Design for six sigma in service*. McGraw-Hill.

Yescombe, E. (2002). *Principles of project finance*. Academic Press.

Yescombe, E. (2007). *Public-private partnerships*. Butterworth-Heinemann.

Young, T. L. (2001). *Successful project management*. Kogan Page.

Young, T. L. (2003). *Handbook of project management*. Kogan Page.

Index

Note: 'Page numbers followed by "*f*" indicate figures and "*t*" indicate tables.'

Notes

Notes

Printed in the United States
by Baker & Taylor Publisher Services